Trends in Developing Metaheuristics, Algorithms, and Optimization Approaches

Peng-Yeng Yin
National Chi Nan University, Taiwan

Managing Director:	Lindsay Johnston
Editorial Director:	Joel Gamon
Book Production Manager:	Jennifer Romanchak
Publishing Systems Analyst:	Adrienne Freeland
Development Editor:	Heather Probst
Assistant Acquisitions Editor:	Kayla Wolfe
Typesetter:	Travis Gundrum
Cover Design:	Nick Newcomer

Published in the United States of America by
Information Science Reference (an imprint of IGI Global)
701 E. Chocolate Avenue
Hershey PA 17033
Tel: 717-533-8845
Fax: 717-533-8661
E-mail: cust@igi-global.com
Web site: http://www.igi-global.com

Library of Congress Cataloging-in-Publication Data

Trends in developing metaheuristics, algorithms, and optimization approaches / Peng-Yeng Yin, editor.
       pages cm
   Summary: "This book provides insight on the latest advances and analysis of technologies in metaheuristics computing, offering widespread coverage on topics such as genetic algorithms, differential evolution, and ant colony optimization"--Provided by publisher.
   Includes bibliographical references and index.
   ISBN 978-1-4666-2145-9 (hardcover) -- ISBN (invalid) 978-1-4666-2146-6 (ebook) -- ISBN (invalid) 978-1-4666-2147-3 (print & perpetual access) 1. Heuristic algorithms. 2. Computer simulation. 3. Combinatorial optimization. 4. Computational intelligence. I. Yin, Peng-Yeng, 1966- editor of compilation.
   T57.84.T74 2012
   006.3--dc23
                                      2012019563

British Cataloguing in Publication Data
A Cataloguing in Publication record for this book is available from the British Library.

The views expressed in this book are those of the authors, but not necessarily of the publisher.

Table of Contents

Detailed Table of Contents

Chapter 1

  Rahul Roy, KIIT University, India
  Satchidananda Dehuri, Fakir Mohan University, India
  Sung Bae Cho, Yonsei University, South Korea

The Combinatorial problems are real world decision making problem with discrete and disjunctive choices. When these decision making problems involve more than one conflicting objective and constraint, it turns the polynomial time problem into NP-hard. Thus, the straight forward approaches to solve multi-objective problems would not give an optimal solution. In such case evolutionary based meta-heuristic approaches are found suitable. In this paper, a novel particle swarm optimization based meta-heuristic algorithm is presented to solve multi-objective combinatorial optimization problems. Here a mapping method is considered to convert the binary and discrete values (solution encoded as particles) to a continuous domain and update it using the velocity and position update equation of particle swarm optimization to find new set of solutions in continuous domain and demap it to discrete values. The performance of the algorithm is compared with other evolutionary strategy like SPEA and NSGA-II on pseudo-Boolean discrete problems and multi-objective 0/1 knapsack problem. The experimental results confirmed the better performance of combinatorial particle swarm optimization algorithm.

Chapter 2

  Masoud Yaghini, Iran University of Science and Technology, Iran
  Mohammad Karimi, Iran University of Science and Technology, Iran
  Mohadeseh Rahbar, Iran University of Science and Technology, Iran
  Rahim Akhavan, Kermanshah University of Technology, Kermanshah, Iran

The fixed-cost Capacitated Multicommodity Network Design (CMND) problem is a well known NP-hard problem. This paper presents a matheuristic algorithm combining Simulated Annealing (SA) metaheuristic and Simplex method for CMND problem. In the proposed algorithm, a binary array is considered as solution representation and the SA algorithm manages open and closed arcs. Several strategies for opening and closing arcs are proposed and evaluated. In this matheuristic approach, for a given design vector, CMND becomes a Capacitated Multicommodity minimum Cost Flow (CMCF) problem. The exact evaluation of the CMCF problem is performed using the Simplex method. The parameter tuning for the proposed algorithm is done by means of design of experiments approach. The performance of the proposed algorithm is evaluated by solving different benchmark instances. The results of the proposed algorithm show that it is able to obtain better solutions in comparison with previous methods in the literature.

Chapter 3

Hernan X. Cordova, Free University of Brussels, Belgium
Leo Van Biesen, Free University of Brussels, Belgium

One of the major sources of performance degradation of current Digital Subscriber Line systems is the electromagnetic coupling among different twisted pairs within the same cable bundle (crosstalk). Several algorithms for Dynamic Spectrum management have been proposed to counteract the crosstalk effect but their complexity remains a challenge in practice. Optimal Spectrum Balancing (OSB) is a centralized algorithm that optimally allocates the available transmit power over the tones making use of a Dual decomposition approach where Lagrange multipliers are used to enforce the constraints and decouple the problem over the tones. However, the overall complexity of this algorithm remains a challenge for practical DSL environments. The authors propose a low-complex algorithm based on a combination of simulated annealing and non-linear simplex to find local (almost global) optimum spectra for multiuser DSL systems, whilst significantly reducing the prohibitive complexity of traditional OSB. The algorithm assumes a Spectrum Management Center (at the cabinet side) but it neither relies on own end-user modem calculations nor on messaging-passing for achieving its performance objective. The approach allows furthering reducing the number of function evaluations achieving further reduction on the convergence time (up to ~27% gain) at reasonable payoff (weighted data rate sum).

Chapter 4

Yannis Marinakis, Technical University of Crete, Greece
Magdalene Marinaki, Technical University of Crete, Greece
Nikolaos Matsatsinis, Technical University of Crete, Greece
Constantin Zopounidis, Technical University of Crete, Greece

Nature-inspired methods are used in various fields for solving a number of problems. This study uses a nature-inspired method, artificial bee colony optimization that is based on the foraging behaviour of bees, for a financial classification problem. Financial decisions are often based on classification models, which are used to assign a set of observations into predefined groups. One important step toward the development of accurate financial classification models involves the selection of the appropriate independent variables (features) that are relevant to the problem. The proposed method uses a discrete version of the artificial bee colony algorithm for the feature selection step while nearest neighbour based classifiers are used for the classification step. The performance of the method is tested using various benchmark datasets from UCI Machine Learning Repository and in a financial classification task involving credit risk assessment. Its results are compared with the results of other nature-inspired methods.

Chapter 5

S. N. Omkar, Indian Institute of Science, India
G. Narayana Naik, Indian Institute of Science, India
Kiran Patil, Indian Institute of Science, India
Mrunmaya Mudigere, Indian Institute of Science, India

In this paper, a generic methodology based on swarm algorithms using Artificial Bee Colony (ABC) algorithm is proposed for combined cost and weight optimization of laminated composite structures. Two approaches, namely Vector Evaluated Design Optimization (VEDO) and Objective Switching Design

Optimization (OSDO), have been used for solving constrained multi-objective optimization problems. The ply orientations, number of layers, and thickness of each lamina are chosen as the primary optimization variables. Classical lamination theory is used to obtain the global and local stresses for a plate subjected to transverse loading configurations, such as line load and hydrostatic load. Strength of the composite plate is validated using different failure criteria—Failure Mechanism based failure criterion, Maximum stress failure criterion, Tsai-Hill Failure criterion and the Tsai-Wu failure criterion. The design optimization is carried for both variable stacking sequences as well as standard stacking schemes and a comparative study of the different design configurations evolved is presented. Performance of Artificial Bee Colony (ABC) is compared with Genetic Algorithm (GA) and Particle Swarm Optimization (PSO) for both VEDO and OSDO approaches. The results show ABC yielding a better optimal design than PSO and GA.

### Chapter 6

An integrated ant colony optimization algorithm (IACS-HFS) is proposed for a multistage hybrid flow-shop scheduling problem. The objective of scheduling is the minimization of the makespan. To solve this NP-hard problem, the IACS-HFS considers the assignment and sequencing sub-problems simultaneously in the construction procedures. The performance of the algorithm is evaluated by numerical experiments on benchmark problems taken from the literature. The results show that the proposed ant colony optimization algorithm gives promising and good results and outperforms some current approaches in the quality of schedules.

This paper presents a Discrete Particle Swarm Optimization (DPSO) approach for the Multi-Level Lot-Sizing Problem (MLLP), which is an uncapacitated lot sizing problem dedicated to materials requirements planning (MRP) systems. The proposed DPSO approach is based on cost modification and uses PSO in its original form with continuous velocity equations. Each particle of the swarm is represented by a matrix of logistic costs. A sequential approach heuristic, using Wagner-Whitin algorithm, is used to determine the associated production planning. The authors demonstrate that any solution of the MLLP can be reached by particles. The sequential heuristic is a subjective function from the particles space to the set of the production plans, which meet the customer's demand. The authors test the DPSO Scheme on benchmarks found in literature, more specifically the unique DPSO that has been developed to solve the MLLP.

### Chapter 8

The metaheuristic approach has become an important tool for the optimization of design in engineering. In that way, its application to the development of the plasmonic based biosensor is apparent. Plasmonics represents a rapidly expanding interdisciplinary field with numerous transducers for physical, biological and medicine applications. Specific problems are related to this domain. The plasmonic structures design depends on a large number of parameters. Second, the way of their fabrication is complex and industrial aspects are in their infancy. In this study, the authors propose a non-uniform adapted Particle Swarm Optimization (PSO) for rapid resolution of plasmonic problem. The method is tested and compared to the standard PSO, the meta-PSO (Veenhuis, 2006) and the ANUHEM (Barchiesi, 2009).These approaches are applied to the specific problem of the optimization of Surface Plasmon Resonance (SPR) Biosensors design. Results show great efficiency of the introduced method.

Chapter 9

Jalel Euchi, University of Sfax, Tunisia
Habib Chabchoub, University of Sfax, Tunisia
Adnan Yassine, University of Le Havre, France

Mismanagement of routing and deliveries between sites of the same company or toward external sites leads to consequences in the cost of transport. When shipping alternatives exist, the selection of the appropriate shipping alternative (mode) for each shipment may result in significant cost savings. In this paper, the authors examine a class of vehicle routing in which a fixed internal fleet is available at the warehouse in the presence of an external transporter. The authors describe hybrid Iterated Density Estimation Evolutionary Algorithm with 2-opt local search to determine the specific assignment of each tour to a private vehicle (internal fleet) or an outside carrier (external fleet). Experimental results show that this method is effective, allowing the discovery of new best solutions for well-known benchmarks.

Chapter 10

Gladys Maquera, Universidad Peruana Unión, Peru
Manuel Laguna, University of Colorado, USA
Dan Abensur Gandelman, Universidade Federal do Rio de Janeiro, Brasil
Annibal Parracho Sant'Anna, Universidade Federal Fluminense, Brasil

Though its origins can be traced back to 1977, the development and application of the metaheuristic Scatter Search (SS) has stayed dormant for 20 years. However, in the last 10 years, research interest has positioned SS as one of the recognizable methodologies within the umbrella of evolutionary search. This paper presents an application of SS to the problem of routing vehicles that are required both to deliver and pickup goods (VRPSDP). This specialized version of the vehicle routing problem is particularly relevant to organizations that are concerned with sustainable and environmentally-friendly business practices. In this work, the efficiency of SS is evaluated when applied to this problem. Computational results of the application to instances in the literature are presented.

Chapter 11

Nashat Mansour, Lebanese American University, Lebanon
Ghia Sleiman-Haidar, Lebanese American University, Lebanon

University exam timetabling refers to scheduling exams into predefined days, time periods and rooms, given a set of constraints. Exam timetabling is a computationally intractable optimization problem, which requires heuristic techniques for producing adequate solutions within reasonable execution time.

For large numbers of exams and students, sequential algorithms are likely to be time consuming. This paper presents parallel scatter search meta-heuristic algorithms for producing good sub-optimal exam timetables in a reasonable time. Scatter search is a population-based approach that generates solutions over a number of iterations and aims to combine diversification and search intensification. The authors propose parallel scatter search algorithms that are based on distributing the population of candidate solutions over a number of processors in a PC cluster environment. The main components of scatter search are computed in parallel and efficient communication techniques are employed. Empirical results show that the proposed parallel scatter search algorithms yield good speed-up. Also, they show that parallel scatter search algorithms improve solution quality because they explore larger parts of the search space within reasonable time, in contrast with the sequential algorithm.

Chapter 12

Fred Glover, OptTek Systems, Inc., USA

Leon Lasdon, The University of Texas at Austin, USA

John Plummer, Texas State University, USA

Abraham Duarte, Universidad Rey Juan Carlos, Spain

Rafael Marti, Universidad de Valencia, Spain

Manuel Laguna, University of Colorado, USA

Cesar Rego, University of Mississippi, USA

Motivated by the successful use of a pseudo-cut strategy within the setting of constrained nonlinear and nonconvex optimization in Lasdon et al. (2010), we propose a framework for general pseudo-cut strategies in global optimization that provides a broader and more comprehensive range of methods. The fundamental idea is to introduce linear cutting planes that provide temporary, possibly invalid, restrictions on the space of feasible solutions, as proposed in the setting of the tabu search metaheuristic in Glover (1989), in order to guide a solution process toward a global optimum, where the cutting planes can be discarded and replaced by others as the process continues. These strategies can be used separately or in combination, and can also be used to supplement other approaches to nonlinear global optimization. Our strategies also provide mechanisms for generating trial solutions that can be used with or without the temporary enforcement of the pseudo-cuts.

Chapter 13

Walid Moudani, Lebanese University, Lebanon

Félix Mora-Camino, Ecole Nationale de l'Aviation Civile, France

This paper presents Bus Driver Allocation Problem (BDAP) which deals with the assignment of the drivers to the scheduled duties so that operations costs are minimized while its solution meets hard constraints resulting from the safety regulations of public transportation authorities as well as from the internal company's agreements. Another concern in this study is the overall satisfaction of the drivers since it has important consequences on the quality and economic return of the company operations. This study proposes a non-dominated bi-criteria allocation approach of bus drivers to trips while minimizing the operations cost and maximizing the level of satisfaction of drivers. This paper proposes a new mathematical formulation to model the allocation problem. Its complexity has lead to solving the BDAP by using techniques such as: Genetic algorithms in order to assign the drivers with minimal operations cost and Fuzzy Logic and Rough Set techniques in order to fuzzyfied some parameters required for the satisfaction module followed by applying the Rough Set technique to evaluate the degree of satisfaction for the drivers. The application of the proposed approach to a medium size Transport Company Bus Driver Allocation Problem is evaluated.

Madhabananda Das, KIIT University, India
Rahul Roy, KIIT University, India
Satchidananda Dehuri, Fakir Mohan University, India
Sung-Bae Cho, Yonsei University, Korea

Associative classification rule mining (ACRM) methods operate by association rule mining (ARM) to
obtain classification rules from a previously classified data. In ACRM, classifiers are designed through
two phases: rule extraction and rule selection. In this paper, the ACRM problem is treated as a multi-
objective problem rather than a single objective one. As the problem is a discrete combinatorial optimization problem, it was necessary to develop a binary multi-objective particle swarm optimization
(BMOPSO) to optimize the measure like coverage and confidence of association rule mining (ARM) to
extract classification rules in rule extraction phase. In rule selection phase, a small number of rules are
targeted from the extracted rules by BMOPSO to design an accurate and compact classifier which can
maximize the accuracy of the rule sets and minimize their complexity simultaneously. Experiments are
conducted on some of the University of California, Irvine (UCI) repository datasets. The comparative
result of the proposed method with other standard classifiers confirms that the new proposed approach
can be a suitable method for classification.

R. Rathipriya, Periyar University, India
K. Thangavel, Periyar University, India
J. Bagyamani, Government Arts College, India

Data mining extracts hidden information from a database that the user did not know existed. Biclustering is one of the data mining technique which helps marketing user to target marketing campaigns
more accurately and to align campaigns more closely with the needs, wants, and attitudes of customers
and prospects. The biclustering results can be tuned to find users' browsing patterns relevant to current business problems. This paper presents a new application of biclustering to web usage data using
a combination of heuristics and meta-heuristics algorithms. Two-way K-means clustering is used to
generate the seeds from preprocessed web usage data, Greedy Heuristic is used iteratively to refine a set
of seeds, which is fast but often yield local optimal solutions. In this paper, Genetic Algorithm is used
as a global optimizer that can be coupled with greedy method to identify the global optimal target user
groups based on their coherent browsing pattern. The performance of the proposed work is evaluated by
conducting experiment on the msnbc, a clickstream dataset from UCI repository. Results show that the
proposed work performs well in extracting optimal target users groups from the web usage data which
can be used for focalized marketing campaigns.

Walid Moudani, Lebanese University, Lebanon
Félix Mora-Camino, Ecole Nationale de l'Aviation Civile (ENAC–DGAC), France

The Crew Reserve Assignment Problem (CRAP) considers the assignment of the crew members to a
set of reserve activities covering all the scheduled flights in order to ensure a continuous plan so that
operations costs are minimized while its solution must meet hard constraints resulting from the safety

regulations of Civil Aviation as well as from the airlines internal agreements. The problem considered in this study is of highest interest for airlines and may have important consequences on the service quality and on the economic return of the operations. A new mathematical formulation for the CRAP is proposed which takes into account the regulations and the internal agreements. While current solutions make use of Artificial Intelligence techniques run on main frame computers, a low cost approach is proposed to provide on-line efficient solutions to face perturbed operating conditions. The proposed solution method uses a dynamic programming approach for the duties scheduling problem and when applied to the case of a medium airline while providing efficient solutions, shows good potential acceptability by the operations staff. This optimization scheme can then be considered as the core of an on-line Decision Support System for crew reserve assignment operations management.

Chapter 17

Masoud Yaghini, Iran University of Science and Technology, Iran

Mohsen Momeni, Iran University of Science and Technology, Iran

Mohammadreza Sarmadi, Iran University of Science and Technology, Iran

A Hamiltonian path is a path in an undirected graph, which visits each node exactly once and returns to the starting node. Finding such paths in graphs is the Hamiltonian path problem, which is NP-complete. In this paper, for the first time, a comparative study on metaheuristic algorithms for finding the shortest Hamiltonian path for 1071 Iranian cities is conducted. These are the main cities of Iran based on social-economic characteristics. For solving this problem, four hybrid efficient and effective metaheuristics, consisting of simulated annealing, ant colony optimization, genetic algorithm, and tabu search algorithms, are combined with the local search methods. The algorithms' parameters are tuned by sequential design of experiments (DOE) approach, and the most appropriate values for the parameters are adjusted. To evaluate the proposed algorithms, the standard problems with different sizes are used. The performance of the proposed algorithms is analyzed by the quality of solution and CPU time measures. The results are compared based on efficiency and effectiveness of the algorithms.

Preface

MOMO: Multi-Objective Metaheuristic Optimization

INTRODUCTION

Optimization techniques have played an important role in engineering and business domains where many complex problems we face in real-world are mathematically modeled as a parametric-form objective function and the optimal parameter setting for obtaining the best objective value is sought for. Recently, metaheuristic computing has shown significant competence against classic optimization methods. The first two volumes of the *International Journal of Applied Metaheuristic Computing* (IJAMC) have disclosed many theoretic breakthroughs and successful applications in metaheuristic computing for optimization with one single objective. We particularly solicit in future volumes of IJAMC more research articles addressing Multi-Objective Optimization Problems (MOOP) using metaheuristic computation.

Multi-Objective Metaheuristic Optimization (MOMO) is increasingly important due to two observations. First, many real-world problems involve multiple objectives that are conflicting with one another. Typical examples are cost-benefit analysis, engine performance and fuel consumption, weight of a mechanical part, and its strength, to name just a few. Secondly, literature findings have shown that MOMO is more beneficial than classic multi-objective optimization approaches. MOMO is able to find a set of non-dominated solutions in a single run, while classic multi-objective optimization approaches, such as weighted-sum, ε-constraint programming, and compromise programming, need repetitive runs with various parameter settings.

Fred Glover (1986) first coined the term *metaheuristic*, which employs a master heuristic to guide the search course of a low-level heuristic to look beyond the local optimality. The master heuristic ranges from phylogenetic evolution, sociocognition, gestalt psychology, social insects foraging, to strategic level problem-solving rules. From a broader perspective, metaheuristic approaches can be clasified as evolutionary algorithms (EA) and adaptive memory programming (AMP) techniques. The EA focuses on intelligent mechanisms inspired by natural metaphors, while the AMP relies on strategic level rules and memory manipulation. MOMO based on EA or AMP has received great attention from the MOOP community, but there are no or only few attempts that were devoted to MOMO using a hybrid framework of EA and AMP.

The previous volume of this series book has disclosed that researchers and practitioners intend to identify the primitive components contained in metaheuristic computing and to develop the so-called hybrid metaheuristics towards more effective metaheuristic computing (Yin, 2012). We particulary emphasize on the innovation named *Cyber-heuristic*, which combines useful notions found in EA and AMP and creates a more effective form of metaheuristic computing. This preface introduces two MOMO methods that illustrate the strengths of Cyber-heuristic.

The remainder of this preface is organized as follows. Section 2 reviews principal classic multi-objective optimization approaches. In Section 3 we introduce important features of MOEAs. Section 4 presents two notions for establishing a MOMO method that synergizes the strengths of EA and AMP. Finally, conclusions are made in Section 5.

CLASSIC MULTI-OBJECTIVE OPTIMIZATION

A widely accepted notion for multi-objective optimization is to search for the Pareto-optimal solutions which are not dominated by any other solution. A solution x *dominates* another solution y if x is strictly better than y in at least one objective and x is no worse than y in the others. Formally, given k minimization objective functions, $f_i(x)$, $i = 1, 2, ..., k$, solution x dominates solution y, denoted as $x \succ y$, if $f_i(x) \leq f_i(y), \forall i = 1, 2, ..., k$ and $f_j(x) < f_j(y), \exists j \in \{1, 2, ..., k\}$.

Figure 1 illustrates a simple case for a bi-objective optimization problem. Solution x dominates solution y because $f_1(x) < f_1(y)$ and $f_2(x) < f_2(y)$ (see Figure 1(a)). Therefore, those solutions whose plots locate in the dark region are dominated by solution x. The plots of objective values for all Pareto-optimal (non-dominated) solutions form a Pareto front in the objective space as shown in Figure 1(b). We aim to find representative members of Pareto-optimal solutions whose plots are desirable to be equally distanced on the Pareto front.

Decision makers are clearly comfortable in seeking Pareto-optimal solutions because if the final solution is not Pareto-optimal it can be improved in at least one objective without deteriorating the solution quality in other objectives. However, it is sometimes difficult to find the true Pareto-optimal solutions due to the high complexity of the problem nature, and an approximate Pareto front is instead sought for. The quality of this front is measured in two aspects: (1) The convergence metric indicates how closely the approximate Pareto front is converging to the true Pareto front, and (2) the diversity metric is in favour of the approximate Pareto front whose plots are most evenly spread. The classic multi-objective optimization approaches include weighted-sum scheme (Chang et al., 2008; Ismail et al., 2011),

Figure 1. Dominance relationship and Pareto front in a bi-objecitve optimization case

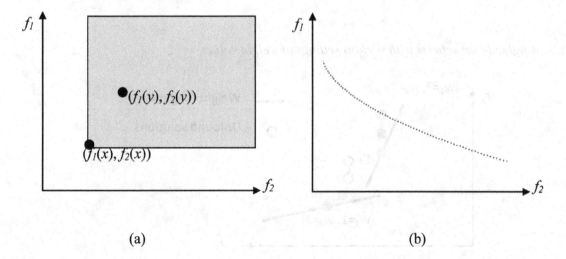

(a)   (b)

goal programming (Schniederjans, 1995) and ε-constraint programming (Yokoyama, 1996). However, all of these approaches decompose the given multi-objective problem into a number of single-objective sub-problems, and multiple executions of the program are needed to obtain the final approximate Pareto front. In the following subsections, the major classic multi-objective optimization approaches are presented.

Weighted-Sum Scheme

The weighted-sum scheme transforms an MOOP into an SOOP (single-objective optimization problem). A linear aggregating function $\sum_{i=1}^{m} w_i f_i(x)$, where w_i is the weight for the ith objective, is employed to combine the m objectives as a weighted sum. Many executions with various settings of the weight values are needed to identify multiple non-dominated solutions on the front. Figure 2 shows an example. A given setting of weight values defines a tangent to the front, and the plot at the intersection is the optimal weighted sum corresponding to this weight setting.

The weighted-sum method, however, has the following drawbacks. (1) Multiple settings with evenly distributed weight values are needed to perform repetitive executions to locate representative plots on the front. (2) Some non-dominated solutions may have never been found using any setting of weights if the Pareto front is concave (as shown in Figure 2). (3) The weight value for each objective is difficult to determine, even the decision makers cannot precisely state and quantify the importance degree of each objective.

ε-Constraint Programming

The ε-constraint programming approach retains one objective and transforms the remaining objectives to constraints. Formally, the MOOP is converted to the following form:

$$\min_{X \in S} f_i(X)$$

subject to

$$f_j(X) < \varepsilon_j \qquad j = 1, 2, ..., m; \quad j \neq i$$

Figure 2. Weighted-sum scheme with various settings of weight values

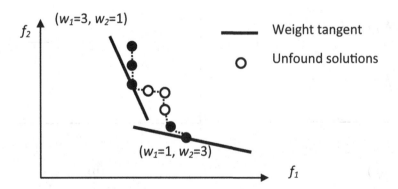

*Figure 3. Illustration of **ε**-constraint programming*

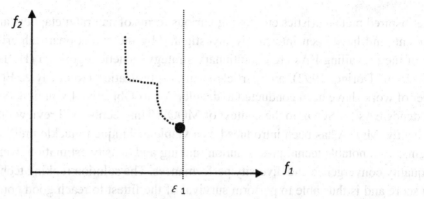

It can be seen in Figure 3 that **ε**-constraint programming can identify a point on the Pareto front by minimizing objective *f2* with respect to the constraint that $f1 \le \varepsilon1$. With multiple executions of various constraint settings which gradually decrease the $\varepsilon1$ value, multiple points on the Pareto front can be located.

Compromise Programming

The compromise programming technique uses the ideal point y as a reference point to estimate how closely the objective plot approaches to the Pareto front. The ideal point is a pseudo plot whose individual objective value is obtained by deriving the optimal value for this single objective. The compromise programming intends to find the point that has the minimum value for the weighted-sum distance, $\sum_{i=1}^{m} w_i \left| f_i(x) - f_i(y) \right|$, or the Chebyshef distance, $\max_i \left(w_i \left| f_i(x) - f_i(y) \right| \right)$. As shown in Figure 4, the objective plot for solution x is optimal with respect to the weighted-sum distance with equal weights.

Figure 4. Illustration of compromise programming

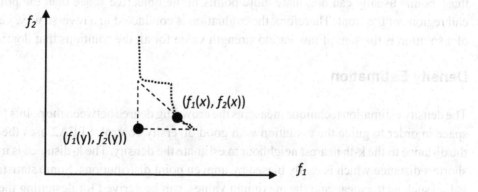

MULTI-OBJECTIVE EVOLUTIONARY ALGORITHMS

Nature-inspired metaheuristics employing various forms of natural metaphors are easy to describe and implement, and have been intensively investigated by the evolutionary algorithm (EA) community. Some of the prevailing EAs are evolutionary strategy, genetic algorithm (Holland, 1975), ant colony optimization (Dorigo, 1992), and particle swarm optimization (Kennedy & Eberhart, 1995). A great number of works have been conducted to develop Multi-Objective Evolutionary Algorithms (MOEAs) that extends EAs for SOOP to the context of MOOP. This section will review the features of MOEAs.

Recently, MOEA has been introduced as a viable technique to tackle multi-objective optimization problems. Two notable techniques, solution ranking and density estimation, were introduced to obtain high-quality convergence and diversity performances. The solution ranking technique gives each solution a score and is thus able to perform survival of the fittest to reach good convergence based on the rank of competing solutions. The density estimation technique measures the degree of crowding between the plot points in the objective space in order to guide the evolution with good diversity control. Another interesting notion for diversity control is via objective decomposition, which decomposes a multi-objective optimization problem into a number of sub-problems and optimizes them simultaneously. Each sub-problem is defined by a weighted-sum of objectives and the weight vector is well separated from the weight vectors for other sub-problems. Thus decomposition-based MOEA can generate a set of evenly distributed points on the front.

Solution Ranking

In order to maneuver the evolution, the quality of the solutions should be quantatively measured and differentiated. One of the useful criteria is the dominance relationship among solutions. Here, we present the non-dominated sorting and Pareto strength broadly used in the literature.

The non-dominated sorting technique was originally proposed in NSGA (Deb et al., 2002). It gives the highest score to the solutions non-dominated by any other solutions in the population. The solutions with the highest score are then removed from the population, and the strategy proceeds to give the second-highest score to the non-dominated solutions in the population. The process is repeated until every solution in the population has been assigned a score.

The notion of Pareto strength was presented in SPEA (Zitzler et al., 2001). Each found solution is given a raw Pareto strength value defined as the ratio of the number of the remaining solutions that are dominatd by this solution. However, this raw value is preferring points on the front central region because these points usually can dominate more points in the objective space than the points located near the end regions of the front. Therefore, the evaluation is conducted in a reverse way, i.e., the quality fitness of a solution is the sum of raw Pareto strength value for all the solutions that dominate this solution.

Density Estimation

The density estimation technique measures the crowding degree between the points found in the objective space in order to guide the evolution with good diversity control. SPEA2 uses the k-distance which is the distance to the k-th nearest neighbour to estimate the density. The k-distance is more reliable than the shortest distance which is easily biased by uneven point distributions. Some statistics on the k-distance value, such as the mean and the maximum values, can be derived for designing the diversity strategy.

As defined in NSGA-II, the crowding distance for a point is the average distance of two points on either side of this point along each of the objectives. The crowding distance is compromise between the k-distance and the shortest distance and does not require any density parameters. When the competing solutions have the same ranking based on dominance relationship, the solution with larger crowding distance is prevailing.

The grid-based technique has been used in the Pareto Archived Evolution Strategy (PAES) (Knowles & Corne, 2000), Fard et al. (2011) and the Multi-Objective Particle Swarm Optimization (MOPSO) (Coello Coello et al., 2004). The objective space is divided into regions called grids, and the number of points located in each grid cell is used as an estimate for the density. When there is a need for replacing a solution in the grid archive, a solution in the densest grid is selected at random for removal.

Decomposition-Based MOEA

Another interesting notion for diversity control is via objective decomposition. The MOEA/D algorithm (Zhang & Li, 2007) decomposes a multiobjective optimization problem into a number of sub-problems and optimizes them simultaneously. As shown in Figure 5, each sub-problem is defined by a weight vector λ and the weighted-sum of objectives is considered as the global optimization goal of this sub-problem. The weight vectors are generated with uniformly distributed slopes and the weight vector of each sub-problem is well separated from the weight vectors for the other sub-problems.

Figure 6 shows the flow chart of the MOEA/D algorithm. The External Population (EP) is a memory space to store the non-dominated solutions that are found during the whole execution. At the initialization step, a population of solutions is randomly generated and each solution is designated to a weight vector that defines a particular sub-problem. The weight vectors are systematically generated such that their slop values are uniformly distributed. Each solution is seeking to optimize a sub-problem with the designated weight vector. For each solution, two neighbors of this solution are chosen randomly and

Figure 5. Illustration of MOEA/D decomposition

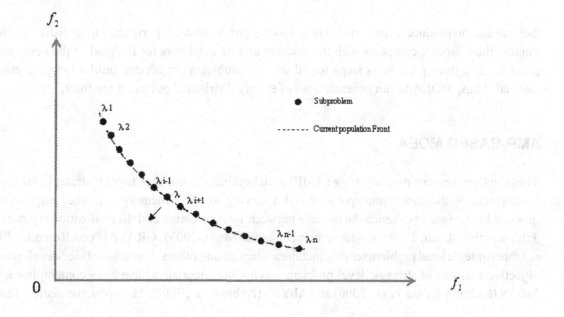

Figure 6. Flow chart of MOEA/D

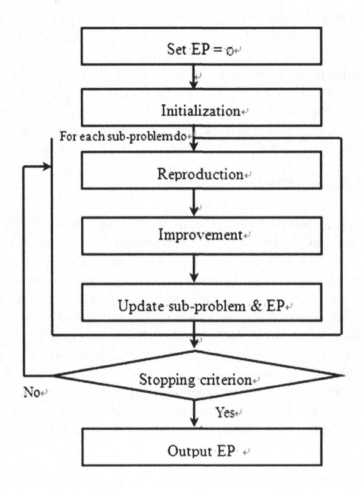

they are used to produce an offspring. Then, a local searh heuristic is performed to improve the offspring. Finally, the offspring competes with the solution and its neighbors for the goal of the designated sub-problem. The three processing steps for all the sub-problems are iterated until a stopping crietrion is reached. Thus, MOEA/D can generate a set of evenly distributed points on the front.

AMP-BASED MOEA

The adaptive memory programming (AMP) metaheuristic approaches takes a strategic level problem solving rules with memory manipulation. With varying forms of memory structure, purposeful strategies can be devised to enhance the balance between intensification and diversification types of search. Tabu search (Glover, 1989), scatter search (Laguna&Marti, 2003), GRASP (Feo&Resende, 1995) are notable strategic level problem solving metaheuristics among others. Few attempts for developing multi-objective versions of strategic level problem solving metaheuristics have been contemplated, such as MOTS (Kulturel-Konak et al., 2006) and AbYSS (Nebro et al., 2008). However, the distinct features of adaptive memory and responsive strategy may be overlooked and there is a promising research area for seeking effective ways of combining AMP with EA to create a powerful MOMO algorithm.

As previously noted, the advantages of AMP have not been fully explored in assistance of MOEA. This section proposes two notions in this vein and presents preliminary results which are very promising. The first notion marries AMP with multi-objective particle swarm optimization (MOPSO), and the second notion uses AMP to enhance the performance of MOEA/D.

AMP-Based MOPSO

We previously introduced the Cyber Swarm Algorithm (Yin et al., 2010) which enhances PSO by emphasizing to generate new solutions using high-level problem-solving strategies. These strategies may involve complex neighborhood concepts, memory structure, and adaptive search prinicples, mainly drawn from the field of Tabu Search. The adjective "Cyber" indicates the connection between the nature-inspired PSO and the high-level problem-solving strategies provided by the Scatter Search/Path Relinking (SS/PR) suite (Glover, 1998). Omran (2011) edited a special issue on Scatter Search and Path Relinking methods which addresses the contributions of SS/PR template for swarm intelligence. Hence, we further propose an extension of Cyber Swarm Algorithm (CSA) for solving multi-objective optimization problems.

The proposed method, named MOCSA, adds new features to CSA for generating non-dominated solutions with good convergence and diversity performances. MOCSA consists of four memory components and responsive strategies. The swarm memory component is the working memory where the population of swarm particles evolves to improve their solution quality based on guided moving by reference to strategically selected solution guides. The individual memory reserves a separate space for each particle and stores the pseudo non-dominated solutions by reference to all the solutions found by the designated particle only. Note that the pseudo non-dominated solutions could be dominated by the solutions found by other particles, but we propose to store the pseudo non-dominated solutions because our preliminary results show that these solutions contain important diversity information along the individual search trajectory and they assist in finding influential solution guides that are otherwise overlooked by just using global non-dominated solutions. The global memory tallies the non-dominated solutions that are not dominated by any other solutions found by all the particles. The solutions stored in the global memory will be output as the approximate Pareto-optimal solutions as the program terminates. Finally, the reference memory selects the most influential solutions based on convergence and diversity measures. Moreover, to arouse the power of MOCSA when the search loses its efficacy, two responsive strategies are performed upon the detection of critical events which disclose the stagnation of the search power. It should be noted that the manipulation of memory in MOCSA is very different from that used in the original CSA. MOCSA determines the ranking of solutions by the dominance power and diversity relationship in the multi-objective space while the CSA considers the single-objective fitness and the diversity in the solution space. The selection of solution guides is also different in the two versions. In CSA, the best solution leader can be uniquely identified due to the single-objective context. In MOCSA, however, there exist multiple non-dominated solutions in each level of memory and alternative strategies may be applied.

The pseudo code of the MOCSA is summarized in Figure 7. In the initialization (Step 1) the initial values for particle positions, velocities, and experience memory are given. In the evolutionary iterations (Step 2), the guided moving, memory update, and responsive strategies are performed. In Step 2.1, each

Figure 7. Pseudo codes for the Multi-Objective CSA (MOCSA)

1 Initialization
  1.1 Randomly generate U particle solutions, $P_i = \{p_{ij}\}$, $0 \le i < U$, $0 \le j < d$
  1.2 Randomly generate U velocity vectors, $V_i = \{v_{ij}\}$, $1 \le i \le U$, $0 \le j < d$
  1.3 Evaluate multiple fitness values for each particle. Update experience (individual, global, and reference) memory
2 Repeat until a stopping criterion is met
  2.1 For each particle Pi, i =1, ..., U, Do
  2.1.1 **Guided moving with selected solution guides:**

$$v_{ij}^m \leftarrow K\left(v_{ij} + (\phi_1 + \phi_2 + \phi_3)\left(\frac{\omega_1\phi_1 pbest_{ij} + \omega_2\phi_2 gbest_{ij} + \omega_3\phi_3\,\mathrm{Re}\,fSol[m]_j}{\omega_1\phi_1 + \omega_2\phi_2 + \omega_3\phi_3} - p_{ij} \right) \right)$$

  Pi←constrained-dominance $\left\{ P_i + v_i^m \mid m \in [1, RS] \right\}$

  2.1.2 Update experience memory if necessary
  2.2 **Convergence PR strategy:** If the global memory has not been updated for t1 iterations, restart every particle by exploiting the region between the particle's two closest gbest (say $gbest_j$ and $gbest_k$) by
          $P_i \leftarrow$ Convergence_PR($gbest_j$, $gbest_k$), i =1, ..., U
  Diversity PR strategy: Else if a particular particle's individual memory has not been updated for t2 iterations, restart this particle by
          $P_i \leftarrow$ Diversity_PR($pbest_i$, P_i)
  2.3 Update experience memory if necessary
3 Output feasible non-dominated solutions in the global memory

particle moves with the guidance information provided by three carefully selected solution guides ($pbest_i$, gbest and RefSol[m], m = 1, ..., RS). The guided moving is repeatedly conducted and each instance uses a different member from the reference memory. All the constrained non-dominated solutions identified in the multiple move trials are used for experience updating of all types of memory. The memory updating is conducted in a hierarchical fashion. The lowest level is the swarm memory that records the current position for all the particles. The new position of each particle is used for updating of the personal best experience in the individual memory for this particle. The personal best experiences of all particles compete to enter the global memory. The highest level is the reference memory that selects its members from individual memory and global memory in terms of the quality and diversity performances. In Step 2.2, responsive PR strategies are employed upon critical events for improving the convergence and diversity of the currently produced front. When the stopping criterion of MOCSA is met, all the feasible non-dominated solutions in the global memory are used to produce the approximate Pareto front.

The advantages of the MOCSA algorithm compared with previous MOEA algorithms for improving convergence and diversity performance are as follows. (1) MOCSA adopts multi-level memory to facilitate solution ranking and density estimation mechanisms. The experience updating is conducted in the order of swarm, individual, global and reference memory. The best experience obtained in a low level memory is used for updating of the next level memory. Thus, the multi-level memory realizes the solution ranking in guiding the evolution. Moreover, as each type of memory performs a different updating strategy which will preserve unique features of its maintained solutions, sustaining good diversity

Figure 8. Illustration of parental updating and within-section updating by using the weighted-sum scheme

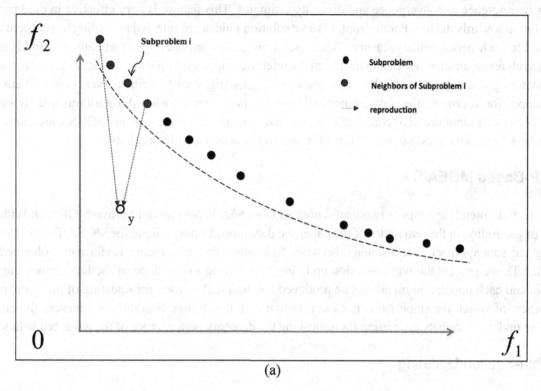

(a)

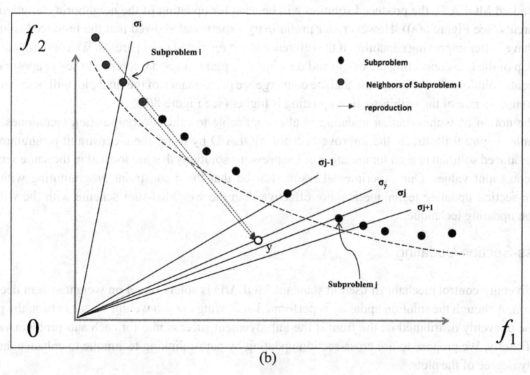

(b)

in the whole population. (2) Reference set is an adaptive memory which stores the most influential solutions by reference to convergence and diversity estimates. This feature is very effective in guiding the revolution towards the true Pareto front. (3) The solution guides are selected according to competitions related to both performance measures. These solution guides are used systematically in combination with a reference solution selected in turn from the reference memory, imposing a dynamic social network for fostering a new particle. (4) Two responsive strategies triggered by critical events are particularly developed for improving the convergence and diversity performance. MOEA algorithms usually suffer the barrier of premature convergence. The responsive strategies employed in MOCSA are useful in detecting these critical events and redirect the search towards uncharted regions.

AMP-Based MOEA/D

This section intends to propose two useful notions from SS/PR domain to improve MOEA/D. Without loss of generality, in the remainder of the paper the decomposition technique for MOEA/D refers to the weighted-sum approach unless stated otherwise. To improve the convergence performance obtained by MOEA/D, we present the within-section updating by referring to the slope of the line connecting the origin and each population member. The produced solution will be used for updating of the population members of which the slope value is closest to that of this solution. Secondly, we present the cross-section update by density estimation for improving the diversity performance of the produced solutions.

Within-Section Updating

In standard MOEA/D, the produced solution will be used for updating of the neighboring solutions of its parents (see Figure 8(a)). However, our preliminary experiment showed that the produced solution may have better improving capability if the reference is not restricted to its parents. We contemplate that the slop of the line connecting the origin and the objective plot is a good reference to select appropriate candidate solutions for updating, and that the convergence performance of the final plot will be enhanced if the success rate of the within-section updating is higher (see Figure 8(b)).

The notion of within-section updating is also applicable to other decomposition techniques. For example, Figure 9 illustrates the improvement for MOEA/D by using the ε-constraint programming. The produced solution is used for updating of the previous solutions that are located in the same section of ε-constraint values. Our experimental result showed that the ε-constraint programming with the within-section updating technique is more effective than the weighted-sum scheme with the within-section updating technique.

Cross-Section Updating

The diversity control mechanism used in standard MOEA/D is solely based on weighted-sum decomposition. Although the solution updating is performed according to each weight vector in turn, the plots will be unevenly distributed on the front if the improvement success rate for each sub-problem varies significantly. We propose to use cross-section updating by path-relinking technique to enhance the diversity degree of the plots.

The cross-section updating proceeds as follows. The objective space is divided to a number of equal-size sections. We estimate the section density by the number of plots locating in this section zone. The

Figure 9. Illustration of within-section updating by using the ε-constraint programming

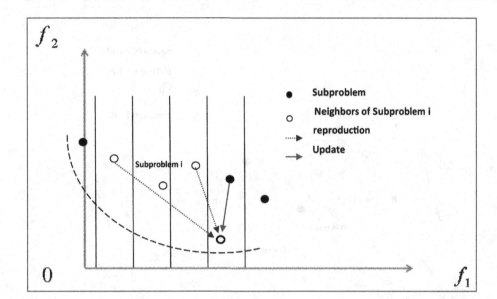

secion zone *i* is bounded by the two radii with slop values σ_i and σ_{i+1}. The section with the lowest density is targeted for cross-section updating for improving the front diversity. The path-relinking is performed by taking an external-archived solution from each of the two neighboring sections of the lowest-density section. One solution is referred to as the initiating solution, the other solution is used as the guiding solution. Path-relinkng undertakes to explore the trajectory space by constructing a path that transforms the initiating solution into the guiding solution by generating a succession of moves that introduce attributes from the guiding solution into the initiating solution (see Figure 10(a)). The cross-section updating technique can be also applied to improve the diversity performance of ε-constaint programming as shown in Figure 10(b), but the section is referring to the zone within two consecutive ε-constaint values. Our preliminary experiment showed that path-relinking technique can identify new non-dominated solutions in the lowest-density section and the cross-section updating is useful in improving the diversity performance of the final solution front.

CONCLUSION

Many real-world problems involve multiple optimization objectives that are conflicting to one another. A compromise to seek a trade-off among these objectives is to find a set of non-dominated solutions that define a Pareto front. However, it is usually too computationally prohibitive to find these non-dominated solutions. Instead, the approximate Pareto front is sought for and the quality of this front is measured in terms of convergence and diversity. The convergence performance metric indicates how closely the approximate Pareto front is converging to the true Pareto front. The diversity performance metric is in favour of the approximate Pareto front whose plots are most evenly spread.

Figure 10. Illustration of cross-section updating

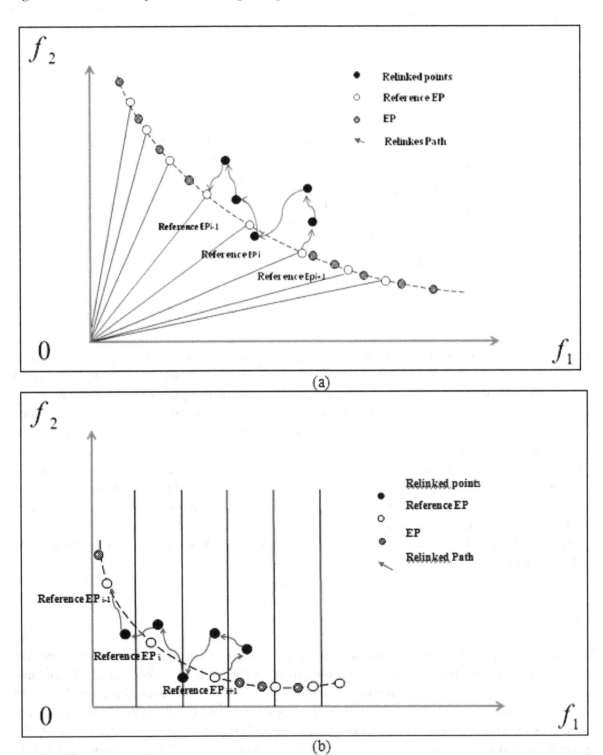

This chapter reviews the major classic multi-objective optimization approaches including the weighted-sum scheme, ε-constraint programming, and compromise programming. The salient features of Multi-Objective Evolutionary Algorithms (MOEAs) are then presented. Finally, we propose two novel Adaptive Memory Programming (AMP)-based MOEA methods. The first method named MOCSA combines PSO with SS and creates more benefits that are not obtainable by the previous version. The second method improves MOEA/D by using two notions. The within-section updating drives the search towards the true Pareto front more effectively than the parental-relationship updating that is employed by the original MOEA/D. The cross-section updating technique detects the section with the lowest density value, and applies path-relinking procedure to construct a path between its two neighboring sections. The cross-section updating is able to produce more non-dominated solutions in the lowest-density section, and the diversity degree of the solution front is thus improved. The chapter has shown the great potentials of AMP for improving MOEA methods and we invite more future studies aiming to this promising research domain.

Peng-Yeng Yin
National Chi Nan University, Taiwan

REFERENCES

Chang, P. C., Chen, S. H., & Mani, V. (2008). Parametric analysis of bi-criterion single machine scheduling with a learning effect. *International Journal of Innovative Computing . Information and Control, 4*(8), 2033–2043.

Coello Coello, A. C., Pulido, G. T., & Lechuga, M. S. (2004). Handling multiple objectives with particle swarm optimization. *IEEE Transactions on Evolutionary Computation, 8*, 256–279.

Deb, K., Pratap, A., Agarwal, S., & Meyarivan, T. (2002). A fast and elitist multiobjective genetic algorithm: NSGA-II. *IEEE Transactions on Evolutionary Computation, 6*, 42–50.

Dorigo, M. (1992). *Optimization, learning, and natural algorithms*. Ph.D. Thesis, Dip. Elettronica e Informazione, Politecnico di Milano, Italy.

Fard, S. M., Hamzeh, A., & Ziarati, K. (2011). A new cooperative co-evolutionary multi-objective algorithm for function optimization. *International Journal of Innovative Computing . Information and Control, 7*(5), 2529–2542.

Feo, T. A., & Resende, M. G. C. (1995). Greedy randomized adaptive search procedures. *Journal of Global Optimization, 6*, 109–133.

Glover, F. (1986). Future paths for integer programming and links to artificial intelligence. *Computers & Operations Research, 13*, 533–549.

Glover, F. (1989). Tabu search - Part I. *ORSA Journal of Computing, 1*, 190-206.

Glover, F. (1998). A template for scatter search and path relinking, Artificial Evolution. *Lecture Notes in Computer Science, 1363*, 13–54.

Holland, J. (1975). *Adaptation in natural and artificial systems: An introductory analysis with applications to biology, control, and artificial intelligence* (pp. 175–177). Ann Arbor: University of Michigan.

Ismail, F. S., Yusof, R., & Khalid, M. (2011). Self organizing multi-objective optimization problem. *International Journal of Innovative Computing . Information and Control, 7*(1), 301–314.

Kennedy, J., & Eberhart, R. C. (1995). Particle swarm optimization. *Proceedings of the IEEE International Conference on Neural Networks, IV,* 1942–1948.

Knowles, J. D., & Corne, D. W. (2000). Approximating the nondominated front using the Pareto archived evolution strategy. *Evolutionary Computation, 8,* 149–172.

Kulturel-Konak, S., Smith, A. E., & Norman, B. A. (2006). Multi-objective tabu search using a multinomial probability mass function. *European Journal of Operational Research, 169,* 918–931.

Laguna, M., & Marti, R. (2003). *Scatter search.* Boston, MA: Kluwer Academic Publishers.

Nebro, A. J., Luna, F., Alba, E., Dorronsoro, B., Durillo, J. J., & Beham, A. (2008). AbYSS: Adapting scatter search to multiobjective optimization. *IEEE Transactions on Evolutionary Computation, 12*(4), 439–457.

Omran, M. (2011). Special Issue on Scatter Search and Path Relinking Methods . *International Journal of Swarm Intelligence Research, 2*(2).

Schniederjans, M. J. (1995). *Goal programming: Methodology and applications.* Norwell, NJ: Kluwer Academic Publishers.

Yin, P. Y. (2012). *Modeling, analysis, and applications in metaheuristic computing: Advancements and trends.* Hershey, PA: IGI-Global Publishing.

Yin, P. Y., Glover, F., Laguna, M., & Zhu, J. X. (2010). Cyber swarm algorithms – Improving particle swarm optimization using adaptive memory strategies. *European Journal of Operational Research, 201*(2), 377–389.

Yokoyama, K. (1996). Epsilon approximate solutions for multiobjective programming problems. *Journal of Mathematical Analysis and Applications, 203,* 142–149.

Zhang, Q., & Li, H. (2007). MOEA/D: A multiobjective evolutionary algorithm based on decomposition. *IEEE Transactions on Evolutionary Computation, 11*(6), 712–731.

Zitzler, E., Laumanns, M., & Thiele, L. (2001). *SPEA2: Improving the strength Pareto evolutionary algorithm.* Technical Report 103, Swiss Federal Institute of Technology (ETH), Switzerland.

Chapter 1
A Novel Particle Swarm Optimization Algorithm for Multi-Objective Combinatorial Optimization Problem

Rahul Roy
KIIT University, India

Satchidananda Dehuri
Fakir Mohan University, India

Sung Bae Cho
Yonsei University, South Korea

ABSTRACT

The Combinatorial problems are real world decision making problem with discrete and disjunctive choices. When these decision making problems involve more than one conflicting objective and constraint, it turns the polynomial time problem into NP-hard. Thus, the straight forward approaches to solve multi-objective problems would not give an optimal solution. In such case evolutionary based meta-heuristic approaches are found suitable. In this paper, a novel particle swarm optimization based meta-heuristic algorithm is presented to solve multi-objective combinatorial optimization problems. Here a mapping method is considered to convert the binary and discrete values (solution encoded as particles) to a continuous domain and update it using the velocity and position update equation of particle swarm optimization to find new set of solutions in continuous domain and demap it to discrete values. The performance of the algorithm is compared with other evolutionary strategy like SPEA and NSGA-II on pseudo-Boolean discrete problems and multi-objective 0/1 knapsack problem. The experimental results confirmed the better performance of combinatorial particle swarm optimization algorithm.

DOI: 10.4018/978-1-4666-2145-9.ch001

INTRODUCTION

Many real world decision making problems involve discrete and disjunctive choices, i.e., such problems' solution can be either represented by 0/1 (represented as yes/ no) or an integer number. These problems are known as combinatorial problems. When these problems are modelled with more than one conflicting objectives and constraints, which need to be satisfied for finding optimal solutions, such problems are called multi-objective combinatorial optimization (MOCO) problems. Garey and Johnson (1990) showed that the decision making problems are intractable in nature. In case of such problems, the use of exact methods that guarantee generation of exact Pareto-optimal solutions may be not possible because of computational requirements of the methods. To solve such problems, many meta-heuristic algorithms have been found in the specialized literature (Zitzler & Thiele, 1998; Deb, Pratap, Agrawal, & Meyarivan, 2002; Jaszkiewicz, 2000). These algorithms aim at effective generation of subset of Pareto-optimal solutions which is an approximate representation of the whole Pareto set. Laumanns et al. (2002) compared the performance of multi-objective evolutionary algorithms on the pseudo-Boolean functions. In this comparative study, they analyzed the running time of the evolutionary based algorithms. Jaszkiewicz (2000) proposed multi-objective genetic local search algorithm and evaluated the performance with the multi-objective 0/1 knapsack problem. Zitzler et al. (1999) implemented some of the state-of-the-art combinatorial optimization problems with multi-objective genetic algorithm. Later Grosan et al. (2003) proposed a multi-objective evolutionary algorithm based on Є-relation for solving multi-objective 0/1 knapsack problem.

In recent days, multi-objective particle swarm optimization (MOPSO) based meta-heuristics algorithm have attained a great attention among the researchers. After the pioneering work by Coello and Lechuga (2002), it has been used in various domain to solve multi-objective problems. But the algorithm is designed to solve real valued problems. However, the algorithm can be used to solve combinatorial problems by truncating the real valued solution to integer. But it would incur a truncation error and will create a hindrance in convergence of the solutions to the true Pareto front.

Thus, an attempt has been made by the authors to solve a combinatorial optimization problems by designing a binary MOPSO (BMOPSO)(Das, Roy, Dehuri, & Cho, 2011).

This algorithm works well for problems where the solution can be represented as strings of 0/1. However, the solutions which are represented as integers, this approach would need to convert the integer values to binary form before they are used for updation. This would incur an updation cost overhead based on number of bits required to represent the solution in each dimension. This motivated us to design a combinatorial MOPSO (CMOPSO) algorithm that can overcome both drawbacks stated above for using real valued or binary MOPSO for solving combinatorial optimization problem whose solution can be represented as integers. During the study of different MOPSO algorithms, we have seen that only the position updation equation, which generates the new solutions, depends on the solutions space. However, all other components, like the gbest selection, repository updation, etc., depend on the objective space which is real valued. Thus we can conclude from this observation that to design a combinatorial optimization algorithm, we need to concentrate more on position update equation whereas the rest of the strategies presented in real valued MOPSO literature can be adapted directly in the algorithm. This motivated us to present a new position update strategy that can adapt to the integer valued solution space without truncating of transforming to binary form for solving the integer valued combinatorial optimization problems. An empirical comparison of the algorithm with binary MOPSO, NSGA-II (Deb et al., 2002) and SPEA (Zitzler & Thiele, 1998) is done to evaluate the performance on the pseudo-Boolean function and multi-objective 0/1 knapsack problem.

The rest of the section is organized as follows. First we provide a brief description of the fundamental concepts of multi-objective optimization. We describe the binary MOPSO algorithm. Then we describe the CMOPSO algorithm approach. We present a theoretically comparison of the computational complexity of the two proposed algorithms. We present the experimental study of the algorithms and draw inferences based on the experiments. Finally, we conclude the paper.

FUNDAMENTAL CONCEPTS OF MULTI-OBJECTIVE OPTIMIZATION

A multi-objective optimization problem can be stated in the general form (Equation 1):

$$Minimize \ / \ maximize \ f_m(x), \ m = 1, 2, ... M$$
$$subject \quad to$$
$$g_j(x) \geq 0 \quad j = 1, 2, ... J;$$
$$h_k(x) = 0 \quad k = 1, 2, ..., K;$$
$$x_i^{(L)} \leq x_i \leq x_i^{(U)} \quad i = 1, 2, ..., n$$

$$(1)$$

A solution x is a vector of n decision variables: $x = (x_1 x_2, .., x_n)^T$. The last sets of constraints are called variable bounds, restricting each decision variable x_i to take a value within a lower $x_i^{(L)}$ and an upper $x_i^{(U)}$ bounds. These bounds constitute a decision variable space D, or simply the decision space. Associated with the problem are J inequality and K equality constraints and the terms $g_j(x)$ and $h_k(x)$ are called constraint functions. Although the inequality constraints are

treated as $\geq$ types, the $\leq$ constraints can also be considered in the above formulation by converting those to $\geq$ types simply by multiplying the constraint function by -1 (Deb, 1995).

In multi-objective optimization, the m objective functions $f(x) = (f_1(x), f_2(x), ..., f_m(x))^T$ can be either minimized or maximized or both. Many optimization algorithms are developed to solve only one type of optimization problems, such as e.g., minimization problems. When an objective is required to be maximized by using such an algorithm, the duality principle (Deb, 1995) can be used to transform the original objective for maximization into an objective for minimization by multiplying objective function by -1. It is to be noted that for each solution x in the decision variable space, there exists a point in the objective space, denoted by $f(x) = z = (z_1, z_2, ..., z_M)$. There are two goals in a multi-objective optimization: firstly, to find a set of solutions as close as possible to the Pareto-optimal front; secondly, to find a set of solutions as diverse as possible. Multi-objective optimization involves two search spaces i.e., the decision variable space and the objective space. Although these two spaces are related by an unique mapping between them, often the mapping is non-linear and the properties of the two search spaces are not similar. In any optimization algorithm, the search is performed in the decision variable space. However, the proceedings of an algorithm in the decision variable space can be traced in the objective space. In some algorithms, the resulting proceedings in the objective space are used to steer the search in the decision variable space. When this happens, the proceedings in both spaces must be coordinated in such a way that the creation of new solutions in the decision variable space is complementary to the diversity needed in the objective space. Most multi-objective optimization algorithms use the concept of domination. In these algorithms, two solutions are compared on the basis of whether one dominates the other solution or not. The concept of domination is described in the following definitions (assuming, without loss of generality, the objective functions to be minimized).

- **Definition 1:** Given two decision or solution vectors x and y, we say that decision vector x weakly dominates (or simply dominates) the decision vector y (denoted by x $\leq$ y) if and only if $f_i(x) \leq f_i(y) \forall$ i = 1, ..., M (i.e., the solution x is no worse than y in all objectives) and $fi(x) \prec fi(y)$ for at least one $i \in$ 1, 2, ..., M (i.e., the solution x is strictly better than y in at least one objective).

- **Definition 2:** A solution x strongly dominates a solution y (denoted by $x \prec y$), if solution x is strictly better than solution y in all M objectives.

  Figure 1 illustrates a particular case of the dominance relation in the presence of two objective functions. However, if a solution x strongly dominates a solution y, the solution x also weakly dominates solution y, but not vice versa.

- **Definition 3:** The decision vector x $\in$ P (where P is the set of solution or decision vectors is non-dominated with respect to set P, if there does not exit another x$\in$ P such that $f'(x) \leq f(x)$.

- **Definition 4:** Among a set of solution or decision vectors P, the non-dominated set of solution or decision vectors P are those that are not dominated by any member of the set P.

- **Definition 5:** A decision variable vector $x \in P$ where P is the entire feasible region or simply the search space is Pareto-Optimal if it is non-dominated with respect to P.

- **Definition 6:** When the set P is the entire search space, the resulting non-dominated set P' is called the Pareto-Optimal set. In other words, $P'= \{x \in P \mid x \text{ is Pareto } -$

Optimal}.The non-dominated set P' of the entire feasible search space P is the globally Pareto-Optimal set.

- **Definition 7:** All Pareto-Optimal solutions in a search space can be joined with a curve (in two-objective space) or with a surface (in more than two-objective space). This curve or surface is termed as Pareto optimal front or simply Pareto front. In other words, $PF = \{f(x) \mid x \in P'\}$.

Figure 2 illustrates a particular case of the Pareto front in the presence of two objective functions. We thus wish to determine the Pareto optimal set form the set P of all the feasible decision variable vectors that satisfy (1). It is to be noted that in practice, the complete Pareto Optimal set is not normally desirable (e.g., it may not be desirable to have different solutions that map to the same values in objective function space) or achievable. Thus a preferred set of Pareto optimal solutions should be obtained from practical point of view.

Figure 1. Illustration of Dominance relation in bi-objective functions

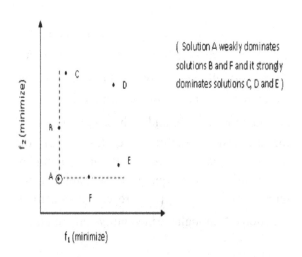

BINARY MULTI-OBJECTIVE PARTICLE SWARM OPTIMIZATION

The design of BMOPSO algorithm to solve combinatorial optimization problem is presented in Das et al. (2011). Here we present a brief description of the BMOPSO algorithm. To change the real valued MOPSO algorithm to binary MOPSO we need to change the concept of velocity as a probability of change of position from rate of change of position. In this method, we represent the particle as a string of bits. The swarm is initialized with m binary strings of fixed length n which are generated randomly. Each particle P_i, $i = 1, ..., m$ is generated independently. For every position p_{ij}, $i = 1, ..., m, j = 1, ..., n$ in P_i, a standard Gaussian random number τ is drawn on the interval (0, 1). The particle representation of the BMOPSO algorithm for integer and binary valued solution is shown in Figure 3 and Figure 4.

Every particle is associated to a unique vector of velocities $v_i = (v_{i1}, v_{i2}, ..., v_{in})$, $i = 1, ..., m$. The elements v_{ij}, $i = 1, ..., m, j = 1, ..., n$ in v_i determine the rate of change of each respective coordinate p_{ij} of P_i. Each element $v_{ij} \in v_i$ is updated according to the equation

$$V_{ij}(t+1) = wv_{ij}(t) + r_1 c_1 \left(PB_{ij} - p_{ij}(t) \right)$$
$$+ r_2 c_2 \left(PG_j - p_{ij}(t) \right)$$

(2)

where w, $(0 < w < 1)$, called the inertia weight, is a constant value chosen by the user and $j = 1, 2, .., n$. Equation 2 is used in PSO algorithms to update the velocities. The factors $r1$ and $r2$ are

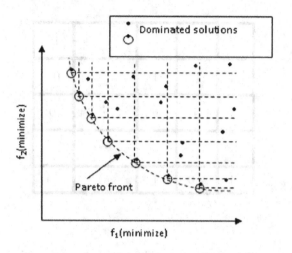

Figure 2. Pareto front of a set of solutions in a bi-objective Problem

Figure 2. Pareto front of a set of solutions in a bi-objective Problem

uniform random numbers independently generated in the interval (0, 1). c_1 and c_2 represents the self cognition factor and global cognition factor. PB_i and PG_i represent the particle from *pbest* repository (PB) and *gbest* repository (PG) respectively. Here we use two different archives to store the *pbest* and *gbest* solutions. The selection of PB_i and PG_i is different from normal PSO.

For each particle Pi and each dimension j, the value of the new co-ordinate $p_{ij} \in Pi$ can be either 0 or 1. The decision of whether p_{ij} will be 0 or 1 is based on its respective velocity $v_{ij} \in v_i$ and is given by $p_{ij} = 1$ if $rand < S\,ig(v_{ij})$, otherwise $p_{ij} = 0$, where $0 < rand < 1$ is a random number, and

$$sig(v_{ij}) = \frac{1}{1 + \exp(-v_{ij})}$$

(3)

Figure 3. Particle representation of solution encoded as binary string

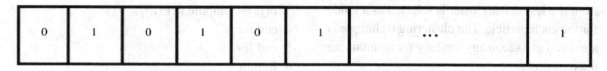

0	1	0	1	0	1	...	1

Figure 4. Particle representation of integer valued solution encoded as binary matrix

0	1	0		1
1	0	1		0
1	1	0	⋮	0
			⋯	
1	0	1		1

$Sig(v_{ij})$ is the sigmoid function. Equation 3 is a standard equation used to generate new particle positions in the binary MOPSO. Note that the lower the value of v_{ij} is, the more likely the value of p_{ij} will be 0. By contrast, the higher the value of v_{ij} is, the more likely the value of p_{ij} will be 1. The motivation to use the sigmoid function is to map v_{ij} into the interval (0, 1) which is equivalent to the interval of a probability function.

At the beginning, the previous best position of P_i, denoted by PB_i stored in an external repository is empty. Note that a list of external repository has to be maintained in order to store the non-dominated solutions. Therefore, once the initial particle P_i is generated, $PB_i = P_i$. After that every time that P_i is updated, PB_i is updated based on whether it is a dominated or non-dominated solution. A similar process is used to update the global best position i.e., PG. As there is not a single global optimal solution, so we need to take an External Archive for maintaining the non-dominated list of PG.

The guide for a particle, to direct the search, is selected by applying the k-medoid clustering technique on the external archive (Dehuri & Cho, 2009). One of the non-dominated particles from the less dense cluster is selected as a guide for the each particle. The clustering technique is also used as a secondary strategy for maintaining the repository.

Algorithm 1. BMOPSO

1. Initialize the swarm
   a. For j=1: max /*Maximum size of the swarm*/
   b. Initialize SWARM[j];
2. Evaluate each of the particle in SWARM by invoking the Fitness-particle()
3. Find the non-dominated particle from the SWARM and store them in external archive called EX-ARCHIVE.
4. Initialize the memory for each particle.(This memory serves as a guide to travel through search space).
   a. For j=1: max
   b. PB[j]=SWARM[j];
5. While (I < I_{Max}) /* I_{Max} is the maximum number of iteration*/
6. Update the velocity of particle with the Equation 5
7. Update the position of the particle using Equation 2
8. Evaluate the fitness of the updated particle in the SWARM.
9. Update the PB[j].
10. Update the content of EX-ARCHIVE. Here we need to determine whether the new solution should be added to the archive or not. This is determined by performing a non-domination test on the external archive. As a secondary strategy we can apply clustering when the archive size exceeds pre-specified value.
11. End

Fitness-Particle(PB)

1: for j = 1: m do
2: for i= 1: k do
3: F(j,i)=Compute fi(PB[j])
4: end for
5: end for
6: Return F

In this algorithm, we need to update each bit position of the particle to generate new particle position. Thus as size of the particle increases number of steps required to update a particle position increases.

COMBINATORIAL MULTI-OBJECTIVE PARTICLE SWARM OPTIMIZATION

Jarboui et al. (2007) proposed a particle swarm optimization algorithm to combinatorial optimization problem with integer values. It considers the mapping of the particle in the {-1, 0, 1} and then update the position of the particle and demap it to the corresponding integer values. We use the same notion to design the CMOPSO. However, the restrictive mapping of the particle position to {-1, 0, 1} is relaxed by using a symmetric function $f(x)$.

The intuition behind the design of this algorithm is that when the particle approaches the global best combinatorial state, the particle velocity should be reduced so that it does not deviate away from the gbest combinatorial state. For this, we define a function that uses a particle combinatorial state to generate a real value which is subtracted from the value generated by the function using the particle gbest and pbest combinatorial state, thereby reducing the velocity. Also, we allow the particle to exploit the search space region without not allowing it to explore the search space because the exploration of the search space occurs due to the *gbest* component. However if the particle combinatorial state is close to pbest combinatorial state, the we need to induce more velocity so that the particle can better explore the search space rather that getting stuck to the local optima. If the particle is not close to both pbest and gbest combinatorial state, then the particle should be allowed to explore and exploit the search space simultaneously under the influence of a value determined by both pbest and

gbest combinatorial state. In case, the particle is in both pbest and gbest combinatorial state, the particle should be allowed to either explore or exploit the search space for finding some other optimal solution which might be present in the search space. The detailed mathematical description of the algorithm is given below in following subsections.

Swarm Initialization of the CMOPSO

The particle of the swarm is encoded as vector of integer (or binary) values of fixed length m to represent the solution of the combinatorial optimization problem, which are randomly generated. The solutions generated follow a uniform distribution. Each particle P_i, $i = 1, ..., m$ is generated independently. The particle representation of binary and integer valued solution of a combinatorial problem for CMOPSO algorithm is shown in Figures 3 and 5. To implement the above mention strategy for position update, we first determine the $f(x)$ value for a particle combinatorial state based on its equality to *pbest* or *gbest* combinatorial state. This procedure is known as the mapping from combinatorial state to continuous state.

Velocity and Position Updation

Every particle is associated to a unique vector of velocities $v_i = (v_{i1}, v_{i2}, ..., v_{in})$, $i = 1, ..., m$. Before updating the particle vector to generate a new position, it is mapped from combinatorial state to continuous state using the Equation 4

$$y_i = \begin{cases} f(x_i) & x_i = p_i^g \\ -f(x_i) & x_i = p_i^t \\ f(x)\,or -f(x) & x_j^t = P_i^t = P_i^g \\ 0 & otherwise \end{cases}$$

(4)

where $f(x) = |x^2 - b|$ where b is a prime number and $b >> x$. The velocity of the particle is update using Equation 5

$$v_{ij}^{t+1} = w \times v_{i,j}^t + c_i r_i \left(f\left(P_{i,j}^t\right) - \left(x_{i,j}\right)\right)$$
$$+ c_2 r_2 \left(f\left(P_{i.j}^g\right) - f\left(x_{i,j}\right)\right)$$

(5)

Here, the value of c_1, c_2, r_1 and r_2 has the same meaning as the one described in binary MOPSO section. When $x_i = P_i^g$, it imposes the particle to fly in positive sense. When $x_i = P_i^t$, the particles are imposed to fly in negative sense. When $x_i = P_i^g = P_i^t$, the particle fly in opposite direction to direction of y_i^t. When $x_i \neq P_i^g \neq P_i^t$, the direction of the particle flight is determined by $r_1, r2, c_1$ and c_2. The position of the particle is calculated by using the Equations 6-8.

$$\lambda_{ij}^{t+1} - y_{ij}^t + v_{ij}^t$$

(6)

The value of y_{ij}^t is adjusted using Equation 7

$$y_{ij}^t = \begin{cases} f(x) & \lambda_{ij}^{t+1} >= \alpha * f(x) \\ -f(x) & \lambda_{ij}^{t+1} < -\alpha * f(x) \\ 0 & otherwise \end{cases}$$

(7)

where α is known as the intensification (or diversification) parameter. Smaller value of α leads to diversification of the Pareto front and large value of α causes intensification of the Pareto front.

The demapping of the y_{ij}^{t+1} with Equation 8 generate the new set of the particle representing new position in the search space.

$$x_{ij}^{t+1} = \begin{cases} P_{ij}^g & if \quad y_{ij} = f(x) \\ P_{ij}^t & if \quad y_{ij} = -f(x) \\ random \quad number \quad otherwise \end{cases}$$

(8)

Selection of P_{ij}^t and P_i^G

At the beginning, each particle position is generated randomly. These initial particle position are considered as the *pbest* (p_{ij}^t) position for the first generation. After every generation, the *pbest* for each particle is selected using the sigma method proposed in Mostaghim and Teich (2003). The selection of *pbest* is done from the external local memory which stores the non dominated solution over the generation.

As there is not a single global optimal solution, so we need to take an External Archive for maintaining the non-dominated list of P_i^G. The guides are stored in external repository and for each particle, a guide is selected by using the guide selection strategy proposed in Coello and Lechuga (2002). In this method, the objective space is divided into adaptive hypercube. Each hypercube is assigned a fitness value which is calculated as

$$\frac{a \quad l\arg e \quad number}{number \quad of \quad particles \quad in \quad each \quad hypercube}$$

Then a guide is selected for each particle from

Figure 5. Particle representation of CMOPSO algorithm for integer valued solution of combinatorial problem

1	1	2	1	2	2	...	1

the cube which has the highest fitness value to guide the particle in the search space.

Repository Updation

As we see in the previous Subsection, we need to maintain two archives to store the global best position and the personal best position. So need a strategy to maintain both the archives because we cannot allow the archive to grow too large and also we cannot randomly remove particles from the archive as they are the bests representation of the true Pareto front.

For maintaining the local memory, we use the non domination test. After every generation, the elements are stored in the local memory if they are not dominated by any members of the archive. Also those members of the archive which are dominated by the new members, which are about to enter the local memory, are removed from the archive. The external repository, storing the global guides, is also maintained using the non domination test. But along with it there is secondary strategy to maintain the repository. We use the crowding sort technique to maintain the elements of the repository when the repository size grows beyond a fixed size. Here we sort the members of the repository based on the objective values in descending order, based on their crowding distance values. Then the elements which have least crowding distance value are removed from the archives. The algorithm for the CMOPSO is

Algorithm 2. CMOPSO

1. Initialize the swarm
   a. For j=1: max /*Maximum size of the swarm*/
   b. Initialize SWARM[j];
2. Evaluate each of the particle in SWARM by invoking the *Fitness-particle()*
3. Find the non-dominated particle from the SWARM and store them in external archive called EX-ARCHIVE.
4. Initialize the memory for each particle.(This memory serves as a guide to travel through search space).
   a. For j=1: max
   b. PB[j]=SWARM[j];
5. While $(I < I_{Max})$ /* I_{Max} is the maximum number of iteration*/
6. Select the *gbest* P_i^g
   a. () for each particle from EX-ARCHIVE
7. Map the particle to continuous state using Equation 4
8. Update the velocity of particle with the Equation 2
9. Update the position of the particle using Equation 6-8
10. Evaluate the fitness of the updated particle in the SWARM.
11. Update the PB[j].
12. Select the *pbest* P_i^t
    a. () for each particle using the sigma method from PB[j]
13. Update the content of EX-ARCHIVE. Here we need to determine whether the new solution should be added to the archive or not. This is determined by performing a non-domination test on the external archive. As a secondary strategy we can apply crowding sort when the archive size exceeds pre-specified value.
14. End

COMPUTATIONAL COMPLEXITY

When BMOPSO and CMOPSO algorithms are used to solve decision making problems, where solutions are represented with 0/1, we encode the particle as vector of fixed length *n*. And let us consider the swarm size as *m*. Now to generate a new set of solutions with velocity and position update equation, we need $O(mn)$ steps. This is the same for both the algorithms.

But when solutions of decision making problems require solutions to be encoded as integer values, we need to convert the integer values to binary values. The each particle representation becomes a matrix of $k \times n$. Thus to update each particle, we require $O(kn)$ steps. Thus to update the entire swarm, we need $O(mkn)$ steps. However, in CMOPSO algorithm, the particle can encode the solution as integer values, thus we can represent solution as a vector of $n \times 1$. Hence, to update the swarm we would need the same number of steps as requires for updating particles when they were encoded as a string of 0 or 1.

From the discussion, we can say the CMOPSO takes $O(mn)$ steps in both representation of the particle whereas in BMOPSO the number of steps increases by factor of k, when it is used to solve combinatorial problems with integer values as solutions. The guide selection and the repository update strategy would take same number of steps, if the same strategy is used in both the algorithms.

EXPERIMENTAL STUDY

This section describes the test problems used for testing the algorithms, the experimental environment, the performance metrics based on which the algorithms are compared and the results and analysis of the experiments.

Test Problems

To compare the algorithms, two pseudo-Boolean function are used, namely leading ones and trailing zeros (LOTZ) and counting ones and counting zeros (COCZ) function. Also we compare the performance of the algorithms on a real world problem, i.e., multi-objective 0/1 knapsack problem.

LOTZ: The pseudo Boolean function $LOT\,Z$: $(0,1)^n \rightarrow N^2$ is defined as:

$$\bar{d} = \sum\nolimits_{i=1}^{|Q|} d_i / \mid Q \mid$$

The objective of this function is to maximize the leading ones and leading zeros simultaneously in a bit string (Laumanns, Thiele, Zitzler, Welzl, & Deb, 2002). As there is no single search point that maximizes both components simultaneously, we want to find the whole set of non-dominated points based on the concept of Pareto optimality, here defined for binary decision variables. The objective space for N=8 is shown in Figure 6.

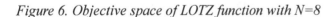

Figure 6. Objective space of LOTZ function with N=8

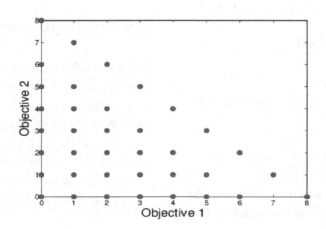

COCZ: The pseudo Boolean function *COCZ:* $\{0,1\}^n \rightarrow N^2$ is defined as:

$$d_i = \min_{k \in Q \wedge k \neq i} \sum\nolimits_{m=1}^{M} \left| f_m^i - f_m^k \right|$$

where $n = 2 \cdot k$ and $k \in N$. The objective space of function is shown in Figure 7.

Here, we maximize the count of ones and zeros simultaneously (Laumanns et al., 2002).

Multi-Objective 0/1 Knapsack Problem:

Generally, a 0/1 knapsack problem consists of a set of items, weight and profit associated with each item, and an upper bound for the capacity of the knapsack. The task is to find a subset of items which maximizes the total profit in the subset, yet al w_{ij} = weight of item j according to knapsack i, c_i = capacity of knapsack i,

Find a vector $x = (x_1, x_2, ..., x_m) \in \{0,1\}^m$

such that

$$\forall i \in \{1, 2, ..., n\} : \sum_{j=1}^{m} w_{ij\,x_j} \leq c_i \quad (9)$$

and for which $f(x) = (f1(x), f2(x), ..., fn(x))$ is maximum, where

$$f_i(x) = \sum_{j=1}^{m} p_{ij} \cdot x_j$$

and $x_j = 1$ iff item j is selected. This typical multi-objective 0/1 knapsack problem assumes that an item is placed either in all knapsacks at one time, if that item is selected (i.e., entry is 1 in the solution vector x for that item index) or in none of the knapsacks, if that item is not selected (i.e., entry is 0 in the solution vector x for that item index). However, as a variation from the formal

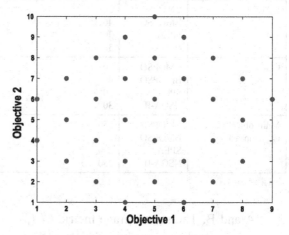

Figure 7. Objective space of LOTZ function with N=10

knapsack problem and for ease of understanding and implementation, we have made an assumption that all items to be distinct from each other (i.e., there is no multiple number of same item). In other words, once an item is placed in one knapsack, it cannot be placed in other knapsacks.

EXPERIMENTAL SETUP

The experiment is performed in Core2Duo processor, 1GB RAM and Matlab (2009b). The parameter settings for the experiments are shown in Table 1.

For multi-objective 0/1 knapsack problem, we consider two synthetic data set of size 250 and 500 respectively. The parameter setting is same for both the dataset.

Performance Metrics

The algorithms are compared on the basis of following performance metrics:

- **Set Coverage:** Zitzler et al. (1999) suggested a set coverage metric which is used to get an idea of the relative spread of solutions between two sets of solutions vectors

Table 1. Parameter setting

Test Problem	Algorithm	swarm size/ population size	iteration	w	c_1	c_2	P_c	P_c
LOTZ	CMOPSO	30	15	0.4	1.8	2.5	-	-
	BMOPSO	40	20	1	1	2.5	-	-
	SPEA	30	18	-	-	-	0.8	0.01
	NSGA-II	30	15	-	-	-	0.8	0.03
COCZ	CMOPSO	30	20	0.4	1.8	2.5	-	-
	BMOPSO	40	20	1	1	2.5	-	-
	SPEA	30	20	-	-	-	0.8	0.01
	NSGA-II	30	20	-	-	-	0.8	0.03
Multi-objective 0/1 knapsack	CMOPSO	250	80	0.4	1.76	2.5	-	-
	BMOPSO	250	80	1	1	2.5	-	-
	SPEA	250	80	-	-	-	0.8	0.01
	NSGA-II	250	80	-	-	-	0.8	0.03

A and B. The set coverage metric $C(A,B)$ calculates the proportion of solution in B, which are weakly dominated by solution of A.

$$C(A,B) = \frac{\left|\{b \in B \mid \exists a \in A : a \leq b\}\right|}{|B|}$$

The metric value $C(A, B)=1$ means all members of B are weakly dominated by A. On the other hand, $C(A, B)=0$ means that no members of B is weakly dominated by A.

- **Maximum Spread:** Zitzler et al. (1999) defined a metric measuring the length of the diagonal of a hyper box formed by the extreme function values observed by the extreme f. $d_i = \min_{k \in Q \wedge k \neq i} \sum_{m=1}^{M} |f_m^i - f_m^k|$ and $\bar{d}$ is the mean value of the above dis-

Table 2. Comparison of different algorithms based on performance metrics

Metrics	Algorithms	LOTZ	COCZ	0/1 knapsack	
				250	500
Set Coverage	C(CMOPSO,BMOPSO)	0	0	0.256	1
	C(BMOPSO,CMOPSO)	0	0	0.250	0
	C(CMOPSO,NSGA-II)	0	0	0.346	1
	C(NSGA-II,CMOPSO)	0	0	0.344	0
	C(CMOPSO,SPEA)	0	0	0.016	0.96
	C(SPEA,CMOPSO)	0	0	0.014	0.0003
Maximum Spread	CMOPSO	11.31	8.48	7784.527	7841.913
	BMOPSO	11.31	8.48	4593.49	4178.56
	SPEA	11.31	8.48	6303.043	7306.552
	NSGA-II	11.31	8.48	5633.977	5524.63
Spacing	CMOPSO	0	0	5.5354	5.2539
	BMOPSO	0	0	18.867	17.9253
	SPEA	0	0	19.654	17.6349
	NSGA-II	0	0	13.4967	7.4364
CPU Time(in sec)	CMOPSO	1.45	1.625	560.32	940.68
	BMOPSO	10.23	10.91	1039.71	4686.34
	SPEA	2.82	2.89	738.95	3984.0
	NSGA-II	2.16	2.21	662.98	2038.0

Figure 8. Implementation result of different algorithms on 0/1 knapsack problem with different size of dataset: Data set size=250

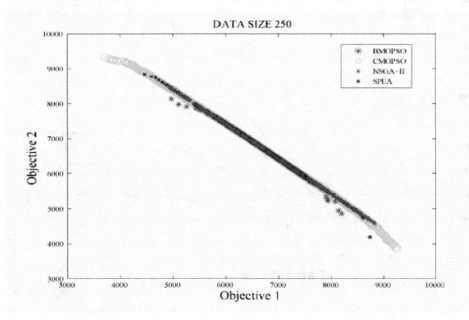

Figure 9. Implementation result of different algorithms on 0/1 knapsack problem with different size of dataset: Data set size 500

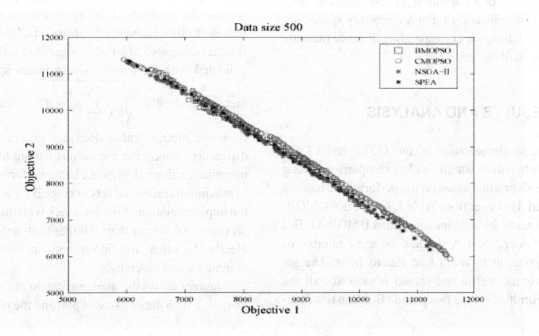

Figure 10. CPU time of different algorithms

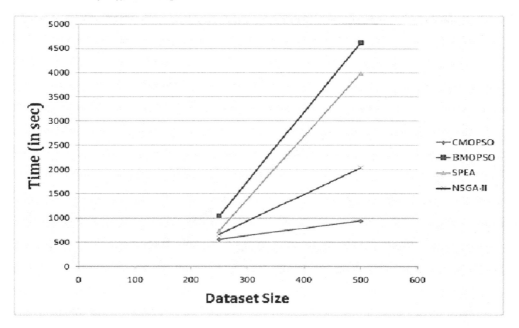

tance measure $\bar{d} = \sum_{i=1}^{|Q|} di / |Q|$. The distance measure is the minimum value of the sum of the absolute difference in objective function values between the ith solution and any other solution in the obtained non-dominated set. For a uniformly spaced solution, the corresponding distance measure will be zero.

RESULTS AND ANALYSIS

The implementation of the LOTZ and COCZ function does not give a clear comparison among the algorithms based on the performance metrics used. However from Table 1, we see that CMOPSO takes lesser population than BMOPSO. But SPEA and NSGA-II takes the same number of population to reach true Pareto front. The set coverage metric and spread is same for all the algorithms for the two pseudo Boolean test func-

tions. The results for the implementation are shown in Table 2. The clear distinction in performance among the algorithms is seen from the implementation of multi-objective 0/1 knapsack. Figure 8 shows the implementation of the multi-objective 0/1 knapsack with 250 data set. In the figure we see that the point of all the algorithms overlaps with each other. However the spread of CMOPSO is better compared to all other algorithms. This is validated with values of the maximum spread metric value $S = \sqrt{\dfrac{1}{|Q|} \sum_{i=1}^{|Q|} (d_i - \bar{d})^2}$. The set coverage metrics value does not give a clear distinction among the algorithms having higher magnitude values. However, a better performance of maximum spread and set coverage is clear from the implementation of 0/1 knapsack with the data set size of 500. We see that CMOPSO out-performs clearly the other algorithms both in terms of magnitude and diversity.

Figure 9 shows the implementation of the 0/1 knapsack with dataset size of 500 and the results of the performance are tabulated in column 5 of Table 2. The spacing metric compares the relative spacing of the non- dominated solutions obtained from different algorithms. It is evident from the Knapsack results that the CMPOSO has the smallest relative distance in non-dominated set. However for the LOTZ and COCZ algorithm the spacing of the solution is uniform for the entire algorithm due to which the spacing metric turns out to be 0. Thus it can be concluded that the CMOPSO provides a better spacing in solutions compared to other algorithms.

Also we have a plot of CPU time vs. dataset size in Figure 10. We see that the time requirement of the CMOPSO grows less as compared to other algorithms, as the data set size increases. These empirical results validate the argument about the computational complexity between the CMOPSO and BMOPSO.

CONCLUSION

In this paper, we present a combinatorial multi-objective particle swarm optimization algorithm to solve combinatorial problems. We showed theoretically that the proposed algorithm requires $O(mn)$ steps for updating the swarm at each generation in case particle are encoded as both integer of binary value. Also we compared the algorithm with BMOPSO, NSGA-II, and SPEA and found it to outperform them both in terms of diversity and magnitude and spacing. Also we see that the algorithm takes lesser time to complete its computation even with the scaling of the dataset size.

However, the algorithms need to be evaluated with other multi-objective problems to confirm its performance. Also, we need to see the algorithms with real world combinatorial optimization problems like task scheduling, task allocation etc.

REFERENCES

Coello, C. A. C., & Lechuga, M. (2002). MOPSO: a proposal for multi-objective particle swarm optimization. In *Proceedings of the 9th IEEE, World Congress on Computational Intelligence*, Honolulu, HI (pp. 1051-1056).

Das, M., Roy, R., Dehuri, S., & Cho, S.-B. (2011). A new approach to associative classification based on binary multi-objective particle swarm optimization. *International Journal of Applied Metaheuristic Computing*, 2(2), 51–73. doi:doi:10.4018/jamc.2011040103

Deb, K. (1995). *Optimization for engineering design: algorithms and examples*. New Delhi, India: Prentice Hall.

Deb, K., Pratap, A., Agrawal, S., & Meyarivan, T. (2002). A fast and elitist multi-objective genetic algorithm: NSGA-II. *IEEE Transactions on Evolutionary Computation*, 6(2), 182–197. doi:doi:10.1109/4235.996017

Dehuri, S., & Cho, S. B. (2009). Multi-criterion Pareto based particle swarm optimized polynomial neural network for classification: a review and state-of-the-art. *Journal of Computer Science Review*, 3, 19–40. doi:doi:10.1016/j.cosrev.2008.11.002

Garey, M. R., & Johnson, D. S. (1990). *Computers and intractability: A guide to the theory of NP-completeness*. Murray Hill, NJ: Bell Telephone Laboratories.

Grosan, C., Oltean, M., & Dumitrescu, D. (2003). Performance metrics for multi-objective optimization evolutionary algorithms. In *Proceedings of the Conference on Applied and Industrial Mathematics*, Oradea, Romania.

Jarboui, B., Damak, N., Siarry, P., & Rebai, A. (2007). A combinatorial particle swarm optimization for solving multi-mode resource-constrained project scheduling problems. *Journal of Applied Mathematics and Computation, 195*, 299–308. doi:doi:10.1016/j.amc.2007.04.096

Jaszkiewicz, A. (2000). *On the performance of multiple objective genetic local search on the 0/1 knapsack problem: A comparative experiment* (Working Paper No. RA002/2000). Poznan, Poland: Institute of Computer Science, Poznan University of Technology.

Laumanns, M., Thiele, L., Zitzler, E., Welzl, E., & Deb, K. (2002). *Running time analysis of a multi-objective evolutionary algorithm on a simple discrete optimization problem* (Tech. Rep. No. TIK-123). Zurich, Switzerland: Swiss Federal Institute of Technology.

Laumanns, M., Thiele, L., Zitzler, E., Welzl, E., & Deb, K. (2002). *Running time analysis of algorithm on vector-valued pseudo Boolean functions* (Tech. Rep. No. TIK-165). Zurich, Switzerland: Swiss Federal Institute of Technology.

Martello, S., & Toth, P. (1990). *Knapsack problem: algorithms and computer implementation.* Chichester, UK: John Wiley & Sons.

Mostaghim, S., & Teich, J. (2003). Strategies for finding good local guides in multi-objective particle swarm optimization. In *Proceedings of the IEEE Symposium on Swarm Intelligence* (pp. 26-33).

Zitzler, E., Deb, K., & Thiele, L. (1999). *Comparison of multiobjective evolutionary algorithms: empirical results.* Zurich, Switzerland: Swiss Federal Institute of Technology.

Zitzler, E., & Thiele, L. (1998). *An evolutionary algorithm for multi-objective optimization: the strength pareto approach* (Tech. Rep. No. TIK-43). Zurich, Switzerland: Swiss Federal Institute of Technology.

This work was previously published in the International Journal of Applied Metaheuristic Computing, Volume 2, Issue 4, edited by Peng-Yeng Yin, pp. 41-57, copyright 2011 by IGI Publishing (an imprint of IGI Global).

Chapter 2
A Hybrid Simulated Annealing and Simplex Method for Fixed–Cost Capacitated Multicommodity Network Design

Masoud Yaghini
Iran University of Science and Technology, Iran

Mohammad Karimi
Iran University of Science and Technology, Iran

Mohadeseh Rahbar
Iran University of Science and Technology, Iran

Rahim Akhavan
Kermanshah University of Technology, Kermanshah, Iran

ABSTRACT

The fixed-cost Capacitated Multicommodity Network Design (CMND) problem is a well known NP-hard problem. This paper presents a matheuristic algorithm combining Simulated Annealing (SA) metaheuristic and Simplex method for CMND problem. In the proposed algorithm, a binary array is considered as solution representation and the SA algorithm manages open and closed arcs. Several strategies for opening and closing arcs are proposed and evaluated. In this matheuristic approach, for a given design vector, CMND becomes a Capacitated Multicommodity minimum Cost Flow (CMCF) problem. The exact evaluation of the CMCF problem is performed using the Simplex method. The parameter tuning for the proposed algorithm is done by means of design of experiments approach. The performance of the proposed algorithm is evaluated by solving different benchmark instances. The results of the proposed algorithm show that it is able to obtain better solutions in comparison with previous methods in the literature.

DOI: 10.4018/978-1-4666-2145-9.ch002

INTRODUCTION

The Network Design Problem (NDP) is one of the important problems in combinatorial optimization. The objective of NDP is to find a minimum cost network on the available arcs for each commodity which satisfy flows of commodities. Network design models have numerous applications in various fields such as transportation (Crainic et al., 1989; Magnanti & Wong, 1984) telecommunication (Gendron et al., 1998; Minoux, 1989, 2001) and distribution planning (Jayaraman & Ross, 2002).

The fixed-cost Capacitated Multicommodity Network Design (CMND) problem is one type of the network design models. In this model, multiple commodities such as goods, data, people, etc., must be routed between different points of origin and destination on the limited capacity arcs. In this problem, in addition to a unit cost (variable cost), a fixed cost is usually added due to the opening cost for the first time the arc is used (Ahuja et al., 1993). CMND problem seeks a network with minimum cost which satisfies demands of commodities. This minimum cost is the sum of the fixed and variable costs (Crainic, 2000).

Network design problems can easily be stated but solving them is too difficult (Balakrishnan et al., 1997). There are effective exact solution methods for uncapacitated network design problem. In exact algorithms, finding optimal solution is guaranteed. Benders decomposition and branch-and-bound methods have been the two most effective ones for this type of problem (Magnanti & Wong, 1984). Several surveys on network design models and their exact solution methods can be found in Balakrishnan et al. (1997), Crainic (2000), Frangioni and Gendron (2008), Magnanti and Wong (1984), and Minoux (1989, 2004). In addition, in Minoux (1989), a good survey on models and solution methods for multicommodity capacitated network design problems, especially simplex-based cutting plane and Lagrangian relaxation solution methods has been presented.

Adding capacity to the arcs of network design problem creates more complexity of this problem (Balakrishnan et al., 1997). Furthermore, large-scale network design problems which occur in real world applications are very difficult to solve (Magnanti & Wong, 1984). There are theoretical and empirical evidence that the capacitated network design problems are NP-hard (Balakrishnan et al., 1997; Magnanti & Wong, 1984). That is to say, there is no efficient algorithm which can solve them in polynomial time. As a result, the researchers proposed approximation methods to solve them. However, this type of solution methods cannot guarantee the optimality of the solutions. Metaheuristics that are a kind of approximation methods deal with these problems by introducing systematic rules to escape from local optima (Glover & Kochenberger, 2003).

To deal with CMND problems, several matheuristic algorithms have been proposed in the literature. Matheuristics are heuristic algorithms made by the interoperation of metaheuristics and mathematic programming techniques. This approach has begun to appear regularly in the metaheuristics literature (Blum et al., 2011; Boschetti et al., 2009; Caserta & Voß, 2009).

In Crainic et al. (2000), a simplex-based Tabu Search (TS) for solving path-based capacitated network design was presented. Their TS method was based on combining pivot moves of the space of path variables with column generation. Crainic and Gendreau (2002) proposed cooperative parallel TS for CMND problem. They implemented several strategies and showed that parallel strategies can result better solution qualities. Ghamlouche et al. (2003) proposed a cycle-based neighborhood for multicommodity capacitated network design. Their proposed method attained the better results than (Crainic et al., 2000). Ghamlouche et al. (2004) proposed a new metaheuristic based on cycle-based neighborhood structure for CMND that outperforms the other previous methods. Crainic et al. (2004) proposed a matheuristic approach to solve CMND that combined characteristics of TS metaheuristic, Lagrangian relaxation, and slope-scaling methods. The proposed method presented better results than pervious methods in several instances. Alvarez et al. (2005) proposed a new method based on Scatter Search framework for undirected CMND problem. The proposed method was the first solution method for undirected CMND problem that could obtain near optimal solution for large-scale problems. Crainic et al. (2006) proposed a multilevel cooperative TS method. The results of the proposed method were compared with those of pervious methods that showed it could obtain better solutions especially for large problems with numerous commodities. Crainic and Gendreau (2007) implemented the Scatter Search metaheuristic for CMND problems. Although Scatter Search was not able to obtain to the results of the best existing metaheuristics for the problem, they showed their success in finding better solutions in some instances. Katayama et al. (2009) proposed a capacity scaling heuristic using a column generation and row generation technique to solve CMND problem. They compared their proposed method with pervious methods by using the same instances. The comparisons showed that the method results were better than those of pervious methods in almost all instances. The aim of this paper is to continue this line of research and propose a new matheuristic for the CMND problem.

This paper presents a matheuristic method for the arc-based formulation of CMND. In the proposed matheuristic, a binary array is considered as a solution representation and the SA algorithm manages open and closed arcs. Several strategies for opening and closing arcs are proposed and evaluated. In this matheuristic approach, for a given design vector, the arc-based formulation of CMND becomes a Capacitated Multicommodity minimum Cost Flow (CMCF) problem. The exact evaluation of the CMCF problems is performed using the Simplex method by an LP solver. To the best of available knowledge, such approach has

not been proposed in the literature as a solution method for CMND. The parameter tuning for the proposed algorithm is done by means of Design of Experiments (DOE) approach. The performance of the proposed algorithm is evaluated by solving benchmark instances of different dimensions from OR-Library (Beasley, 1990).

The paper is organized as follows. In the next section, mathematical formulation of CMND model is presented. Then, the proposed solution method to solve CMND is presented, and its parameters tuning and computational results are given in further sections. Finally, the conclusions are presented.

The CMND Mathematical Formulation

In CMND problem, multiple commodities must be routed between different points of origin and destination on the arcs with limited capacity. Suppose a directed network $G = (N, A)$ in which N is the set of nodes and A is the set of directed arcs. Let K be the set of commodities, then the amount of each commodity k which must flow from its origin $O(k)$ to its destination $D(k)$ is d_k. Let c_{ij}^k and f_{ij} be the per unit arc routing cost of commodity k on arc (i, j), and fixed arc designing cost of arc (i, j), respectively. CMND problem can be formulated as follows (Crainic, 2000).

$$\min \sum_{k \in K} \sum_{(i,j) \in A} c_{ij}^k x_{ij}^k + \sum_{(i,j) \in A} f_{ij} y_{ij} \qquad (1)$$

subject to

$$\sum_{j \in N} x_{ij}^k - \sum_{j \in N} x_{ji}^k = \begin{cases} d_k & if \quad i = O(k) \\ -d_k & if \quad i = D(k) \quad for \ all \quad k \in K \\ 0 & otherwise \end{cases} \qquad (2)$$

$$\sum_{k \in K} x_{ij}^k \leq u_{ij} y_{ij} \qquad for \ all \quad (i, j) \in A \qquad (3)$$

$$x_{ij}^k \geq 0 \ , \quad y_{ij} = 0 \ or \ 1 \qquad for \ all \ (i, j) \in A, \ k \in K \tag{4}$$

where, y_{ij} and x_{ij}^k are decision variables. y_{ij} is 0, if arc (i, j) is closed, and 1, if it is open. x_{ij}^k is the amount of commodity k that is shipped on arc (i, j). The objective function (1) is to minimize variable and fixed costs. Constraints (2) are the usual balancing equations of network flow problem which insure that each demand is starting from origin node, passing intermediate nodes, and reaching destination node. Constraints (3) demonstrate that the sum of flows on each arc (i, j) must not exceed the capacity u_{ij} of the arc. The CMND model which is presented above is arc-based formulation.

The Proposed Matheuristic Method for CMND

The Simulated Annealing (SA) algorithm was proposed by Kirkpatrick et al. (1983). The overall procedure of SA is as follows. The SA starts with initial solution and moves iteratively to neighborhood solutions. If the neighborhood solution is better than current solution, it is set as a current solution, otherwise it accepted by the probability $\exp(-\Delta E / T)$ in which ΔE is the dif-

ference between current and neighborhood objective value, and T is a parameter called temperature. At each temperature, several iterations are executed and the temperature is gradually decreased. At the first step of the search, temperature is set at high value (more probability of accepting worst solutions), and gradually it is decreased to a little value at the end of the search (less probability of accepting worst solutions). In the following subsections, the overall structure of the proposed SA and its components are described.

The Overall Structure of the Proposed Method

The solution method proposed in this paper is a matheuristic method that combines SA metaheuristic and the Simplex method as an exact solution method. In the proposed matheuristic method, the SA specifies open and closed arcs (when each y_{ij} variable must either be 0 or 1), and the Simplex method determines the amount of flows (x_{ij}^k) on the open arcs that are determined by the SA algorithm. With this process, the arc-based formulation of CMND becomes a CMCF problem (Formulation (5)-(8)) that is extremely easier than CMND problem to be solved.

Figure 1. Overall structure of the proposed matheuristic algorithm

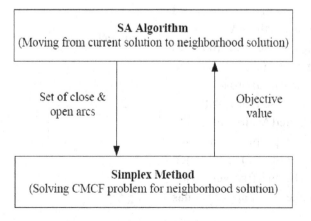

$$\min \sum_{k \in K} \sum_{(i,j) \in A} c_{ij}^{k} x_{ij}^{k} \qquad (5)$$

subject to

$$\sum_{j \in N} x_{ij}^{k} - \sum_{j \in N} x_{ji}^{k} = \begin{cases} d_k & if \quad i = O(k) \\ -d_k & if \quad i = D(k) \quad for \ all \quad k \in K \\ 0 & otherwise \end{cases}$$

$$(6)$$

$$\sum_{k \in K} x_{ij}^{k} \leq u_{ij} \qquad for \ all \quad (i,j) \in A \qquad (7)$$

$$x_{ij}^{k} \geq 0 , \qquad for \ all \quad (i,j) \in A, \quad k \in K \qquad (8)$$

Indeed, CMCF is a linear programming problem that can be solved easily by Simplex method. After solving this problem with CPLEX solver, the value of the objective function of this problem is added to the cost of open arcs, and the obtained value is passed to the SA algorithm. In the SA algorithm, acceptance of the solution and moving to new neighborhood solution is determined by this value (Figure 1).

Figure 2 shows a pseudocode for the proposed matheuristic algorithm for CMND problem. In this pseudocode, *currentSolution*, *initialSolution*, and *bestSolution* are current, initial, and best solutions, respectively. *Cycle* is a counter for the number of opening and closing moves. When this number reaches a predefined number (*MAX_CY-CLE*), the inner loop of the algorithm is ended, and this loop is iterated in a new temperature. Acceptance criterion is checked in opening and closing moves. In the following sections, components of this pseudocode are described.

The Solution Representation

As mentioned in the proposed algorithm, the SA algorithm manages open and closed arcs, and to do this, a binary array is considered as a solution representation (Figure 3). The length of the array

Figure 2. The pseudocode of the proposed matheuristic algorithm

```
// The pseudocode of the proposed algorithm
Input: parameters
Generate initial solution, x₀ ;
Set T = Tₘₐₓ ; /*setting initial temperature*/
Set currentSolution = initialSolution,
Set bestSolution = currentSolution;
Repeat
    Set Cycle = 0;
    Repeat
        Do closing move;
        Do opening move;
        Set Cycle = Cycle + 1;
    Until Cycle < MAX_CYCLE
    Update T;
Until stopping criteria satisfied
Output: the best found solution.
```

Figure 3. Solution representation

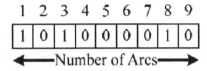

is the number of arcs. If an arc is open then the value of the relevant element is 1; otherwise, it is 0. For instance, the array of Figure 3 is related to a problem with nine arcs in which the value of 1 for the first element indicates the arc being open, and the value of 0 for the second element indicates the arc being closed.

Initial Solution Generation

As it is illustrated in Figure 2, generating initial solution is the first step of the proposed algorithm. To generate an initial solution in the proposed algorithm, seven procedures are tested and evaluated, and, finally, the best one is selected. The procedures are as follows.

1. Start with an array of arcs with zero values and sequentially open the closed arcs randomly until a feasible solution is found.
2. Start with an array of arcs with zero values and sequentially open the closed arcs based on the minimum combined cost until a feasible solution is found.
3. Open all arcs and allocate the demands to arcs according to the variable cost.
4. Open all arcs and allocate the demands to arcs according to the combined cost.
5. Start with an array of arcs with zero values and sequentially open the closed arcs based on the per unit fixed cost until a feasible solution is found.

6. Start with an array of arcs with zero values and sequentially open the closed arcs based on the combined cost until a feasible solution is found.
7. Start with an array of arcs with zero/one values that are indicated by a sub-MIP model.

In seventh procedure, an LP model by relaxing binary constraints of the CMND model is generated and solved. Then, a new sub-MIP model with an additional constraint is generated. This constraint (constraint 9) fixes the value of relative decision variables of those arcs which have no commodity for shipping to zero, in where A_c^{LP} is subset of A that are closed arcs in the LP solution. And a set of binary constraints is used for the relative decision variables of those arcs with commodity for shipping in the LP solution. Then the created small sub-MIP model is solved

$$\sum_{(i,j)\in A_c^{LP}} y_{ij} = 0 \qquad (9)$$

After using each of these procedures, the arcs with no assigned demands are closed. In the above procedures, the combined cost is per unit fixed cost plus variable cost of the arc.

Figure 4 illustrates the pseudocode of the first procedure for generating an initial solution. In this procedure, the first step randomly selects a closed arc to open. Then, by the Simplex method the model is solved according to the open arcs. If the obtained solution is feasible, then it is used as the initial solution; otherwise, another arc is selected for opening, and the procedure is iterated until a feasible solution is found. Finally, all the arcs with no assigned demands need to be closed. The fifth, second and sixth procedures are similar to the first one, with the criteria of selecting an arc for opening being different. In fourth procedure, all the arcs are assumed to be open and then the model is solved with the Simplex algorithm, and

Figure 4. The first procedure to generate an initial solution

```
// The first procedure to generate an initial solution
Generate an array of arcs with 0 values;
Repeat
    Select an arc for randomly opening;
    Solve CMCF problem with open arcs;
Until a feasible solution found;
Close idle arcs;
Output: the generated initial solution.
```

the combined cost as the multiplication of flow variables to specify the demand allocation to each arc. After allocating demands, the idle arcs must be closed.

To select one of these procedures, they are implemented and run several times. The seventh procedure gains better result than all the other ones. Therefore, this method is used for generating initial solution.

Moving to Neighborhood Solutions

Following generation of an initial solution, a move must be performed for searching the solution space. In the proposed algorithm, opening and closing moves are presented.

In the closing move, an open arc is selected to close. This type of move reduces the cost of objective function, but it may cause the infeasibility of the solution. In the proposed algorithm, eight strategies are considered and tested for closing move as here below.

1. Close an arc randomly.
2. Close an arc randomly, when the solution becomes infeasible, a shortest path is replaced by the closed arc to ensure solution feasibility.
3. Close an arc with the maximum per unit fixed cost.
4. Close an arc with the maximum variable cost.
5. Close an arc with the maximum unused capacity utilization.
6. Start from a random point and close the open arcs.
7. Close an arc with the maximum combined cost.
8. Close an arc with the maximum per unit combined cost.

As an example, Figure 5 provides a pseudocode for the second method of closing move. In this method, after closing an open arc, if the solution becomes infeasible, the algorithm opens the arcs of a path replace of the closed arc. The path is randomly selected from k shortest paths by a k-shortest path algorithm. This path is from the origin of the closed arc to its destination. Twenty shortest paths are generated (i.e., $k = 20$) for each arc as input to the proposed method. In the other closing strategies, after closing an open arc, if the solution becomes infeasible, the algorithm cancels this closing move. In the first and second methods, the termination condition is the maximum number of closing in which the solution is not accepted (*MAX_CLOSING*).

In the opening move, a closed arc is selected to become open. This move is performed to search the solution space more comprehensively with no

Figure 5. The pseudocode of the second closing move procedure

```
// The pseudocode of the second closing move procedure
Set countClosing = 0;
Repeat
  Close an open arc randomly;
  Solve LP model for finding the neighborhood solution;
  If (neighbrSolution is feasible) {
      If (neighbrObjective < currentObjective) { // accepting the closed arc
        Set currentSolution = neighbrSolution;
        Set currentObjective = neighbrObjective;
        If (neighbrObjective < bestObjective) { // updating the best solution
          Set bestSolution = neighbrSolution;
          Set bestObjective = neighbrObjective;
        }
      }Else {    // neighbrObjective ≥ currentObjective
        If (acceptance criteria of SA met) { // Accept the closed arc
          Set currentSolution = neighbrSolution;
          Set currentObjective = neighbrObjective;
        }Else {    // if acceptance criteria of SA not met, cancel the closed arc
          Set countClosing = countClosing +1;
        }
      }
  }
  Else {  // solution is infeasible
      Replace the closed arc with a shortest path randomly;
      Solve LP model for the the neighborhood solution;
      If (neighbrSolution is feasible) {
          If (neighbrObjective < currentObjective) { // Accept the closed arc
            Set currentSolution = neighbrSolution;
            Set currentObjective = neighbrObjective;
            If (neighbrObjective < bestObjective) {
              Set bestSolution = neighbrSolution;
              Set bestObjective = neighbrObjective;
            }
          }Else {  // neighbrObjective ≥ currentObjective
            If (acceptance criteria of SA met) { // Accept the closed arc
              Set currentSolution = neighbrSolution;
              Set currentObjective = neighbrObjective;
              }Else {  // if acceptance criteria of SA not met, cancel the closed arc
              Set countClosing = countClosing +1;
            }
          }
      }Else {  // if solution is infeasible, cancel the closed arc
          Set countClosing = countClosing +1;
      }
  }
Until countClosing < MAX_CLOSING;
```

effect on the solution feasibility. In the proposed algorithm, for opening move following procedures are evaluated.

1. Open an arc randomly.
2. Open an arc with the minimum per unit fixed cost.
3. Open an arc with the minimum variable cost.
4. Start from an arc randomly in the solution array, and the next closed arcs are opened one by one.
5. Open an arc with the minimum combined cost.
6. Open an arc with the minimum per unit combined cost.

As an instance, a pseudocode of the fifth procedure is illustrated in Figure 6. In this method, the termination condition of the opening move is a pre-specified number of arcs opening (*MAX_OPENING*) when the SA acceptance function does not satisfy.

Stopping Criteria

In the proposed algorithm, two stopping criteria are evaluated, namely, time limitation and *m* number of reducing temperature with no improvement.

Figure 6. The pseudocode of the fifth opening move procedure

```
// The pseudocode of the fifth opening move procedure
Set countOpening = 0;
Repeat
    Open a closed arc with the minimum combined cost;
    Solve LP model for finding the neighborhood solution;
    If (neighbrObjective < currentObjective) { // Accept the closed arc
        Set currentSolution = neighbrSolution;
        Set currentObjective = neighbrObjective;
        If (neighbrObjective < bestObjective) { // updating the best solution
            Set bestSolution = neighbrSolution;
            Set bestObjective = neighbrObjective;
        }
    }Else { // neighbrObjective ≥ currentObjective
        If (acceptance criteria of SA met) { // Accept the closed arc
            Set currentSolution = neighbrSolution;
            Set currentObjective = neighbrObjective;
        }Else { // if acceptance criteria of SA not met, cancel the closed arc
            Close the open arc
            Set countOpening = countOpening +1;
        }
    }
Until countOpening< MAX_OPENING;
```

Table 1. The characteristics of the benchmark instances

Problem No.	No. of nodes	No. of arcs	No. of Commodity	F/V	L/T
1	20	230	40	V	L
2	20	230	40	V	T
3	20	230	40	F	T
4	20	230	200	V	L
5	20	230	200	F	L
6	20	230	200	V	T
7	20	230	200	F	T
8	20	300	40	V	L
9	20	300	40	F	L
10	20	300	40	V	T
11	20	300	40	F	T
12	20	300	200	V	L
13	20	300	200	F	L
14	20	300	200	V	T
15	20	300	200	F	T
16	25	100	10	V	L
17	25	100	10	F	L
18	25	100	10	F	T
19	25	100	30	V	T
20	25	100	30	F	L
21	25	100	30	F	T
22	30	520	100	V	L
23	30	520	100	F	L
24	30	520	100	V	T
25	30	520	100	F	T
26	30	520	400	V	L
27	30	520	400	F	L
28	30	520	400	V	T
29	30	520	400	F	T
30	30	700	100	V	L
31	30	700	100	F	L
32	30	700	100	V	T
33	30	700	100	F	T
34	100	400	10	V	L
35	100	400	10	F	L
36	100	400	10	F	L
37	100	400	30	V	T
38	100	400	30	F	L
39	100	400	30	F	T

Parameters Tuning

The parameters of the proposed algorithm are tuned using the Design of Experiments (DOE) approach and Design-Expert statistical software. One can define an experiment as a test or series of tests in which purposeful changes are made to the input variables of a process or system so that to observe and identify the reasons for changes that may be observed in the output response. DOE refers to the process of planning the experiments so that appropriate data that can be analyzed by statistical methods, will be collected, resulting in valid and objective conclusions (Biratteri, 2009).

The important parameters in DOE approach are response variable, factor, level, treatment and effect. The response variable is the measured variable of interest. In the analysis of metaheuristics, the typical measures are the solution quality and computation time (Adenso-Díaz & Laguna, 2006). A factor is an independent variable manipulated in an experiment because it is thought to affect one or more of the response variables. The various values at which the factor is set are known as levels. In metaheuristic performance analysis, the factors include both the metaheuristic tuning parameters and the most important problem characteristics (Biratteri, 2009). A treatment is a specific combination of factor levels. The particular treatments will depend on the particular experiment design and on the ranges over which factors are varied. An effect is a change in the response variable due to a change in one or more factors (Ridge, 2007). DOE is a method that can be used to determine important parameters and interactions between them. Four-stages of DOE consist of screening and diagnosis of important factors, modeling, optimization and assessment (Montgomery, 2005). This methodology is known sequential experimentation, which is used to set the parameters in the DOE approach and has been used in this paper for the proposed algorithm.

To tune parameters for the proposed algorithm four instances with different characteristics are

Table 2. The results of implementing algorithm for the benchmark instances

Prob No.	TC		PR		MLEVEL	Proposed algorithm	
	ObjVal	Time	ObjVal	Time	ObjVal	ObjVal	Time
1	430,628	214.12	424,385	148.82	426,702	423,484	101.74
2	372,522	241.44	371,811	156.92	371,475	371,475	142.21
3	652,775	259.53	645,548	172.16	652,894	645,412	113.34
4	100,001	2585.94	100,404	2494.92	98,582	95,638	4827.24
5	148,066	3141.91	147,988	2878.27	143,150	140,263	7241.68
6	106,868	2729.57	104,689	2210.86	102,030	99,589	4911.42
7	147,212	3634.1	147,554	3385.75	141,188	137,729	6729.39
8	432,007	304.6	429,398	224.91	429,837	429,417	91.57
9	602,180	335.59	590,427	228.33	593,544	589,081	428.09
10	466,115	378.7	464,509	247.89	466,004	464,509	231.47
11	615,426	349.48	609,990	214.43	619,203	610,243	371.53
12	81,367	4085.89	78,184	3565.98	78,209.50	76,112	18000
13	122,262	4210.14	123,484	4012.64	121,951	116,346	14391.61
14	80,344	4203.84	78,866.80	3924.21	77,251	76,095	5182.44
15	113,947	4854.61	113,584	3857.14	111,173	108,388	6719.88
16	14,712	19.48	14,712	12.54	14,712	14,712	9.14
17	14,941	22.33	14,941	14.13	14,941	14,941	7.26
18	50,529	33.84	49,899	24.06	49,937	49,899	8.68
19	365,385	141.51	365,385	101.44	365,385	365,272	94.29
20	37,515	112.48	37,654	75.15	37,607	37,500	142.41
21	87,325	132.92	86,428	96.98	86,461.30	85,591	82.52
22	56,603	2260.69	54,904	1194.12	55,754	54,243	93.25
23	103,657	2683.74	102,054	1459.99	99,817	98,678	8401.33
24	54,454	2715.76	53,017	1513.66	53,512	52,480	2396.11
25	105,130	2891.93	106,130	1522.68	102,477	101,513	11429.29
26	122,673	55,771.2	119,416	27477.40	115,671	112,973	18000
27	164,140	40070.40	163,112	36669.30	156,601	154,233	18000
28	122,655	4678.8	120,170	23089.10	120,980	116,181	16281.92
29	169,508	49886.80	163,675	52173.20	160,217	146,295	18000
30	50,041	2959.51	48,723	1860.61	48,869	48,156	5196.48
31	64,581	3181.72	63,091	1837.5	63,756	62,574	17271.51
32	48,176	3745.82	47,209	1894.08	47,457	46,255	4269.26
33	57,628	3547.09	56,575.50	1706.06	56,910	56,591	7195.39
34	28,786	252.15	28,485	89.21	28,553	28,443	52.88
35	24,022	196.5	24,022	82.86	24,022	25,336	34.77
36	67,184	451.28	65,278	209.93	66,284	65,131	1423.98
37	385,508	1199.88	384,926	492.76	385,282	384,934	1702.76
38	51,831	717.42	51,325	314.97	50,456	50,281	1629.27
39	147,193	1300.79	141,359	480.86	145,721	142,893	2192.71

generated. In the proposed algorithm, solution quality and CPU time are considered as the response variables. As a solution quality indicator, for each instance the average deviation from the solution of the other algorithm is calculated (Equation 10).

$$Relative\ Gap =$$

$$\frac{\text{Obtained solution - Other algorithm solution}}{\text{Other algorithm solution}} \times 100 \qquad (10)$$

Several parameters exist in the proposed algorithm. *CLOSING_MOVE_TYPE* and *OPENING_MOVE_TYPE* are the type of closing and opening moves. *MAX_CYCLE* is the number of cycle in each temperature to reach the equilibrium status. *MAX_OPENING* is the number of opening and *MAX_CLOSING* is the number of closing in each cycle. *MAX_TEMPERATURE* is the initial temperature, and, finally, *COOLING_RATE* is the rate of reducing (updating) the temperature.

CLOSING_MOVE_TYPE and *OPENING_MOVE_TYPE* are the two parameters considered for the first step of tuning parameters. To tune this parameters, the value of *MAX_CYCLE*, *MAX_OPENING*, *MAX_CLOSING*, *MAX_TEMPERATURE*, and *COOLING_RATE* are set at 5, 5, 5, 1000, and 0.9, respectively. Then, the type of opening and closing are obtained by DOE method. The best type of opening move is the fifth type. This move opens an arc with the minimum combined cost. The final obtained type of closing move is the second type. This move closes an arc randomly; when the solution becomes infeasible, a shortest path is replaced by the closed arc to ensure solution feasibility. To tune the other parameters, the types of opening and closing moves are set at the obtained types, and the best values are obtained by DOE approach. The final values for *MAX_CYCLE*, *MAX_OPENING*, *MAX_CLOSING*, *MAX_TEMPERATURE*, and *COOLING_RATE* are 3, 6, 5, 200, and 0.99, respectively. The final termination criteria is two conditions including "100 times of

reducing temperature" and "30 times of reducing temperature with no improvement" are considered simultaneously.

Experimental Results

Several samples with different characteristics are selected from OR-Library to perform evaluation of the proposed algorithm (Beasley, 1990). These are randomly generated benchmark instances (Gendron & Crainic, 1994). The same instances have been used in several other papers such as (Crainic et al., 2000, 2004, 2006; Ghamlouche et al., 2003, 2004; Katayama et al., 2009). Table 1 shows these samples and their characteristics. In the last two columns, F means that the fixed costs are predominate relative to the variables costs, and V means the contrary; L indicates that the capacities are loose while T indicates that the capacities are tight.

The proposed algorithm is implemented with Java programming language. The experiments have been conducted on a computer with Intel Core 2 Duo 2.53 GHz CPU and 4.00 GB RAM.

The algorithm is implemented with tuned parameters for all of the samples. The results are summarized and compared with those of the other pervious methods in Table 2. The last column is related to the proposed algorithm in this paper. The column "TC" is the results of cycle-based TS of Ghamlouche et al. (2003). The column "PR" is related to path relinking method of Ghamlouche et al. (2004). The column "MLEVEL" shows the results of the multilevel cooperative TS of Crainic et al. (2006).

Table 3 illustrates in which samples the proposed algorithm can obtain better solution in comparison with other methods. These improvements can be seen as negative gap values in the table. The proposed algorithm improved 36, 29 and 35 problems in the cycle-based TS (Ghamlouche et al., 2003), path relinking (Ghamlouche et al., 2004), and multilevel cooperative TS (Crainic et al., 2006), respectively.

Table 3. Comparing obtained results with those of other methods by means of calculating gaps

PROB	TC	PR	MLEVEL
1	-1.66	-0.21	-0.75
2	-0.28	-0.09	0.00
3	-1.13	-0.02	-1.15
4	-4.36	-4.75	-2.99
5	-5.27	-5.22	-2.02
6	-6.81	-4.87	-2.39
7	-6.44	-6.66	-2.45
8	-0.60	0.00	-0.10
9	-2.18	-0.23	-0.75
10	-0.34	0.00	-0.32
11	-0.84	0.04	-1.45
12	-6.46	-2.65	-2.68
13	-4.84	-5.78	-4.60
14	-5.29	-3.51	-1.50
15	-4.88	-4.57	-2.50
16	0.00	0.00	0.00
17	0.00	0.00	0.00
18	-1.25	0.00	-0.08
19	-0.03	-0.03	-0.03
20	-0.04	-0.41	-0.28
21	-1.99	-0.97	-1.01
22	-4.17	-1.20	-2.71
23	-4.80	-3.31	-1.14
24	-3.63	-1.01	-1.93
25	-3.44	-4.35	-0.94
26	-7.91	-5.40	-2.33
27	-6.04	-5.44	-1.51
28	-5.28	-3.32	-3.97
29	-13.69	-10.62	-8.69
30	-3.77	-1.16	-1.46
31	-3.11	-0.82	-1.85
32	-3.99	-2.02	-2.53
33	-1.80	0.03	-0.56
34	-1.19	-0.15	-0.39
35	5.47	5.47	5.47
36	-3.06	-0.23	-1.74
37	-0.15	0.00	-0.09
38	-2.99	-2.03	-0.35
39	-2.92	1.09	-1.94

Table 4. The comparison of the proposed algorithm with other methods

	Cycle Based	Path Relinking	Multilevel
Average GAPs (%)	-3.11	-1.91	-1.43
Number of problems improved	36	29	35
Number of problems not improved	1	6	1
Number of problems with same solutions	2	4	3

Table 4 illustrates the comparison of the proposed algorithm with other methods. The first row gives averages of gap between the proposed algorithm and other methods, for example the proposed algorithm improved the cycle-based tabu search in Ghamlouche et al. (2003) about -3.11 percent. The second row corresponds to those problems which the proposed algorithm improved the other methods solutions. The third row shows the number of problems which the proposed algorithm failed to improve. And the last row points to the number of problems for which the proposed algorithm and the other methods obtain the same solutions.

CONCLUSION

In this paper, a matheuristic combining Simulated Annealing and Simplex method is proposed for Multicommodity Capacitated Network Design (CMND) problems. The results of the proposed method are compared with the solutions of the other previous methods in the literature. The results show that the proposed matheuristic algorithm is able to obtain better solutions in most of instances than previous methods. It is hoped the proposed approach for CMND problem will draw more researchers to develop and implement this approach for real-life problems. In addition, the proposed representation and neighborhood structure can be used in other metaheuristics such as genetic algorithm and tabu search, so that, to pave the way for future research.

REFERENCES

Adenso-Díaz, B., & Laguna, M. (2006). Fine-tuning of algorithms using fractional experimental design and local search. *Operations Research, 54*, 99–114. doi:10.1287/opre.1050.0243

Ahuja, R. K., Magnanti, T. L., & Oril, J. B. (1993). *Theory, algorithms, and application.* Upper Saddle River, NJ: Prentice Hall.

Alvarez, A. M., Lez-Velarde, J. L. G., & De-alba, K. (2005). Scatter search for network design problem. *Annals of Operations Research, 138*, 159–178. doi:10.1007/s10479-005-2451-4

Balakrishnan, A., Magnanti, T. L., & Mirchandani, P. (1997). Annotated bibliographies in combinatorial optimization . In Dell'Amico, M., Maffioli, F., & Martello, S. (Eds.), *Network design* (pp. 311–334). New York, NY: John Wiley & Sons.

Beasley, J. E. (1990). OR-Library: Distributing test problems by electronic mail. *The Journal of the Operational Research Society, 41*, 1069–1072.

Biratteri, M. (2009). *Tuning Metaheuristics: A machine learning perspective.* Heidelberg, Germany: Springer-Verlag.

Blum, C., Puchinger, J., Raidl, G. R., & Roli, A. (2011). Hybrid metaheuristics in combinatorial optimization: A survey. *Applied Soft Computing, 11*, 4135–4151. doi:10.1016/j.asoc.2011.02.032

Boschetti, M., Maniezzo, V., Roffilli, M., & Rohler, A. B. (2009). Mateheuristics: Optimization, simulation and control. In *Proceedings of the 6th International Workshop on Hybrid Metaheuristics* (pp. 171-177).

Caserta, M., & Voß, S. (2009). Metaheuristics: Intelligent problem solving . In Maniezzo, V., Stützle, T., & Voß, S. (Eds.), *Matheuristics: Hybridizing metaheuristics and mathematical programming* (pp. 1–38). Berlin, Germany: Springer-Verlag.

Crainic, T. G. (2000). Service network design in freight transportation. *European Journal of Operational Research, 122*, 272–288. doi:10.1016/S0377-2217(99)00233-7

Crainic, T. G., Dejax, P., & Delorme, L. (1989). Models for multimode location problem with interdepot balancing requirement. *Annals of Operations Research, 18*, 277–302. doi:10.1007/BF02097809

Crainic, T. G., & Gendreau, M. (2002). Cooperative parallel tabu search for capacitated network design. *Journal of Heuristics, 8*, 601–627. doi:10.1023/A:1020325926188

Crainic, T. G., & Gendreau, M. (2007). A scatter search heuristic for the fixed-charge capacitated network design problem. *Metaheuristics, Operations Research/Computer Science Interfaces Series, Part I, 39*, 25-40.

Crainic, T. G., Gendreau, M., & Farvolden, J. M. (2000). A Simplex-based tabu search method for capacitated network design. *INFORMS Journal on Computing, 12*, 223–236. doi:10.1287/ijoc.12.3.223.12638

Crainic, T. G., Gendreau, M., & Hernu, G. (2004). A slope scaling/Lagrangian perturbation heuristic with long-term memory for multicommodity capacitated fixed-charge network design. *Journal of Heuristics, 10*, 525–545. doi:10.1023/B:HEUR.0000045323.83583.bd

Crainic, T. G., Li, Y., & Toulouse, M. (2006). A first multilevel cooperative algorithm for capacitated multicommodity network design. *Computers & Operations Research, 33*, 2602–2622. doi:10.1016/j.cor.2005.07.015

Frangioni, A., & Gendron, B. (2008). 0-1 reformulation of the multicommodity capacitated network design problem. *Discrete Applied Mathematics*, *157*, 1229–1241. doi:10.1016/j.dam.2008.04.022

Gendron, B., & Crainic, T. G. (1994). *Relaxations for multicommodity capacitated network design problems* (Publication No. CRT-945). Montreal, QC, Canada: Centre de recherche sur les transports, Université de Montréal.

Gendron, B., Crainic, T. G., & Frangioni, A. (1998). Multicommodity capacitated network design . In Sanso, B., & Soriono, P. (Eds.), *Telecommunication network planning* (pp. 1–19). Boston, MA: Kluwer Academic.

Ghamlouche, I., Crainic, T. G., & Gendreau, M. (2003). Cycle-based neighbourhoods for fixed-charge capacitated multicommodity network design. *Operations Research*, *51*, 655–667. doi:10.1287/opre.51.4.655.16098

Ghamlouche, I., Crainic, T. G., & Gendreau, M. (2004). Path relinking, cycle-based neighborhoods and capacitated multicommodity network design. *Annals of Operations Research*, *131*, 109–133. doi:10.1023/B:ANOR.0000039515.90453.1d

Glover, F., & Kochenberger, G. A. (2003). *Handbook of metaheuristics*. Boston, MA: Kluwer Academic.

Jayaraman, V., & Ross, A. (2002). A simulated annealing methodology to distribution network design and management. *European Journal of Operational Research*, *144*, 629–645. doi:10.1016/S0377-2217(02)00153-4

Katayama, N., Chen, M., & Kubo, M. (2009). Capacity scaling heuristic for the multicommodity capacitated network design problem. *Journal of Computational and Applied Mathematics*, *232*, 90–101. doi:10.1016/j.cam.2008.10.055

Kirkpatrick, S., Gelatt, C., & Vecchi, M. (1983). Optimization by simulated annealing. *Science*, *220*, 671–680. doi:10.1126/science.220.4598.671

Magnanti, T. L., & Wong, R. T. (1984). Network design and transportation planning: models and algorithms. *Transportation Science*, *18*, 1–55. doi:10.1287/trsc.18.1.1

Minoux, M. (1989). Network synthesis and optimum network design problems: Models, solution methods and applications. *Networks*, *19*, 313–360. doi:10.1002/net.3230190305

Minoux, M. (2001). Discrete cost multicommodity network optimization problems and exact solution methods. *Annals of Operations Research*, *106*, 19–46. doi:10.1023/A:1014554606793

Minoux, M. (2004). Polynomial approximation schemes and exact algorithms for optimum curve segmentation problems. *Discrete Applied Mathematics*, *144*, 158–172. doi:10.1016/j.dam.2004.05.003

Montgomery, D. (2005). *Design and analysis of experiments*. New York, NY: John Wiley & Sons.

Ridge, E. (2007). *Design of experiments for the tuning of optimization algorithms* (Unpublished doctoral dissertation). University of York, York, UK.

This work was previously published in the International Journal of Applied Metaheuristic Computing, Volume 2, Issue 4, edited by Peng-Yeng Yin, pp. 13-28, copyright 2011 by IGI Publishing (an imprint of IGI Global).

Chapter 3
A Hybrid Meta-Heuristic Algorithm for Dynamic Spectrum Management in Multiuser Systems:
Combining Simulated Annealing and Non-Linear Simplex Nelder-Mead

Hernan X. Cordova
Free University of Brussels, Belgium

Leo Van Biesen
Free University of Brussels, Belgium

ABSTRACT

One of the major sources of performance degradation of current Digital Subscriber Line systems is the electromagnetic coupling among different twisted pairs within the same cable bundle (crosstalk). Several algorithms for Dynamic Spectrum management have been proposed to counteract the crosstalk effect but their complexity remains a challenge in practice. Optimal Spectrum Balancing (OSB) is a centralized algorithm that optimally allocates the available transmit power over the tones making use of a Dual decomposition approach where Lagrange multipliers are used to enforce the constraints and decouple the problem over the tones. However, the overall complexity of this algorithm remains a challenge for practical DSL environments. The authors propose a low-complex algorithm based on a combination of simulated annealing and non-linear simplex to find local (almost global) optimum spectra for multiuser DSL systems, whilst significantly reducing the prohibitive complexity of traditional OSB. The algorithm assumes a Spectrum Management Center (at the cabinet side) but it neither relies on own end-user modem calculations nor on messaging-passing for achieving its performance objective. The approach allows furthering reducing the number of function evaluations achieving further reduction on the convergence time (up to ~27% gain) at reasonable payoff (weighted data rate sum).

DOI: 10.4018/978-1-4666-2145-9.ch003

INTRODUCTION

Ideally, Optical Fiber should be the technology to be deployed to reaching the end-users (Fiber to the Home -FTTH-). However, the investment and maintenance costs are still prohibitive for some countries as indicated by Cioffi (2007).

Digital subscriber line (DSL) technology is currently the most widely deployed broadband access technology and will continue to play an important role during the coming years. In a previous report from Point-Topic by Vanier (2009), it is mentioned that the number of DSL subscribers adds up to ~65% of all worldwide broadband access technologies (being followed by ~21% of cable modem subscribers).

However, to cope with the bandwidth-intensive and a mixed set of quality-of-service (QoS) and quality-of-experience (QoE) requirements of the many emerging broadband applications and services (i.e., VoIP, triple-play services including HDTV, IPTV, video-conferences, etc.), it is essential to further improve DSL technology.

One of the major sources of performance degradation of current DSL systems is the electromagnetic coupling among different twisted pairs within the same cable bundle (also referred to as crosstalk). The presence of crosstalk transforms DSL systems into a very challenging multi-user multicarrier interference environment, where different users, i.e., modems, can significantly impact each other's datarate transmission and the quality of the line; thus, having a direct impact on the final end-user experience.

Spectrum coordination management consists of coordinating the transmit spectrum of each modem so as to prevent (or substantially mitigate) the impact of crosstalk; nowadays, relatively new techniques like the different levels of DSM (single-user coordination, multi-user, vectoring, etc.) as proposed in the literature (Cioffi, 2007; Cioffi & Mohseni, 2004; Song et al., 2002; Ginis & Cioffi, 2002; Cioffi et al., 2006; Yu, Ginis, & Cioffi, 2002; Cendrillon, Moonen, Verlinden, Bostoen, & Yu, 2004; Cendrillon et al., 2006; Yu et al., 2004; Tsiaflakis et al., 2005; Papandriopoulus & Evans, 2009; Cendrillon & Moonen, 2005; Tsiaflakis et al., 2008) allow increasing dramatically the bitrates over copper lines, up to (or even higher than) fiber capacities.

One of the first proposed DSM algorithms (Yu, Ginis, & Cioffi, 2002) was iterative waterfilling (IWF) where it was demonstrated a significant performance improvement over traditional (let say, static spectrum management) multiuser DSL lines. This algorithm iteratively measures the aggregate interference received from all other users, greedily water-pouring their own power allocation. However, this approach disregards the impact on other users.

Cendrillon et al. (2004, 2006), proposed an Optimum Spectrum Balancing (centralized) algorithm, therefore assuming centralized control in a Spectrum Management Center (SMC). OSB applies dual decomposition (Yu, Lui, & Cendrillon, 2004) making use of Lagrange multipliers to enforce constraints that are coupled over the tones to find the optimal power allocation to a predetermined quantum and to lower the complexity from exponential to linear in the number of tones. As this algorithm considers the damage done to other modems, OSB avoids a selfish optimum, thereby significantly improving its performance over IWF. However, OSB remains difficult to implement over practical DSL lines though many attempts to reduce this complexity appear in Cendrillon et al. (2004, 2006).

Tsiaflakis et al. (2005) proposed to exploit Lagrange multipliers properties, making the number of Lagrange multiplier evaluations independent of the number of users and smaller compared to other existing search algorithms. However, the convergence time still remains a challenge.

Papandriopoulos and Evans (2009) proposed a new (semi-centralized/distributed) algorithm whose performance outperforms IWF at similar complexity, whereby applying a series of convex relaxations. Cendrillon and Moonen (2005) proposed an iterative spectrum balancing algorithm to reduce the complexity and the convergence time, leading to suboptimal performance, yet.

We initially explored a different approach by using a simplex bounded search algorithm (Cordova & van Biesen, 2010), to reduce overall OSB complexity obtaining near-optimal results but the convergence time was still about 900s for a two-user case. Extra efforts have been performed since then to drastically reduce the convergence time by using Globalized Bounded Nelder-Mead (GBNM) leading to tractable computational complexity and reducing the convergence time to less than one minute. Other algorithms have been proposed in the literature for distributed multiuser systems (Tsiaflakis, Diehl, & Moonen, 2008; Tsiaflakis & Moonen, 2008). We do not focus on distributed algorithms but we further continue elaborating on the idea to use meta-heuristics algorithms for practical implementations over DSL systems.

Thus, in this paper, we combine simulated annealing and non-linear simplex to find a near-optimal solution. It is proven by simulations (two-user scenario) that the near-optimal solution hereby found corresponds "practically" (negligible differences) to the global optimum found by OSB at enormous reduction in the convergence time.

First we describe the system model and the problem definition. An overview of the combined algorithms we will be using to solve the problem statement will be provided. Our proposed solution and algorithm is revised. Simulations results are presented and discussed. Conclusions are drawn.

SYSTEM MODEL AND OBJECTIVE

System Model

A multiuser (MU) DMT channel environment of users with a maximum number of tones is represented by:

$$y_i^n = H_i^n . X_i^n + Z_i^n \qquad (1)$$

where H_{ii}^n is the diagonal representing the main channel transfer function of user i at tone n H_{ij}^n ($i \neq j$) corresponds to the off-diagonal elements, representing the crosstalk transfer function from user j to user i at tone n; Z_i^n represents the additive white Gaussian noise (AWGN) of user at tone whose power is given by $\tilde{A}_{i,n}^2$.

The vector $x_n = [x_n^1, x_n^2, \ldots, x_n^M]^T$ represents the M transmitted signals on tone n; the vector $z_n = [z_n^1, z_n^2, \ldots, z_n^M]^T$ represents the vector additive noise on tone n, containing thermal noise, alien crosstalk, and radio-frequency interference. The transmit power is defined as $s_n^i \triangleq \Delta f E\{|x_n^i|^2\}$, the noise power is defined as $\tilde{A}_n^i \triangleq \Delta f E\{|z_n^i|^2\}$; the vector containing the transmit power of user i on all tones is given by $s^i = [s_1^i, s_2^i, \ldots, s_N^i]^T$; the vector containing the transmit power of all users on tone n is $s_n = [s_n^1, s_n^2, \ldots, s_n^M]^T$.

The general expression for bitloading in a multiuser environment, assuming that each modem treats interference from other modems as noise (so, leading to a Gaussian distribution) is given by:

$$b_i^n = \log_2(1 + \frac{S_i^n |H_{ii}^n|^2}{(N_i^n)}) \qquad (2)$$

where the total noise power is given by:

$$N_i^n = \tilde{A}_{i,n}^2 + \sum_{\substack{j=1 \\ j \neq i}}^{M} S_j^n |H_{ij}^n|^2 \qquad (3)$$

And depends on the bit-error rate, coding gain and noise margin. The aggregate datarate in a multiuser environment is thus given by:

$$R_{total} = \sum_{i=1}^{M} R_i = f_{sym} . \sum_{i=1}^{M} \sum_{n=1}^{N} b_i^n \qquad (4)$$

where f_{sym} is the symbol rate (a typical value for DMT systems is 4000 symbols/s).

Spectrum Management Problem

The main purpose of management and optimization of the transmit power across all users and tones is to balance the impact of crosstalk; this is based on a set of objectives and/or constraints. Higher data rates and (recently also very important) power minimization (for green purposes), are of common interest. Thus, the spectrum management optimization problem is related to maximize (or minimize the loss given the protection to legacy systems) the data rate of the incumbent modem given that it does not cause additional noise as if it were a disturber from the legacy modem family (in the specific frequency-range). We will start focusing on the rate adaptive problem: that is, to maximize the weighted datarate of all users, under different power constraints (i.e., total power per user is less or equal than the maximum power and/or PSD at every tone should be lower than the power spectral mask).

The achievable region $\mathcal{R}$ features the set of all possible combinations of datarates that can be achieved. This is defined as:

$$\mathcal{R} = \left\{ (R^i : i \in \mathcal{M}) | R^i = f_{sym} \sum_{n \in \mathcal{N}} b_i^n (s^n), \{s_i, \ i \in \mathcal{M}\} \in \mathbb{S} \right\}$$
$$(5)$$

with $\mathbb{S}$ being defined as:

$$\mathbb{S} = \left\{ \left(s_i^n : n \in \mathcal{N}, i \in \mathcal{M} \right) : \sum_{n=1}^{N} S_i^n \leq S_{i,\ total}, \right.$$
$$\left. 0 \leq S_i^n \leq S_i^{n,\ mask} \right\} \quad (6)$$

Where the vector containing the transmit power of user i on all tones is given by $s^i = [s_1^i, s_2^i, ..., s_N^i]^T$; the vector containing the transmit power of all users on tone n is

$s_n = [s_n^1, s_n^2, ..., s_n^M]^T$, and where $\mathbb{S}$ models the set of constraints on the transmit powers. We can also notice a non-convex relation between the transmit powers and datarate, see expressions (2), (3) and (4), hence we are dealing with NP-hard non-convex optimization type of problem.

We will focus on the rate sum optimization that is, maximizing a weighted rate sum. Mathematically this can be expressed as:

$$\max R_{total} = \sum_{i=1}^{M} w_i R_i = f_{sym}. \sum_{i=1}^{M} w_i \sum_{n=1}^{N} b_i^n \quad (7)$$

subject to

$$0 \leq S_i^n \leq S_i^{n,\ mask}; \ n \in \mathcal{N}, i \in \mathcal{M}$$

$$\sum_{n=1}^{N} S_i^n \leq S_{i,\ total}; \ i \in \mathcal{M}$$

$$b_i^n \in Z_0^{b_{max}}$$

where $w_i R_i$ is weighted data rate to account for different class of services; S_i^n is the power in each subchannel n (for $n = 1, ..., \ N$) and for each user i (for $i = 1, ..., \ M$), b_{max} is the maximum number of possible bits to allocate, $Z_0^{b_{max}}$ is a set of elements of $\{0, 1, ..., b_{max}\}$ and $S_i^{n,\ mask}$ is the power mask typically imposed by regulatory entities and Standardization bodies to safeguard other existing systems.

A COMBINED META-HEURISTICS ALGORITHM: SIMULATED ANNEALING AND SIMPLEX NELDER-MEAD

Simulated annealing is a powerful technique for combinatorial optimization, i.e., for the optimization of large-scale functions that may assume several distinct discrete configurations, as indicated by Cardoso and Salcedo (1997). Algorithms based on simulated annealing employ stochastic generation of solution vectors and share similarities between the physical process of annealing and a minimization problem. Annealing is the physical process of melting a solid by heating it, followed by slow cooling and crystallization into a minimum free energy state. During the cooling process transitions are accepted to occur from a low to a high energy level through a Boltzmann probability distribution. For an optimization problem, this corresponds to 'wrong-way' movements as implemented in other optimization algorithms. Thus, simulated annealing may be viewed as a "randomization device" that allows some ascent steps during the course of the optimization, through an adaptive acceptance/rejection criterion (Schoen, 1991).

It is recognized that these are powerful techniques which might help in the location of near optimum solutions, despite the drawback that they may not be rigorous (except in an asymptotic way) and may be computationally expensive.

Simulated annealing algorithms thus have a two-loop structure (Cardoso & Salcedo, 1997) whereby alternative configurations are generated at the inner loop and the temperature is decreased at the outer loop. For unconstrained continuous optimization, a robust algorithm in arriving at the global optimum is the simplex method of Nelder and Mead (1965). This method proceeds by generating a simplex (with N dimensions; i.e., a simplex is a convex hull generated by joining N+1 points which do not lie on one hyperplane, Mangasarian, 1994) which evolves at each iteration through reflections, expansions and contractions in one direction or in all directions, so as to mostly move away from the worst point.

The correct way to use stochastic techniques in global optimization seems to be as an extension to, and not as a substitute for, local optimization (Schoen, 1991). An algorithm based on a proposal by Press and Teukolsky (1991) that combines the non-linear simplex of Nelder and Mead 1965 and simulated annealing (the SIMPSA algorithm) which was developed for global optimization of unconstrained and constrained NLP problems (Press & Tuekolsky, 1991).

This algorithm showed good robustness for a number of difficult NLP problems described in the literature (Cardoso & Salcedo, 1997; Cardoso, Salcedo, & Azevedo, 1996), i.e., insensitivity to the starting point, and reliability in attaining the global optimum. However, it is possible that the simplex moves in the SIMPSA algorithm do not maintain ergodicity of the Markov chain, since the search space is constrained to the evolution of mostly unsymmetric simplexes around their centroids. As a consequence, the algorithm might not reach the global optimum, though this was not apparent in the tested NLP problems (Cardoso, Salcedo, & Azevedo, 1996) nor on our experiments where the difference between the global optimum found by other algorithms and our implementation was of less than 3% with significant convergence time reductions.

Thus, this strategy matches very well our purpose: by using simulated annealing we find the right trajectory and by using the simplex Nelder-Mead we locate and converge quickly to the nearby optimal.

In this specific study, we observed near-optimal results at tractable computation complexity. Results are provided in the next section.

PROPOSED ALGORITHM

For a given $P^{tot} = \{P^{1,tot}, P^{2,tot}\}$, $w = \{w^1, w^2\}$, $\mathfrak{z} = \{\mathfrak{z}_1, \mathfrak{z}_2\}$ and $\mathfrak{w} = \{\mathfrak{w}^1, \mathfrak{w}^2\}$, we propose the following algorithm to reduce the complexity of OSB and making it more practical for typical DSL scenarios. Here the problem can be summarized as follows:

$$\text{fun} = S_1^{opt}, \quad S_2^{opt} =$$
$$\text{argmax}_{S_1,S_2} w_1 R_1 + w_2 R_2 +$$
$$\sum_{i=1}^{M} »_i \left(S_{i,\,total} - \sum_{n=1}^{N} s_i^n \right) + \sum_{i=1}^{M} {}^3_i \left(S_{i,\,mask} - \sum_{n=1}^{N} s_i^n \right),$$
$$\text{for } i = \{1,2\}$$

(8)

Algorithm: Improving Search in OSB in a Two-User Case via the Combination of Simulated Annealing and Non-Linear Simplex (SIMPSA)

```
Get best values of {»¹,»²}, as in
[11]
Define lower (LB) and upper bounds
(UB) for S₁,S₂
Initial guess Èⁱⁿⁱᵗ = S₁ⁱⁿⁱᵗ, S₂ⁱⁿⁱᵗ
Define set of constraints, ℂ (e.g.,
S₁, total, S₂, total; S₁,mask,S₂,mask )
For n=1:N (number of tones)
  Define fun as a function of S₁,S₂
as in (5)
  S₁ᵒᵖᵗ(n), S₂ᵒᵖᵗ(n) = SIMPSA(@fun, Èⁱⁿⁱᵗ,
LB, UB, ℂ)
End for
```

Constraints in step 4 can be further explained as follows: $S_{1,\,total}$ and $S_{2,\,total}$ related to the maximum possible (physical) power supplied by each line; $S_{1,mask}$ and $S_{2,mask}$ correspond to the limits imposed by regulatory bodies (e.g., ETSI). $S_{1,mask}$ and $S_{2,mask}$ can be neglected for simulation purposes without affecting significantly the convergence time results.

SIMULATION RESULTS

Two-User Scenario Definition

The algorithm has been tested in an ADSL (N=256 tones; 224 tones used for the downstream) environment as shown in Figure 1. The design is based on a simplified version of "TP150 "cable which is a Dutch 0.5mm cable, also known as "KPN_L1". Only two users have been considered though it is straightforward to extend the analysis to any number of users. The focus is on the downstream and a maximum power of 20.5dBm is applied to the system (per user). The Γ was chosen to be 12.9dB corresponding to a BER of 10^{-7}, 3dB of coding gain and a noise margin of 6dB. For this study, $S_i^{max} = -40$ and $S_i^{min} = -110$. Our initial weights are $w^{init} = \{w^{1,init}, w^{2,init}\} = \{0.4, 0.6\}$. No power mask constraints have been setup (though it is straightforward to incorporate them).

We start by presenting the results originally gotten when only Global Bounded Nelder-Mead was used, in an experiment run several times, leading to the histogram shown in Figure 2.

Table 1 shows the evolution in terms of the concept we are proposing, which is using a meta-

Figure 1. Two-user Simulation Scenario

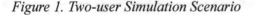
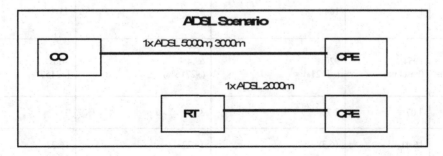

Figure 2. Histogram for GBNM Results

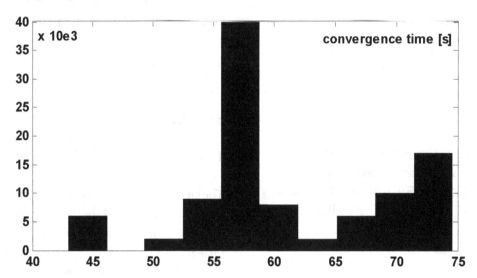

heuristic approach to solve a very complex problem. In the next subsection, we will present the actual results from the proposed algorithm.

Discussion on Simulation Results Obtained from the Proposed Algorithm using SIMPSA

From Table 2, we can observe that our proposed algorithm converges much faster than the algorithm proposed by Cendrillon et al. (2006) and it

Table 1. Initial results when using Global Bounded Nelder-Mead and some variations of it

	A	B	C	D
Description	**Classic OSB with Lagrange update (Cendrillon et al, 2006)**	**Simplex Search with Lagrange (as in Cordova et al, 2010)** $S_1^{init}, S_2^{init} = S_1^{max}, S_2^{max}$	**Proposed GBNM Algorithm with Lagrange** $S_1^{init}, S_2^{init} = S_1^{max}, S_2^{max}$	**Proposed GBNM Algorithm with Lagrange** $S_1^{init}, S_2^{init} = S_1^{max}, S_2^{max}$ **showing a broader range of fluctuation.**
# of λ -evaluations (MAX = 100)	132	104	33	33
Average time to converge [seconds] / Gain [%]	1954.4 (baseline)	920 (~112%)	69.4 (~2716%)	~60s (~3157%)
Rate User 1 [Mbps]	3.07	2.71	3	3
Rate User 2 [Mbps]	16.41	16.38	16.22	16.22

Table 2. Comparing algorithm results

Reference	Classic OSB (Cendrillon et al., 2006)	Initial Results (Cordova & van Biesen, 2010)	Proposed Algorithm	
Description		GBNM starting with S_1^{init}, $S_2^{init} = S_1^{max}$, S_2^{max}	SIMPSA S_1^{init}, $S_2^{init} = S_1^{max}$, S_2^{max}	SIMPSA S_1^{init}, $S_2^{init} = S_1^{max}$, S_2^{max}
Max number of function evaluations	n/s	10	10	5
Average time to converge [seconds]	1954.4 baseline	~60	43.58	31.87
Rate User 1 [Mbps]	3.07	3	3.026	3.37
Rate User 2 [Mbps]	16.41	16.22	16.22	16.02

actually allows the use of smaller range steps to strongly enforce constraints which is prohibitive and time-consuming in the original OSB approach. Furthermore, the current proposed algorithm provides better convergence times than our initial work with only GBNM (Table 1), leading to very tractable times when compared to other algorithms and still benefiting from achieving near-optimal

results. Thus, the combination of simulated annealing with a non-linear simplex in the current spectrum management problem presented here, including the set of constraints, helps reducing the problem complexity and proves to be a valid approach (by reaching near-optimal global results). The simplex search algorithm efficiently searches across the D+1 dimensions of the simplex

Figure 3. Average convergence time when starting with S_1^{init}, $S_2^{init} = S_1^{max}$, S_2^{max}

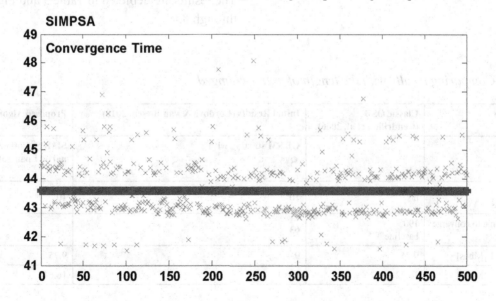

Figure 4. Expected Bitrate for user 1

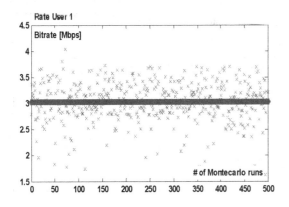

Figure 5. Expected Bitrate for user 2

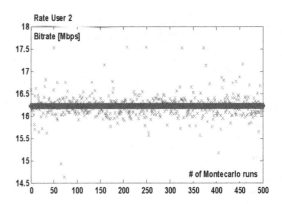

and evaluates the best vertex avoids evaluating all possible PSD levels for each user but only indicating the search space (via the boundaries [S_i^{min}, S_i^{max}]).

We have limited the maximum number of » -evaluations (MAX- ») to 25, following the results found by Tsiaflakis (2005) where less than 40 » -evaluations are suggested for the two-user case. However, we did not make this a strong constraint (and therefore we allow a tolerance of 25% without having a significant performance impact). A typical value for the total number of » -evaluations results to be 33.

Figures 3, 4, and 5 show the convergence time and bitrates, respectively, achieved by user1 and user2 after an extensive Montecarlo simulation. This is also visualized in Table 2 where we can observe that our SIMPSA-based algorithm allows to reduce further the number of function evaluations at reasonable payoff (weight sum datarate), achieving up to 27% reduction related to the convergence time. Clearly, the bit rates for user 1 and 2 remain almost the same but the convergence time is significantly reduced, as desired.

An extra experiment has been conducted using the same scenario as Figure 1 but changing the distance of the first user (from 5000m to 3000m). The results are depicted in Table 3 and Figures 6 through 8.

Table 3. Comparing results with the length of user 1 changed

Reference	Classic OSB (Cendrillon et al, 2006)	Initial Results (Cordova & van Biesen, 2010)	Proposed Algorithm
Description		GBNM starting with $S_1^{init}, S_2^{init} = S_1^{max}, S_2^{max}$	SIMPSA, other scenario, Lpan =3000m
Max number of function evaluations	n/s	10	10
Average time to converge [seconds]	1905 baseline	63	34
Rate User 1 [Mbps]	9.75	9.4	9.75
Rate User 2 [Mbps]	16.72	16.01	16.64

Figure 6. Results from new experiment (convergence time) - Length of user 1 changed to 3000m

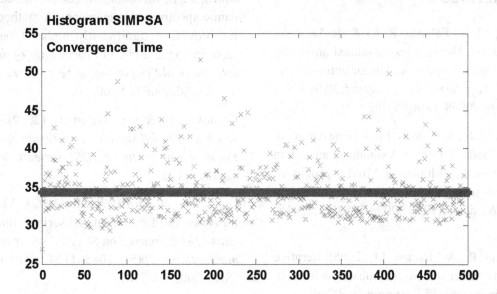

CONCLUSION

A low-complex algorithm based on a combination of simulated annealing and non-linear simplex has been proposed for getting local (almost global) optimum spectra for multiuser DSL systems, whilst significantly reducing the prohibitive complexity of traditional OSB.

Our algorithm assumes a Spectrum Management Center (at the cabinet side) but it neither relies on own end-user modem calculations nor on messaging-passing (as distributed algorithms assumptions) for achieving its performance objective. Extensive simulation results show that the convergence time is better than Cendrillon et al. (2006) and our initial work using only GBNM (Cordova & van Biesen, 2010). Furthermore, our approach allows reducing the number of function evaluations and thus also reducing the convergence time (up to ~27% gain) at reasonable payoff (weighted datarate sum)

The extension to more users though important remains future work and it clearly shows the advantage of pursuing our meta-heuristic approach.

Figure 7. Results from new experiment - Length of user 1 changed to 3000m (Rate User 1)

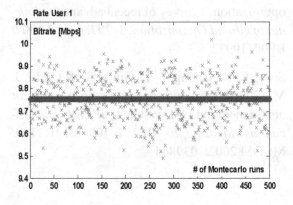

Figure 8. Results from new experiment - Length of user 1 changed to 3000m (Rate User 2)

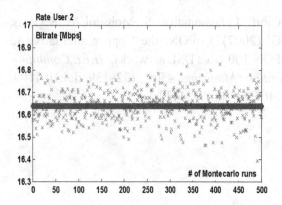

REFERENCES

Cardoso, M. E., Salcedo, R. L., & de Azevedo, S. E. (1996). The simplex-simulated annealing approach to continuous non-linear optimization. *Computers & Chemical Engineering, 20*(9), 1065. doi:10.1016/0098-1354(95)00221-9

Cardoso, M. E., Salcedo, R. L., Feyo de Azevedo, S., & Barbosa, D. (1997). A simulated annealing approach to the solution of MINLP problems. *Computers & Chemical Engineering, 21*(12), 1349–1364. doi:10.1016/S0098-1354(97)00015-X

Cendrillon, R., & Moonen, M. (2005). Iterative spectrum balancing for digital subscriber lines. In *Proceedings of the IEEE International Conference on Communications*.

Cendrillon, R., Moonen, M., Verlinden, J., Bostoen, T., & Yu, W. (2004, June). Optimal multi-user spectrum management for digital subscriber lines. In *Proceedings of the IEEE International Conference on Communications* (Vol. 1, pp. 1-5).

Cendrillon, R., Yu, W., Moonen, M., Verlinden, J., & Bostoen, T. (2006). Optimal multi-user spectrum management for digital subscriber lines. *IEEE Transactions on Communications, 54*(5), 922–933. doi:10.1109/TCOMM.2006.873096

Cioffi, J., Jagannathan, S., et al. (2006, March). *Full Binder Level 3 DSM Capacity and Vectored DSL Reinforcement* (ANSI NIPP-NAI Contribution 2006-041). Las Vegas, NV: ANSI NIPP-NAI.

Cioffi, J., Jagannathan, S., Mohseni, M., & Ginis, G. (2007). CuPON: the Copper alternative to PON 100 Gb/s DSL networks. *IEEE Communications Magazine, 45*(6), 132–139. doi:10.1109/MCOM.2007.374437

Cioffi, J., & Mohseni, M. (2004, March). Dynamic spectrum management - A methodology for providing significantly higher broadband capacity to the users. In *Proceedings of the 15th International Symposium on Services and Local Access*, Edinburgh, Scotland.

Cordova, H., & van Biesen, L. (2010). *Using simplex bounded search for Optimum Spectrum Balancing in multiuser xDSL systems. Internal Report*. VUB.

Ginis, G., & Cioffi, J. M. (2002). Vectored transmission for digital subscriber line systems. *IEEE Journal on Selected Areas in Communications, 20*(5), 1085–1104. doi:10.1109/JSAC.2002.1007389

Mangasarian, O. L. (1994). *Nonlinear programming* (p. 42). Philadelphia, PA: SIAM. doi:10.1137/1.9781611971255

Nelder, J. A., & Mead, R. (1965). A simplex for function minimization. *The Computer Journal, 7*, 308–313.

Papandriopoulos, J., & Evans, J. S. (2009). SCALE: A low-complexity distributed protocol for spectrum balancing in multiuser DSL networks. *IEEE Transactions on Information Theory, 55*(8). doi:10.1109/TIT.2009.2023751

Press, W. H., & Teukolsky, S. A. (1991). Simulated annealing optimization over continuous spaces. *Computers in Physics, 5*(4), 426.

Schoen, F. (1991). Stochastic techniques for global optimization: a survey of recent advances. *Journal of Global Optimization, 1*, 207. doi:10.1007/BF00119932

Song, K. B., Chung, S. T., Ginis, G., & Cioffi, J. M. (2002). Dynamic spectrum management for next-generation DSL systems. *IEEE Communications Magazine, 40*(10), 101–109. doi:10.1109/MCOM.2002.1039864

Tsiaflakis, P., Diehl, M., & Moonen, M. (2008). Distributed spectrum management algorithms for multiuser DSL networks. *IEEE Transactions on Signal Processing, 56*(10). doi:10.1109/TSP.2008.927460

Tsiaflakis, P., & Moonen, M. (2008, April). Low-complexity dynamic spectrum management algorithms for digital subscribre lines. In *Proceedings of the IEEE International Conference on Acoustics, Speech, and Signal Processing*, Las Vegas, NV (pp. 2769-2772).

Tsiaflakis, P., Vangorp, J., Moonen, M., Verlinden, J., Van Acker, K., & Cendrillon, R. (2005). An efficient lagrange multiplier search algorithm for optimal spectrum balancing in crosstalk dominated xDSL systems. *IEEE Journal on Selected Areas in Communications, 4*, 4.

Vanier, F. (2009). *World broadband statistics: Q1 2009*. Retrieved from http://www.point-topic.com

Yu, W., Ginis, G., & Cioffi, J. (2002). Distributed multiuser power control for digital subscriber lines. *IEEE Journal on Selected Areas in Communications, 20*(5), 1105–1115. doi:10.1109/JSAC.2002.1007390

Yu, W., Lui, R., & Cendrillon, R. (2004, December). Dual optimization methods for multiuser orthogonal frequency-division multiplex systems. In *Proceedings of the IEEE International Conference on Global Communications* (Vol. 1, pp. 225-229).

This work was previously published in the International Journal of Applied Metaheuristic Computing, Volume 2, Issue 4, edited by Peng-Yeng Yin, pp. 29-40, copyright 2011 by IGI Publishing (an imprint of IGI Global).

Chapter 4

Discrete Artificial Bee Colony Optimization Algorithm for Financial Classification Problems

Yannis Marinakis
Technical University of Crete, Greece

Nikolaos Matsatsinis
Technical University of Crete, Greece

Magdalene Marinaki
Technical University of Crete, Greece

Constantin Zopounidis
Technical University of Crete, Greece

ABSTRACT

Nature-inspired methods are used in various fields for solving a number of problems. This study uses a nature-inspired method, artificial bee colony optimization that is based on the foraging behaviour of bees, for a financial classification problem. Financial decisions are often based on classification models, which are used to assign a set of observations into predefined groups. One important step toward the development of accurate financial classification models involves the selection of the appropriate independent variables (features) that are relevant to the problem. The proposed method uses a discrete version of the artificial bee colony algorithm for the feature selection step while nearest neighbour based classifiers are used for the classification step. The performance of the method is tested using various benchmark datasets from UCI Machine Learning Repository and in a financial classification task involving credit risk assessment. Its results are compared with the results of other nature-inspired methods.

NTRODUCTION

Artificial Bee Colony (ABC) optimization algorithm is a population-based swarm intelligence algorithm that was originally proposed by Karaboga and Basturk (2007, 2008) and it simulates the foraging behaviour that a swarm of bees perform. In this algorithm, there are three groups of bees, the employed bees (bees that determine the food source (possible solutions) from a prespecified set of food sources and share this information (waggle dance) with the other bees in the hive),

DOI: 10.4018/978-1-4666-2145-9.ch004

the onlookers bees (bees that based on the information that they take from the employed bees they search for a better food source in the neighborhood of the memorized food sources) and the scout bees (employed bees that their food source has been abandoned and they search for a new food source randomly). The initially proposed Artificial Bee Colony optimization algorithm is applied in continuous optimization problems. In our study, as the feature selection problem is a discrete problem, we made some modifications in the initially proposed algorithm in order the algorithm to be suitable for solving this kind of problems.

The development of financial classification models is a complicated process, involving careful data collection and pre-processing, model development, validation and implementation. Focusing on model development, several methods have been used, including statistical methods, artificial intelligence techniques and operations research methodologies. In all cases, the quality of the data is a fundamental point. This is mainly related to the adequacy of the sample data in terms of the number of observation and the relevance of the decision attributes (i.e., independent variables) used in the analysis.

The latter is related to the feature selection problem. Feature selection refers to the identification of the appropriate attributes (features) that should be introduced in the analysis in order to maximize the expected performance of the resulting model. This has significant implications for issues such as (Kira & Rendell, 1992) the noise reduction through the elimination of noisy features, the reduction of the time and cost required to implement an appropriate model, the simplification of the resulting models and the facilitation of the easy use and updating of the models.

The basic feature selection problem is an optimization problem, with a performance measure for each subset of features, which represents expected classification performance of the result-

ing model. The problem is to search through the space of feature subsets in order to identify the optimal or near-optimal one with respect to the performance measure. Unfortunately, finding the optimal feature subset has been proved to be NP-hard (Kira & Rendell, 1992). Many algorithms are, thus, proposed to find the suboptimal solutions in comparably smaller amount of time (Jain & Zongker, 1997). Branch and bound approaches (Narendra & Fukunaga, 1977), sequential forward/backward search (Aha & Bankert, 1996; Cantu-Paz, Newsam & Kamath, 2004) and filters approaches (Cantu-Paz, 2004) deterministically search for the suboptimal solutions. One of the most important of the filter approaches is the Kira and Rendell's Relief algorithm (Kira & Rendell, 1992). Stochastic algorithms, including simulated annealing (Siedlecki & Sklansky, 1988; Lin, Lee, Chen & Tseng, 2008), scatter search (Chen, Lin & Chou, 2010; Lopez, Torres, Batista, Perez & Moreno-Vega, 2006), ant colony optimization (Al-Ani, 2005a, 2005b; Parpinelli, Lopes & Freitas, 2002; Shelokar, Jayaraman & Kulkarni, 2004), GRASP (Yusta, 2009), tabu search (Yusta, 2009), particle swarm optimization (Lin & Chen, 2009; Lin, Ying, Chen & Lee, 2008; Pedrycz, Park & Pizzi, 2009) and genetic algorithms (Cantu-Paz, Newsam & Kamath, 2004; Rokach, 2008; Yusta, 2009) are of great interest recently because they often yield high accuracy and are much faster.

In this paper, a nature inspired intelligent technique, that uses the Discrete Artificial Bee Colony optimization algorithm (DABC) is presented and analyzed in detail for the solution of the feature selection problem. A hybridized version of the Discrete Artificial Bee Colony algorithm has been presented in (Marinakis, Marinaki & Matsatsinis, 2009a). The algorithm is combined with three nearest neighbour based classifiers, the 1-nearest neighbour, the k-nearest neighbour and the weighted k-nearest neighbour classifier. The algorithm is applied to a data set involving financial decision-making problems which involves credit risk assessment. Also, the method is compared with the results of a number of other metaheuristic algorithms for feature selection problem. More precisely, the proposed algorithm is compared with a Tabu Search based algorithm (Glover, 1989, 1990; Marinakis, Marinaki, Doumpos, Matsatsinis & Zopounidis, 2008), a Genetic based algorithm (Goldberg, 1989; Marinaki, Marinakis & Zopounidis, 2007), a Particle Swarm based Optimization algorithm (Kennedy & Eberhart, 1995; Marinakis, Marinaki, Doumpos & Zopounidis, 2009), a Honey Bees Mating based Optization algorithm (Abbass 2001a, 2001b; Marinaki, Marinakis & Zopounidis, 2009) and an Ant Colony based Optimization algorithm (Dorigo, Maniezzo & Colorni, 1996; Dorigo & Stützle, 2004; Marinakis, Marinaki, Doumpos & Zopounidis, 2009).

The rest of this paper is organized as follows: In the next section a short literature review of the bees inspired optimization algorithms is presented. The proposed DABC algorithm is also presented and analyzed in detail. The parameters of the proposed algorithm and the algorithms used for the comparisons are given, as well as the results for the data taken from the UCI Machine Learning Repository. The application context using the aforementioned financial data sets is described and the experimental settings and the analytical computational results for this dataset are given. In the last section, conclusions and future research are presented.

BEES INSPIRED OPTIMIZATION ALGORITHMS

In the recent few years, a number of swarm intelligence algorithms based on the behaviour of the bees have been presented (Baykasoglu, Ozbakor & Tapkan, 2007). These algorithms are divided, mainly, in two categories according to their behaviour in the nature, the foraging behaviour and the mating behaviour. The most important approaches

that simulate the foraging behaviour of the bees are the Artificial Bee Colony (ABC) Algorithm proposed by Karaboga and Basturk (2007, 2008), the Virtual Bee Algorithm proposed by Yang (2005), the Bee Colony Optimization Algorithm proposed by Teodorovic and Dell'Orco (2005), the BeeHive algorithm proposed by Wedde, Farooq and Zhang (2004), the Bee Swarm Optimization Algorithm proposed by Drias, Sadeg and Yahi (2005) and the Bees Algorithm proposed by Pham, Kog, Ghanbarzadeh, Otri, Rahim and Zaidi (2006).

The Artificial Bee Colony algorithm is, mainly, applied in continuous optimization problems and simulates the waggled dance behaviour that a swarm of bees perform during the foraging process of the bees. In this algorithm, there are three groups of bees, the employed bees (bees that determine the food source (possible solutions) from a prespecified set of food sources and share this information (waggle dance) with the other bees in the hive), the onlookers bees (bees that based on the information that they take from the employed bees they search for a better food source in the neighborhood of the memorized food sources) and the scout bees (employed bees that their food source has been abandoned and they search for a new food source randomly). The Virtual Bee Algorithm is, also, applied in continuous optimization problems. In this algorithm, the population of the bees is associated with a memory, a food source, and, then, all the memories communicate between them with a waggle dance procedure. The whole procedure is similar with a genetic algorithm and it has been applied on two function optimization problems with two parameters. In the BeeHive algorithm, a protocol inspired from dance language and foraging behaviour of honey bees is used. In the Bees Swarm Optimization, initially, a bee finds an initial solution (food source) and from this solution the other solutions are produced with certain strategies. Then, every bee is assigned in a solution and when they accomplish their search, the bees communicate between them with a waggle dance strategy and the best solution will become the new reference solution. To avoid cycling the authors use a tabu list. In the Bees Algorithm, a population of initial solutions (food sources) is randomly generated. Then, the bees are assigned to the solutions based on their fitness function. The bees return to the hive and based on their food sources a number of bees are assigned to the same food source in order to find a better neighborhood solution. In the Bee Colony Optimization algorithm, a step by step solution is produced by each forager bee and when the foragers return to the hive a waggle dance is performed by each forager. Then the other bees, based on a probability, follow the foragers. This algorithm looks like the Ant Colony Optimization algorithm but it does not use at all the concept of pheromone trails.

Contrary to the fact that there are many algorithms that are based on the foraging behaviour of the bees, only two algorithms have been proposed that are based on the marriage behavior, the Honey Bees Mating Optimization Algorithm (HBMO), that was presented in Abbass (2001a, 2001b) and the Bumble Bees Mating Optimization Algorithm that was presented in Marinakis, Marinaki and Matsatsinis (2009b). The Honey Bees Mating Optimization algorithm simulates the mating process of the queen of the hive. The mating process of the queen begins when the queen flights away from the nest performing the mating flight during which the drones follow the queen and mate with her in the air. In the Bumble Bees Mating Optimization Algorithm the mating behavior of the Bumble Bees is presented.

DISCRETE ARTIFICIAL BEE COLONY ALGORITHM

Nearest Neighborhood Classifiers

Initially, the classic 1-nearest neighbour (1-nn) Duda and Hart (1973) method is used. The near-est neighbour classifier was selected as it is a method very easy to implement it and it does not need any optimization procedure as for example it is necessary in support vector machines and in neural networks. Assume a training sample of M_{train} vectors $\mathbf{y}_j = (y_{j1},...,y_{jd})$, $j = 1,...,M_{train}$, where d is the number of selected features and y_{jl} is the description of observation j on feature l. In the 1–nn algorithm, the classification of an unknown observation $\mathbf{x}_i = (x_{i1},...,x_{id})$ is based on its Euclidean distance D_{ij} from each training observation j, which is defined as follows:

$$D_{ij} = \sqrt{\sum_{l=1}^{d} |x_{il} - y_{jl}|^2} \tag{1}$$

Then $\mathbf{x}_i$ is classified in the class to which its nearest training observation belongs to.

The previous approach may be extended to the k-nearest neighbour (k-nn) method, where the k-nearest training observations are used to classify the test case through majority voting scheme.

The k-nn algorithm can be extended by assigning weights to each of the nearest neighbours. The weights are defined in terms of the closeness of the corresponding training observations to the unknown case $\mathbf{x}_i$. This extension is called the weighted k-nearest neighbor (wk-nn). In particular, assuming that the neighbours are sorted in descending order according to their distance from $\mathbf{x}_i$, the weight of the ith neighbour is defined as:

$$w_i = \frac{i}{\sum_{i=1}^{k} i} \tag{2}$$

Thus, the following hold:

$$w_k \geq w_{k-1} \geq ... \geq w_1 > 0 \tag{3}$$

$$w_k + w_{k-1} + ... + w_1 = 1 \tag{4}$$

Fitness Function

The fitness function measures the quality of a solution. In this study, the quality is measured with the overall classification accuracy (OCA; see below for a complete description). In the fitness function we would like to maximize the OCA. The accuracy of a C class problem can be described using a $C \times C$ confusion matrix. The element c_{ij} in row i and column j describes the number of samples of true class j classified as class i, *i.e.,* all correctly classified samples are placed in the diagonal and the remaining misclassified cases in the upper and lower triangular parts. Thus, the overall classification accuracy (OCA) can be easily obtained as:

$$OCA = 100 \frac{\sum_{i=1}^{C} c_{ii}}{\sum_{i=1}^{C}\sum_{j=1}^{C} c_{ij}} \quad (5)$$

DABC Algorithm

In this paper, an extended version of the Artificial Bee Colony optimization algorithm (Karaboga & Basturk, 2007, 2008) in the discrete space, the DABC, is proposed for the feature selection problem. In the Artificial Bee Colony optimization algorithm, there are three kind of artificial bees in the colony, the employed bees, the onlooker bees and the scouts. Initially, a set of food source positions (possible solutions) are randomly selected by the employed bees and their nectar amounts (fitness functions) are determined. One of the key issues in designing a successful algorithm for Feature Selection Problem is to find a suitable mapping between Feature Selection Problem solutions and food sources in Artificial Bee Colony Optimization Algorithm. In the Artificial Bee Colony optimization algorithm, the food sources are randomly generated. However in this study, the solutions should have values equal to 0 or to

1, where 0 denotes that the feature is not activated and 1 denotes that the feature is activated. In our proposed algorithm, the food sources are calculated exactly as in the initially proposed algorithm and, then, the values are transformed by using a sigmoid function:

$$sig\left(x_{ij}\right) = \frac{1}{1 + \exp\left(x_{ij}\right)} \quad (6)$$

and then the food sources are calculated by:

$$y_{ij} = \begin{cases} 1, if \ rand1 < sig(x_{ij}) \\ 0, otherwise \end{cases} \quad (7)$$

where x_{ij} is the solution (food source), $i = 1, ..., N$ (N is the number of food sources), $j = 1, ..., d$ (d is the dimension of the problem - the number of features), y_{ij} is the transformed solution and *rand1* is a random number in the interval $(0,1)$. These equations have also been used for the Discrete Particle Swarm Optimization (Shi & Eberhart 1998). Afterwards, the fitness of each food source is calculated and an employed bee is attached to each food source. The employed bees return in the hive and perform the waggle dance in order to inform the other bees (onlooker bees) about the food sources. Then, the onlooker bees choose the food source that will visit based on the nectar information taken from the waggle dance of the employed bees. The probability of choosing a food source is given by (Karaboga & Basturk, 2007, 2008):

$$p_i = \frac{f_i}{\sum_{n=1}^{N} f_n} \quad (8)$$

where f_i is the fitness function of each food source. As it was mentioned previously the nectar information corresponds to the fitness function of each food source. Afterwards, the employed and the onlooker bees are placed in the selected food sources. In order to produce a new food position from the old one the Discrete Artificial Bee Colony algorithm, uses the same equation as in the Artificial Bee Colony Algorithm:

$$x_{ij}' = x_{ij} + rand2(x_{ij} - x_{kj}) \qquad (9)$$

where x_{ij}' is the candidate food source, k is a different from i food source and $rand2$ is a random number in the interval $(0,1)$. As the values of the candidate food sources are not suitable for the feature selection problem they are transformed to the y'_{ij} using the Equations 6 and 7. Afterwards, the fitness of each food source is calculated. It should be noted that if there is a large number of bees in a food source, then, from the local search moves that each bee performs, this food source has larger exploration abilities in each iteration. If a better food source is found in an iteration, this food source replaces the old one. If for a number of iterations a solution is not improved, then this solution is assumed to be abandoned and a scouter bee is placed in a new random position (a new food source).

PARAMETER SELECTION

The algorithms were implemented in Fortran 90 (Lahey f95 compiler) on a Centrino Mobile Intel Pentium M750/1.86GHz, running Suse Linux 9.1. To test the efficiency of the proposed methods a 10-fold cross validation procedure is utilized for the financial classification problem and for the benchmark instances from the UCI Machine Learning Repository. Initially, the data set is divided in 10 disjoint parts containing approximately $M/10$ observations each, where M is the total number of sample observations. Next, each of these parts is systematically removed from the data set, a model is built from the remaining parts (the training set) and, then, the accuracy of

the model is calculated using the excluded parts (the test set).

As it has already been mentioned, three approaches that use different classifiers, the 1-nn, k-nn and the wk-nn, are used. In DABC based classifier, the value of k is changed dynamically depending on the number of iterations. Each generation uses different k. The reason why k does not have a constant value is that we would like to ensure the diversity of the bees in each iteration of DABC. The determination of k is done by using a random number generator with a uniform distribution $(0,1)$ in each iteration. Then, the produced number is converted to an integer k (e.g., if the produced number is in the interval $0.2 - 0.3$, then $k = 3$).

The parameters of the proposed algorithm were selected after thorough testing. A number of different alternative values were tested and the ones selected are those that gave the best computational results concerning both the quality of the solution and the computational time needed to achieve this solution. Thus, the selected parameters are given in the following:

1. The number of employed is set equal to 50.
2. The number of onlookers is set equal to 100.
3. The number of generations is set equal to 50.

For comparison purposes, five other metaheuristic algorithms were used, a tabu based metaheuristic, a genetic (GA) based metaheuristic, an ant colony optimization (ACO), a honey bees mating optimization algorithm (HBMO) and a particle swarm optimization (PSO) based algorithm. For more details of how these methods are applied for the solution of this problem please see (Marinaki, Marinakis & Zopounidis, 2007, 2009; Marinakis, Marinaki, Doumpos, Matsatsinis & Zopounidis, 2008; Marinakis, Marinaki & Matsatsinis, 2009a; Marinakis, Marinaki, Doumpos & Zopounidis, 2009). These methods use in the classification phase the 1-nn, the k-nn and the wk-nn classifiers. In the genetic, ACO, HBMO and PSO based metaheuristics with the k-nn and wk-nn classifier the value of *k* is changed dynamically depending on the number of iterations. Each generation uses different *k*. The maximum number of iterations for the tabu based metaheuristic is equal to 1000 and the size of the tabu list is equal to 10.

The parameter settings for the genetic based metaheuristic are:

1. Population size equal to 200,
2. Number of generations equal to 50,
3. Probability of crossover equal to 0.8, and
4. Probability of mutation equal to 0.25.

The parameter settings for the ant colony optimization based metaheuristic are:

1. The number of ants used is set equal to the number of features (26 for the credit risk assessment case),
2. The number of iterations that each ant constructs a different solution, based on the pheromone trails, is set equal to 50, and
3. $q=0.5$.

The parameter settings for the particle swarm optimization based metaheuristic are:

1. The number of swarms is set equal to 1,
2. The number of particles is set equal to 50,
3. The number of generations is set equal to 50,
4. The coefficients are $c_1 = 2$, $c_2 = 2$, and
5. $w_{max} = 0.9$ and $w_{min} = 0.01$.

The parameter settings for the honey bees mating optimization based metaheuristic are:

1. The number of queens is set equal to 1,
2. The number of drones is set equal to 200,
3. The number of mating flights (M) is set equal to 50,
4. The size of queen's spermatheca is set equal to 50,

5. The number of broods is set equal to 50, and
6. The number of different workers are seven ($w = 7$).

RESULTS FOR THE UCI MACHINE LEARNING REPOSITORY'S DATA

The performance of the proposed methodology is tested, initially, on 9 benchmark instances taken from the UCI Machine Learning Repository. The datasets were chosen to include a wide range of domains and their characteristics are given in Table 1. The data vary in term of the number of observation from very small samples (Hepatitis with 80 observations) up to larger data sets (Spambase with 4601 observations). In two cases (Breast Cancer Wisconsin, Hepatitis) the data sets are appeared with different size of observations. This is performed because in these datasets there is a number of missing values. The problem of missing values was faced with two different ways. In the first way where all the observations are used we took the mean values of all the observations in the corresponding feature while in the second way where we have less values in the observations we did not take into account the observations that they had missing values. Some data sets involve only

Table 1. Data sets characteristics

Data Sets	Observations	Features
Australian Credit (AC)	690	14(8)
Breast Cancer Wisconsin 1 (BCW1)	699	9
Breast Cancer Wisconsin 2 (BCW2)	683	9
German Credit (GC)	1000	24(13)
Heart Disease (HD)	270	13(7)
Hepatitis 1 (Hep1)	155	19(13)
Hepatitis 2 (Hep2)	80	19(13)
Ionosphere (Ion)	351	34
Spambase (spam)	4601	57
Pima Indian Diabetes (PID)	768	8

numerical features and the remaining include both numerical and categorical features. For each data set, Table 1 reports the total number of features and the number of categorical features in parentheses. All the data sets involve 2-class problems and they are analyzed with 10-fold cross validation.

In Table 2 a comparison of the proposed discrete artificial bee colony algorithms with a Genetic algorithm, a Particle Swarm Optimization algorithm and an Ant Colony algorithm is given. As it can be seen from this table the proposed methods give superior classification results from all other methods. In Table 2 the average selected

Table 2. Classification results (OCA(%)) and average selected features for all algorithms

Data Sets	Classifier	DABC		GA		ACO		PSO	
		OCA	Features	OCA	Features	OCA	Features	OCA	Features
AC	1nn	90.21	7.62	86.42	8.12	87.88	8.21	86.95	8.17
	Knn	92.37	7.91	87.12	8.11	89.37	8.05	90.01	7.98
	Wknn	90.24	7.87	86.12	8.07	88.18	8.23	89.39	8.12
BCW1	1nn	99.15	4.34	98.01	4.57	98.65	5.01	99.15	4.98
	Knn	99.01	4.42	97.37	5.12	98.41	5.07	98.85	4.95
	Wknn	99.26	4.18	97.48	5.01	98.08	5.18	99.05	4.84
BCW2	1nn	99.14	5.23	98.55	5.34	99.17	5.21	99.23	5.15
	Knn	99.57	5.18	97.88	5.66	98.79	5.73	99.09	5.43
	Wknn	99.23	5.37	97.45	5.87	98.03	5.49	97.65	5.53
GC	1nn	79.15	14.18	70.96	15.31	75.37	15.52	76.23	15.28
	Knn	79.23	14.23	76.01	14.87	77.57	15.01	78.01	15.31
	Wknn	81.18	13.88	77.35	14.32	78.27	15.12	77.11	15.31
HD	1nn	90.11	6.38	88.45	6.91	90.38	6.92	90.49	6.88
	Knn	92.15	6.54	87.45	6.87	91.78	6.98	91.24	6.91
	Wknn	92.21	6.42	89.39	6.94	91.22	6.88	90.88	6.77
Hep1	1nn	98.25	10.51	94.37	10.78	97.08	11.02	98.15	11.17
	Knn	98.50	10.28	94.32	11.12	96.23	10.88	96.92	10.79
	Wknn	95.00	10.37	95.31	11.23	96.23	11.09	97.12	10.98
Hep2	1nn	100	10.58	98.85	10.88	100	9.85	100	9.88
	Knn	100	10.25	95.23	10.87	97.15	10.85	98.23	10.95
	Wknn	100	10.18	94.29	10.74	96.75	10.82	98.18	10.95
Ion	1nn	96.50	15.67	95.08	16.05	97.12	16.34	98.17	16.21
	Knn	95.50	16.05	95.11	16.45	96.37	16.38	96.14	16.21
	Wknn	95.01	15.42	94.21	15.91	96.44	15.85	97.01	16.02
Spam	1nn	84.43	22.05	82.23	23.12	83.01	22.85	84.28	22.98
	Knn	86.75	22.25	81.32	22.76	83.44	24.65	83.79	24.32
	Wknn	86.25	21.95	81.18	25.01	82.11	24.97	83.18	23.88
PID	1nn	79.05	4.18	71.57	4.21	74.11	4.37	75.18	4.28
	Knn	79.28	4.10	74.19	4.23	76.37	4.26	78.21	4.37
	Wknn	79.77	4.25	74.24	4.28	76.47	4.37	75.39	4.25

features for all algorithms in all runs of the algorithms are presented. The significance of the solution of the feature selection problem using the proposed methods is demonstrated by the fact that with the proposed algorithm the best solutions were found by using fewer features than the other algorithms used in the comparisons.

More precisely, in the most difficult instance, the Spambase instance, the proposed algorithms needed between 21.95 to 22.25 average number of features in order to find their best solutions, while the other three algorithms (Genetic Algorithm, ACO, PSO) needed between 22.76 − 25.01 average number of features to find their best solutions.

A statistical analysis based on the Mann-Whitney U-test for all algorithms using the City Block metric is presented in Table 3. In this table, a value equal to 1 indicates a rejection of the null hypothesis at the 5% (or 10%) significance level, which means that the method is statistically significant different from the other methods. On the other hand, a value equal to 0 indicates a failure to reject the null hypothesis at the 5% (or 10%) significance level, meaning that no statistical significant difference exists between the two methods. As it can be seen from this table, at the 5% significance level there is no statistically significant differences between the algorithms while at the 10% significance level the results of the Discrete Artificial Bee Colony are statistically significant different from the results of the Genetic Algorithms.

The results of the algorithm are, also, compared (Table 4) with the results of a number of meta-heuristic approaches from the literature. In these implementations, the same databases are used as the ones we use in this paper and, thus, comparisons of the results can be performed. More precisely in Table 4 the results of the proposed algorithm are compared with the results of the following algorithms:

Table 3. Results of Mann - Whitney test for all algorithms

5% significance level				
	DABC	GA	ACO	PSO
DABC	-	0	0	0
GA	0	-	0	0
ACO	0	0	-	0
PSO	0	0	0	-
10% significance level				
	DABC	GA	ACO	PSO
DABC	-	1	0	0
GA	1	-	0	0
ACO	0	0	-	0
PSO	0	0	0	-

1. The Parallel Scatter Search algorithm proposed by (Lopez, Torres, Batista, Perez & Moreno-Vega, 2006). In this paper three different versions of Scatter Search are proposed, named Sequential Scatter Search Greedy Combination (SSSGC), Sequential Scatter Search Reduced Greedy Combination (SSSRGC) and Parallel Scatter Search (PSS).

2. The Particle Swarm Optimization – Linear Discriminant Analysis (PSOLDA) proposed by (Lin and Chen, 2009).

3. The Particle Swarm Optimization – Support Vector Machines proposed by (Lin, Ying, Chen & Lee, 2008). In this paper two different version of the algorithm are presented, one with solving the feature selection problem (PSOSVM1) and the other without solving the feature selection problem (PSOSVM2).

4. The Simulated Annealing – Support Vector Machines proposed by (Lin, Lee, Chen & Tseng, 2008). In this paper two different version of the algorithm are presented one with solving the feature selection problem (SASVM1) and the other without solving the feature selection problem (SASVM2).

Table 4. Comparison of the proposed algorithm with other metaheuristic approaches

Method	Data Set							
	AC	BCW	GC	HD	Hep	Ion	Spam	PID
DABC-1nn	90.21	99.15	79.15	90.11	**100**	96.50	84.43	79.05
DABC-knn	**92.37**	**99.57**	79.23	92.15	**100**	95.50	**86.75**	79.28
DABC-wknn	90.24	99.26	81.18	92.21	**100**	95.01	86.25	79.77
SSSGC	-	95.22	-	74.99	-	87.75	-	67.92
SSSRGC	-	94.88	-	74.99	-	87.12	-	67.66
PSS	-	95.11	-	74.91	-	87.35	-	68.10
PSOLDA	84.5	96.5	75.6	84.7	-	92.2	-	76.7
PSOSVM1	91.03	99.18	81.62	92.83	-	99.01	-	82.68
PSOSVM2	88.09	97.95	79.00	88.17	-	97.50	-	80.19
SASVM1	92.19	99.38	-	93.33	-	**99.07**	-	82.22
SASVM2	88.34	97.95	-	87.97	-	97.50	-	80.19
PSONN	-	-	74.7	83.9	-	94.6	-	-
GOV	85.35	97.13	-	-	81.29	-	-	-
SS-ensemble	91.74	99.46	85.49	**96.24**	97.46	-	-	**83.92**
GRASP	-	93	**92.7**	-	-	90.4	84.6	-
Tabu Search	-	92.6	**92.7**	-	-	90.6	82.64	-
Memetic	-	91.8	92.6	-	-	89	79.76	-

5.   A Particle Swarm Optimization algorithm with a Nearest Neighbour (PSONN) classifier proposed by (Pedrycz, Park & Pizzi, 2009).

6.   A Genetic Algorithm using an adjacency matrix-encoding, GWC operator, and fitness function based on the VC dimension of multiple ODTs combined with naive Bayes (GOV - genetic algorithm for ODTs using VC dimension upper bound) proposed by (Rokach, 2008).

7.   A Scatter Search (SS-ensemble) with different classifiers like Support Vector Machines (SVM), Decision Trees (DT) and Back Propagation Networks (BPN) proposed by (Chen, Lin & Chou, 2010).

8.   Three different metaheuristics (GRASP, Tabu Search and a Memetic algorithm) proposed by (Yusta, 2009).

More precisely, the proposed algorithm gives better results in four instances, the Australian Credit (AC), the Breast Cancer Wisconsin (BCW), the spambase (Spam) and the Hepatitis. For the other four instances the algorithms that perform better are: for the German Credit the GRASP and the Tabu Search proposed by (Yusta, 2009), for the Heart Disease and for the Pima Indian Diabetes (PID) the Scatter Search – ensemble proposed by (Chen, Lin & Chou, 2010) and for the ionosphere the Simulated Annealing – Support Vector Machines (SASVM1) proposed by (Lin, Lee, Chen & Tseng, 2008).

FINANCIAL CLASSIFICATION PROBLEM

The Discrete Artificial Bee Colony Optimization algorithm is applied to a financial classification problem. The data, taken from Doumpos and Pasiouras (2005), involve 1330 firm-year observations for UK non-financial firms over the period 1999-2001. The sample observations are classified into five risk groups according to their level of likelihood of default, measured on the basis of their QuiScore, a credit rating assigned by Qui Credit Assessment Ltd. In particular, On the basis of their QuiScore, the firms are classified into five risk groups as follows:

- Secure group, consisting of firms for which failure is very unusual and normally occurs only as a result of exceptional market changes.
- Stable group, consisting of firms for which failure is rare and will only come about if there are major company or market changes.
- Normal group, consisting of firms that do not fail, as well as some that fail.
- Unstable (caution) group, consisting of firms that have a significant risk of failure.
- High-risk group, consisting of firms which are unlikely to be able to continue their operation unless significant remedial action is undertaken.

Such multi-group credit risk rating systems are widely used in practice, mainly by banking institutions. According to Treacy and Carey (2000) the internal reporting process in banking institutions is facilitated by credit risk management systems that are based on a multi-group scheme for rating risk. Internal reports prepared in this context are more informative to senior management and they provide a better representation of the actual risk exposure for the banking institution.

The evaluation criteria used in the analysis involve ratios based on the financial statements of the firms. It should be noted that credit scoring has non-financial aspects as well but obtaining objective, comprehensive, and reliable non-financial information is a cumbersome process, as such

data are rarely available (Doumpos & Pasiouras, 2005). Therefore, within the scope of this current study the analysis was based only on financial information. A total of 26 financial ratios (Table 5) and annual changes (features) were initially considered based on data availability and previous studies on credit assessment and bankruptcy prediction.

Table 5. Financial features for the problem domain

Multigroup credit risk assessement task	
Feature number	**Financial Ratio**
1	Current ratio
2	Quick ratio
3	Shareholders liquidity ratio
4	Solvency ratio
5	Asset cover
6	Annual change in fixed assets
7	Annual change in current assets
8	Annual change in stock
9	Annual change in debtors
10	Annual change in total assets
11	Annual change in current liabilities
12	Annual change creditors
13	Annual change in loans/overdraft
14	Annual change in long-term liabilities
15	Net profit margin
16	Return on total assets
17	Interest cover
18	Stock turnover
19	Debtors turnover
20	Debtors collection (days)
21	Creditors payment (days)
22	Fixed assets turnover
23	Salaries/turnover
24	Gross profit margin
25	Earnings Before Interest and Taxes (EBIT) margin
26	Earnings Before Interest and Taxes, Depreciation and Amortization (EBITDA) margin

The selection of a set of appropriate input feature variables is an important issue in building a good classifier. The purpose of feature variable selection is to find the smallest set of features that can result in satisfactory predictive performance. Because of the curse of dimensionality, it is often necessary and beneficial to limit the number of input features in a classifier in order to have a good predictive and less computationally-intensive model. In the credit risk assessment problem analysed in this paper, there are 2^{26}-1 possible feature combinations. The objective of the computational experiments is to show the performance of the proposed algorithm in searching for a reduced set of features with high accuracy.

Table 6 presents the classification results for the optimal solution of the proposed algorithm, the DABC based metaheuristic, for the financial classification problem. The results of the algorithms used for the comparisons are, also, shown. In the credit risk assessment problem, which involves 5 classes, the nature inspired metaheuristic provides considerably better results compared to the other methods in terms of OCA. DABC 1-nn provides the best results on the evaluation criterion followed by DABC k-nn. In terms of the number of features used in the final model, DABC 1-nn also provides the best results with an average (over the 10 folds of the cross validation analysis) of 9.8 features. The significance of the proposed algorithm is demonstrated by the fact that when the classification problem is solved using 1-nn classifier and without solving the feature selection problem the overall classification accuracy is only 55.78%.

The selection frequencies of the features (Table 7) indicate that the solvency ratio, EBIT margin, net profit margin, shareholders liquidity ratio and the quick ratio are the most important variables. It should be noted that the combination of all the features of the proposed Discrete Artificial Bee Colony algorithm gave the ability in the proposed algorithm to perform better than the other algorithms used in the comparisons.

Table 6. Classification results

Methods	Overall classification accuracy (%)	Average No. of features
DABC 1-nn	**75.48**	**9.8**
DABC k-nn	75.37	10.6
DABC wk-nn	75.18	10.5
HBMO 1-nn	75.21	10.2
HBMO k-nn	74.18	10.3
HBMO wk-nn	73.52	10.6
ACO 1-nn	72.18	11.2
ACO k-nn	72.85	12
ACO wk-nn	71.72	11.8
PSO 1-nn	72.10	11.3
PSO k-nn	73.30	10.4
PSO wk-nn	71.05	11.7
Gen 1-nn	68.12	12.4
Gen k-nn	69.84	13
Gen wk-nn	69.02	11.2
Tabu 1-nn	67.74	12.2
Tabu 5-nn	66.09	12
Tabu w5-nn	67.36	12.4
1-nn	55.78	26

Table 7. Selection frequencies of the features

Features	Frequency		Features	Frequency	
1	15	(50.00)	14	9	(30.00)
2	**19**	**(63.33)**	15	**17**	**(56.66)**
3	**17**	**(56.66)**	16	12	(40.00)
4	**26**	**(86.66)**	17	14	(46.66)
5	11	(36.66)	18	9	(30.00)
6	10	(33.33)	19	6	(20.00)
7	9	(30.00)	20	9	(30.00)
8	7	(23.33)	21	11	(36.67)
9	2	(6.67)	22	12	(40.00)
10	10	(33.33)	23	13	(43.33)
11	8	(26.66)	24	10	(33.33)
12	7	(23.33)	25	**17**	**(56.66)**
13	13	(43.33)	26	16	(53.33)

CONCLUSION

An important issue in building a good classifier is the selection of a set of appropriate input feature variables. The discrete artificial bee colony algorithm has been proposed in this study for solving this feature subset selection problem. Three different classifiers were used for the classification problem, based on the nearest neighbour classification rule. The performance of the proposed algorithm was tested using financial data involving credit risk assessment. The obtained results indicate the high performance of the proposed algorithm in searching for a reduced set of features with high accuracy. DABC 1-nn was found to provide the best results in terms of accuracy rates using less then the half of the available features. Future research will focus on the use of different machine learning classifiers (SVM, neural networks, etc.)

REFERENCES

Abbass, H. A. (2001a). A monogenous MBO approach to satisfiability. In *Proceedings of the International Conference on Computational Intelligence for Modelling, Control and Automation, CIMCA'2001*, Las Vegas, NV.

Abbass, H. A. (2001b). Marriage in honey-bee optimization (MBO): a haplometrosis polygynous swarming approach. In *Proceedings of the Congress on Evolutionary Computation, CEC2001*, Seoul, Korea (pp. 207-214).

Aha, D. W., & Bankert, R. L. (1996). A comparative evaluation of sequential feature selection algorithms . In Fisher, D., & Lenx, J.-H. (Eds.), *Artificial Intelligence and Statistics*. New York, NY: Springer.

Al-Ani, A. (2005a). Feature subset selection using ant colony optimization. *International Journal of Computational Intelligence*, 2(1), 53–58.

Al-Ani, A. (2005b). Ant colony optimization for feature subset selection. *Transactions on Engineering . Computing and Technology*, 4, 35–38.

Baykasoglu, A., Ozbakor, L., & Tapkan, P. (2007). Artificial bee colony algorithm and its application to generalized assignment problem . In Chan, F. T. S., & Tiwari, M. K. (Eds.), *Swarm Intelligence, Focus on Ant and Particle Swarm Optimization* (pp. 113–144). Vienna, Austria: I-Tech Education and Publishing.

Cantu-Paz, E. (2004). Feature subset selection, class separability, and genetic algorithms. In *Proceedings of the Genetic and Evolutionary Computation Conference* (pp. 959-970).

Cantu-Paz, E., Newsam, S., & Kamath, C. (2004). Feature selection in scientific application. In *Proceedings of the 2004 ACM SIGKDD International Conference on Knowledge Discovery and Data Mining* (pp. 788-793).

Chen, S. C., Lin, S. W., & Chou, S. Y. (2010). Enhancing the classification accuracy by scatter-search-based ensemble approach. *Applied Soft Computing*, 11(1).

Dorigo, M., Maniezzo, V., & Colorni, A. (1996). Ant system: Optimization by a colony of cooperating agents. *IEEE Transactions on Systems, Man, and Cybernetics – Part B, 26*(1), 29-41.

Dorigo, M., & Stützle, T. (2004). *Ant Colony Optimization*. Cambridge, MA: MIT Press.

Doumpos, M., & Pasiouras, F. (2005). Developing and testing models for replicating credit ratings: A multicriteria approach. *Computational Economics, 25*, 327–341. doi:10.1007/s10614-005-6412-4

Drias, H., Sadeg, S., & Yahi, S. (2005). Cooperative bees swarm for solving the maximum weighted satisfiability problem. In *Proceedings of the IWAAN International Work Conference on Artificial and Natural Neural Networks* (LNCS 3512, pp. 318-325).

Duda, R. O., & Hart, P. E. (1973). *Pattern classification and scene analysis.* New York, NY: John Wiley & Sons.

Glover, F. (1989). Tabu Search I. *ORSA Journal on Computing, 1*(3), 190-206.

Glover, F. (1990). Tabu Search II. *ORSA Journal on Computing, 2*(1), 4–32.

Goldberg, D. E. (1989). *Genetic algorithms in search, optimization, and machine learning.* Reading, MA: Addison-Wesley.

Jain, A., & Zongker, D. (1997). Feature selection: Evaluation, application, and small sample performance. *IEEE Transactions on Pattern Analysis and Machine Intelligence, 19,* 153–158. doi:10.1109/34.574797

Karaboga, D., & Basturk, B. (2007). A powerful and efficient algorithm for numerical function optimization: artificial bee colony (ABC) algorithm. *Journal of Global Optimization, 39,* 459–471. doi:10.1007/s10898-007-9149-x

Karaboga, D., & Basturk, B. (2008). On the performance of artificial bee colony (ABC) algorithm. *Applied Soft Computing, 8,* 687–697. doi:10.1016/j.asoc.2007.05.007

Kennedy, J., & Eberhart, R. (1995). Particle swarm optimization. In *Proceedings of 1995 IEEE International Conference on Neural Networks* (pp. 1942-1948).

Kira, K., & Rendell, L. (1992). A practical approach to feature selection. In *Proceedings of the Ninth International Conference on Machine Learning,* Aberdeen, UK (pp. 249-256).

Lin, S. W., & Chen, S. C. (2009). PSOLDA: A Particle swarm optimization approach for enhancing classification accurate rate of linear discriminant analysis. *Applied Soft Computing, 9,* 1008–1015. doi:10.1016/j.asoc.2009.01.001

Lin, S. W., Lee, Z. J., Chen, S. C., & Tseng, T. Y. (2008). Parameter determination of support vector machine and feature selection using simulated annealing approach. *Applied Soft Computing, 8,* 1505–1512. doi:10.1016/j.asoc.2007.10.012

Lin, S. W., Ying, K. C., Chen, S. C., & Lee, Z. J. (2008). Particle swarm optimization for parameter determination and feature selection of support vector machines. *Expert Systems with Applications, 35,* 1817–1824. doi:10.1016/j.eswa.2007.08.088

Lopez, F. G., Torres, M. G., Batista, B. M., Perez, J. A. M., & Moreno-Vega, J. M. (2006). Solving feature subset selection problem by a parallel scatter search. *European Journal of Operational Research, 169,* 477–489. doi:10.1016/j.ejor.2004.08.010

Marinaki, M., Marinakis, Y., & Zopounidis, C. (2007). Application of a genetic algorithm for the credit risk assessment problem. *Foundations of Computing and Decisions Sciences, 32*(2), 139–152.

Marinaki, M., Marinakis, Y., & Zopounidis, C. (2009). Honey bees mating optimization algorithm for financial classification problems. *Applied Soft Computing, 10*(3).

Marinakis, Y., Marinaki, M., Doumpos, M., Matsatsinis, N., & Zopounidis, C. (2008). Optimization of nearest neighbor classifiers via metaheuristic algorithms for credit risk assessment. *Journal of Global Optimization, 42,* 279–293. doi:10.1007/s10898-007-9242-1

Marinakis, Y., Marinaki, M., Doumpos, M., & Zopounidis, C. (2009). Ant colony and particle swarm optimization for financial classification problems. *Expert Systems with Applications, 36*(7), 10604–10611. doi:10.1016/j.eswa.2009.02.055

Marinakis, Y., Marinaki, M., & Matsatsinis, N. (2009a). A Hybrid discrete artificial bee colony – GRASP algorithm for clustering. In *Proceedings of the 39th International Conference on Computers and Industrial Engineering*, Troyes, France.

Marinakis, Y., Marinaki, M., & Matsatsinis, N. (2009b). A hybrid bumble bees mating optimization – GRASP algorithm for clustering. In E. Corchado, X. Wu, E. Oja, Á. Herrero, & B. Baruque (Eds.), *Proceedings of HAIS 2009* (LNAI 5572, pp. 549-556).

Narendra, P. M., & Fukunaga, K. (1977). A branch and bound algorithm for feature subset selection. *IEEE Transactions on Computers, 26*(9), 917–922. doi:10.1109/TC.1977.1674939

Parpinelli, R. S., Lopes, H. S., & Freitas, A. A. (2002). An ant colony algorithm for classification rule discovery. In H. Abbas, R. Sarker, & C. Newton (Eds.). *Data mining: A heuristic approach* (pp. 191-208).

Pedrycz, W., Park, B. J., & Pizzi, N. J. (2009). Identifying core sets of discriminatory features using particle swarm optimization. *Expert Systems with Applications, 36*, 4610–4616. doi:10.1016/j.eswa.2008.05.017

Pham, D. T., Kog, E., Ghanbarzadeh, A., Otri, S., Rahim, S., & Zaidi, M. (2006). The bees algorithm - A novel tool for complex optimization problems. In *IPROMS 2006: Proceedings of the 2nd International Virtual Conference on Intelligent Production Machines and Systems*.

Rokach, L. (2008). Genetic algorithm-based feature set partitioning for classification problems. *Pattern Recognition, 41*, 1676–1700. doi:10.1016/j.patcog.2007.10.013

Shelokar, P. S., Jayaraman, V. K., & Kulkarni, B. D. (2004). An ant colony classifier system: application to some process engineering problems. *Computers & Chemical Engineering, 28*, 1577–1584. doi:10.1016/j.compchemeng.2003.12.004

Shi, Y., & Eberhart, R. (1998). A modified particle swarm optimizer. In *Proceedings of 1998 IEEE World Congress on Computational Intelligence* (pp. 69-73).

Siedlecki, W., & Sklansky, J. (1988). On automatic feature selection. *International Journal of Pattern Recognition and Artificial Intelligence, 2*(2), 197–220. doi:10.1142/S0218001488000145

Teodorovic, D., & Dell'Orco, M. (2005). Bee colony optimization - A cooperative learning approach to complex transportation problems. In *Advanced OR and AI Methods in Transportation* (pp. 51-60).

Treacy, W. S., & Cavey, M. (2000). Credit risk rating systems at large US banks. *Journal of Banking & Finance, 24*, 167–201. doi:10.1016/S0378-4266(99)00056-4

Wedde, H. F., Farooq, M., & Zhang, Y. (2004). BeeHive: An efficient fault-tolerant routing algorithm inspired by honey bee behavior. In M. Dorigo (Ed.), *Ant colony optimization and swarm intelligence* (LNCS 3172, pp. 83-94).

Yang, X. S. (2005). Engineering optimizations via nature-inspired virtual bee algorithms. In J. M. Yang & J. R. Alvarez (Eds.), *Proceedings of IWINAC 2005* (LNCS 3562, pp. 317-323).

Yusta, S. C. (2009). Different metaheuristic strategies to solve the feature selection problem. *Pattern Recognition Letters, 30*, 525–534. doi:10.1016/j.patrec.2008.11.012

This work was previously published in the International Journal of Applied Metaheuristic Computing, Volume 2, Issue 1, edited by Peng-Yeng Yin, pp. 1-17, copyright 2011 by IGI Publishing (an imprint of IGI Global).

Chapter 5
Vector Evaluated and Objective Switching Approaches of Artificial Bee Colony Algorithm (ABC) for Multi–Objective Design Optimization of Composite Plate Structures

S. N. Omkar
Indian Institute of Science, India

Kiran Patil
Indian Institute of Science, India

G. Narayana Naik
Indian Institute of Science, India

Mrunmaya Mudigere
Indian Institute of Science, India

ABSTRACT

In this paper, a generic methodology based on swarm algorithms using Artificial Bee Colony (ABC) algorithm is proposed for combined cost and weight optimization of laminated composite structures. Two approaches, namely Vector Evaluated Design Optimization (VEDO) and Objective Switching Design Optimization (OSDO), have been used for solving constrained multi-objective optimization problems. The ply orientations, number of layers, and thickness of each lamina are chosen as the primary optimization variables. Classical lamination theory is used to obtain the global and local stresses for a plate subjected to transverse loading configurations, such as line load and hydrostatic load. Strength of the composite plate is validated using different failure criteria—Failure Mechanism based failure criterion, Maximum stress failure criterion, Tsai-Hill Failure criterion and the Tsai-Wu failure criterion. The design optimization is carried for both variable stacking sequences as well as standard stacking schemes and a comparative study of the different design configurations evolved is presented. Performance of Artificia Bee Colony (ABC) is compared with Genetic Algorithm (GA) and Particle Swarm Optimization (PSO) for both VEDO and OSDO approaches. The results show ABC yielding a better optimal design than PSO and GA. l

DOI: 10.4018/978-1-4666-2145-9.ch005

INTRODUCTION

Extensive Research in behaviour of living systems has provided scientists with powerful methods for designing optimization algorithms. Such algorithms use ideas inspired from the nature. The basic idea behind developing these techniques is to simulate the natural phenomena such as flocking behaviour of birds, ants, etc., seeking a path between their colony and a source of food, simulating living systems and their evolution, immunology and observed immune functions. These algorithms thus operate on a set of individuals rendering population or swarm based methodologies. Swarm intelligence is characterized by some features, where every individual is self-autonomous and can obtain local information, and interact with their geographical neighbours. Complex group behaviour emerges from the interactions of individuals who exhibit simple behaviours by themselves.

Inspired by the behaviour of a bee, Karaboga introduced Artificial Bee Colony (ABC) technique for optimizing numerical functions and real-world problems (Karaboga, 2005). In addition, for achieving good performance on a wide spectrum of problems, such techniques tend to exhibit a high degree of flexibility and robustness (Karaboga & Basturk, 2007). ABC is a population based stochastic optimization technique which incorporates a flexible and well-balanced mechanism to accommodate the global and local exploration and exploitation abilities within a short computation time. Artificial Bee Colony (ABC) optimization technique has been used in variety of scientific and engineering applications over the past few years. A modified ABC algorithm which can solve constrained optimization problems is developed by Karaboga and Basturk (2008). In this study,

performance of ABC is compared with differential evolution (DE) and particle swarm optimization (PSO) algorithms by considering linear, quadratic and non linear objective functions and found that ABC performs better in all the cases considered.

An ABC algorithm to determine the sectionalizing switch to be operated in order to solve the distribution system loss minimization problem is implemented by Srinivasa et al. (2008). They found ABC to be advantageous in terms of its global search capability and computational capability on comparison with GA, DE and simulated annealing (SA). A modified ABC algorithm to adapt to the grid computing environment and applied it to optimize the equilibrium of confined plasma in a nuclear fusion device is developed by (Antonio et al., 2010). Work on optimization aspects of a multi-pass milling operation using Artificial Bee Colony (ABC) algorithm is presented by Venkat Rao and Pawar (2010). The objective considered is minimization of production time (i.e., maximization of production rate) subjected to various constraints of arbor strength, arbor deflection, and cutting power. Study shows that ABC is good in converging to an optimal solution in terms of process parameters like depth of cut, cutting speed, feed and number of passes with good local as well as global search capability.

Clustering analysis has become an important method in identifying homogeneous group of objects which gather as a group (cluster). ABC algorithm for data clustering on classification problems and performance of the algorithm is compared with particle Swarm optimization (PSO) and nine classification techniques are employed by Karaboga and Oztruk (2011). The study depicts that an improved classification is obtained in case of ABC and also it can be applied for multivariate data clustering.

ABC algorithm is successfully adopted for the classification of acoustic emission signal by Omkar et al. (2009). They have presented that an improved classification can be achieved for the complex acoustic emission data set using ABC. Although the above literature studies showcase extensive use of ABC in different engineering domains, not much work is reported on the design optimization of composite structures.

In order to achieve an efficient design that fulfils the design criteria and the difficulty to select the values out of a large set of constrained design variables makes optimization algorithms a natural tool for the design of composite structures. In the previous year an attempt by Omkar et al. (2011) was done in carrying out multi-objective design optimization using the Vector Evaluated Variant of the ABC (VEABC) algorithm for composite structures constituting laminate stacked with various stacking sequences subjected to uni-axial, bi-axial and bending loads.

The motivation for the present paper, which also forms an extension for the previous work (Omkar et al., 2011) are briefly summarized through points listed below.

1. Composite plates are increasingly used in thin-walled structural components of aircrafts, submarines, automobiles, and other high-performance application areas. Their importance is increasing as many structural composite parts can be seen as composition of thin laminated plate elements. Hence, this study focuses on the design of laminated composite plates. Composite laminated plate design optimization typically involves identifying the optimal configuration (minimum weight and cost) in terms of its design variables such as laminate stacking sequence, ply thickness, and number of plies, for specified boundary conditions evaluated under standard failure criteria.

2. The composite plates are subjected to uniform or non-uniform loads depending upon their role. For example, an aircraft wing is one of such kind which experiences different loadings, such as aerodynamic lift, load due to wing structure, weight and load due to fuel contained in the wing. These loads act perpendicular to the wing surface, and their magnitude varies along the length of the wing. In line with this, transverse loading and hydrostatic loading conditions are considered in this study.

3. Multi-objective optimization using nature inspired algorithms have been extensively explored. Two major approaches have emerged - Vector Evaluated design optimization (VEDO) (Schaffer, 1984) and Objective Switching design optimization (OSDO) (Deepti Chafekar et.al., 2003). In this paper, both these approaches are used and a comparative study is carried out.

4. The performance evaluation of ABC on comparison with other popular meta-heuristic algorithms like Genetic Algorithm (GA) and Particle Swarm optimization (PSO) is also performed.

5. Several runs are carried out and best seven optimal values are considered for performance evaluation of ABC. The results in the form of box plot are presented for the statistical evidence.

PROBLEM FORMULATION

In this paper, a simply supported laminated composite plate model is considered as a case study to demonstrate the effectiveness of the design optimization technique using ABC. The further sections give a detailed description of the structural analysis and governing equations considered for the laminated composite plate simply supported on all four edges subjected to two types of transverse loading configurations namely Line Load and Hydrostatic Load evaluated with failure criteria.

Figure 1. Geometry of a simply supported thin laminated plate

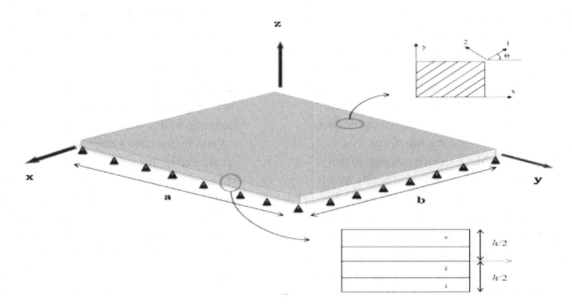

SIMPLY SUPPORTED RECTANGULAR PLATE

A thin rectangular composite laminated plate of length a in x- *direction,* width b in the y -direction, and thickness h in the z-direction is considered in the present work. The plate is assumed to be constructed of arbitrary number, n, of linearly elastic orthotropic layers. Each layer consists of homogeneous fibre reinforced composite material and the plate is simply supported on all four edges. A rectangular Cartesian coordinate system x, y and z is used to describe the infinitesimal deformations of an n-layer laminated composite plate. The laminate consists of n plies with the individual thicknesses h_i and the layer orientation angles $\theta_i \; (i = 1, 2, 3......n)$ as shown in Figure 1. Total thickness of the plate is h and bottom and top surfaces are located at $z = -h/2$ and $z = h/2$ respectively. We assume that the middle surface of the undeformed plate coincides with the *xy-plane*. The principal fibre direction is oriented at an angle ϕ to the x-axis. The layerwise principal material coordinate system, denoted by *1, 2* and

3 are used to find the stresses and strains in order to analyze lamina failure. The 1 and 2 axes are parallel and perpendicular to the x-y plane respectively, and the 3-axis is aligned parallel to the global z-axis.

LAMINATE ANALYSIS

In the present work, classical lamination plate theory is used to develop an analytical solution for orthotropic composite plate. Navier methods are employed for designing of rectangular composite plate with all four edges of the plate as simply supported. Governing equation for static bending in absence of in-plane and thermal forces is expressed as follows (Reddy, 1997).

$$D_{11}\frac{\partial^4 \omega_o}{\partial x^4} + 2(D_{12} + 2D_{66})\frac{\partial^4 \omega_o}{\partial x^2 \partial y^2} + D_{22}\frac{\partial^4 \omega_o}{\partial y^4} = q$$

(1)

where, D_{11}, D_{12}, D_{22} and D_{66} are bending stiffness's

The out of plane displacement of the plate w is assume to be of the form

$$w = \sum_{n=1}^{\infty}\sum_{m=1}^{\infty} W_{mn} \sin\frac{m\pi x}{a} \sin\frac{n\pi y}{b} \qquad (2)$$

where, w – deflection function

W_{mn} – deflection co-efficient.

The transverse distributed load is expanded in double Fourier series as

$$q = \sum_{n=1}^{\infty}\sum_{m=1}^{\infty} Q_{mn} \sin\frac{m\pi x}{a} \sin\frac{n\pi y}{b} \qquad (3)$$

where, q – load function.

Q_{mn} – load co-efficient.

Substituting the equation (2) and equation (3) in equation (1) and on simplification we can obtain the deflection coefficient W_{mn}, as

$$W_{mn} = \frac{Q_{mn} b^4/\pi^4}{\left[D_{11} m^4 s^4 + 2\left(D_{12} + 2D_{66}\right) m^2 n^2 s^2 + D_{22} n^4 \right]} \qquad (4)$$

TRANSVERSE LOADING CONFIGURATIONS

The deflection coefficient W_{mn} for different types of loading, load coefficient Q_{mn} and deflection w are given in separate subsection as follows.

Type I: Line Load

The First type is a line load $q(x,y)$ considered acting along the centre of the plate.

Load coefficient Q_{mn} is given as

$$Q_{mn} = \frac{4q_0}{\pi bm} \sin\left(\frac{n\pi y_0}{b}\right) \qquad (5)$$

where, $q = q_0$, for (m, n=1, 2, 3.....).

Hence for a special orthotropic plate that is subjected to a transverse line load q_0, the deflection w is given by

$$w = \frac{4q_0}{\pi^5}\sum_{n=1}^{\infty}\sum_{m=1}^{\infty} \frac{b^3 \sin\frac{m\pi x}{a} \sin^2\frac{n\pi y}{b}}{m\left[D_{11} m^4 s^4 + 2\left(D_{12} + 2D_{66}\right) m^2 n^2 s^2 + D_{22} n^4 \right]} \qquad (6)$$

Type II: Hydrostatic Load

The Second type is a Hydrostatic load $q(x,y)$.

Load coefficient Q_{mn} is given as

$$Q_{mn} = \frac{8q_0 \cos\left(n\pi\right)}{\pi^2 mn} \qquad (7)$$

where, $q = q_0$, for (m, n=1, 3, 5....) and $q = 0$ for (m, n = 2,4,6....).

$$w = \frac{8q_0}{\pi^6}\sum_{n=1}^{\infty}\sum_{m=1}^{\infty} \frac{b^4 \sin\frac{m\pi x}{a} \sin\frac{n\pi y}{b} \cos\left(n\pi\right)}{mn\left[D_{11} m^4 s^4 + 2\left(D_{12} + 2D_{66}\right) m^2 n^2 s^2 + D_{22} n^4 \right]} \qquad (8)$$

The consecutive equation relating stress and the curvature for a simply supported rectangular laminated plate is expressed as

$$\begin{Bmatrix} \sigma_{xx} \\ \sigma_{yy} \\ \sigma_{xy} \end{Bmatrix}_k = Z_k \begin{bmatrix} \bar{Q}_{11} & \bar{Q}_{12} & \bar{Q}_{16} \\ \bar{Q}_{21} & \bar{Q}_{22} & \bar{Q}_{26} \\ \bar{Q}_{61} & \bar{Q}_{62} & \bar{Q}_{66} \end{bmatrix}_k \begin{Bmatrix} -\dfrac{\partial^2 \omega_0}{\partial x^2} \\ -\dfrac{\partial^2 \omega_0}{\partial y^2} \\ -2\dfrac{\partial^2 \omega_0}{\partial x \partial y} \end{Bmatrix} \qquad (9)$$

where Q_{ij} are the transformed reduced stiffnesses, which are expressed in terms of the orientation angle and the engineering constant of the material.

THE OPTIMIZATION PROCESS

The multi-objective design optimization has been carried out on a laminated composite plate

Table 1. Material properties of carbon/epoxy composites

Elastic Moduli of laminate (GPa)			Lamina Poisson's rations			Rigidity Moduli of laminate (GPa)		
E_{XX}	E_{YY}	E_{ZZ}	v_{XZ}	v_{XY}	v_{YZ}	G_{XY}	G_{YZ}	G_{XZ}
126	11	11	0.28	0.28	0.4	6.6	3.93	6.6

for two approaches namely VEDO and OSDO based ABC algorithm for the composite plate design problem. A carbon/epoxy laminated composite plate with dimensions of $a = 800mm$, $b = 800mm$ is considered in the present work. The twelve discrete variables $[n_{\theta 1}/ n_{\theta 2}/, ..., /n_{\theta 12}]$ corresponding to the number of layers at each of the twelve different fiber orientation angles and the lamina thickness – t which is also considered to be a discrete variable, capable of taking values between the specified ceiling and floor limits with a least count of 0.001mm are considered for the present work. The material properties considered for the study are given in Table 1. A Line loading configuration of the plate is illustrated in Figure 2 and a Hydrostatic loading configuration is illustrated in Figure 3. As discussed earlier, four

different failure theories have been considered to obtain optimal design configurations for two different loading conditions.

OBJECTIVE FUNCTIONS

In this present case, design optimization of the composite plate involves two objectives namely minimizing the weight and to minimize the total cost of the composite plate which includes the material cost and manufacturing cost.

Figure 2. Composite Laminated Plate with line load q(x,y)

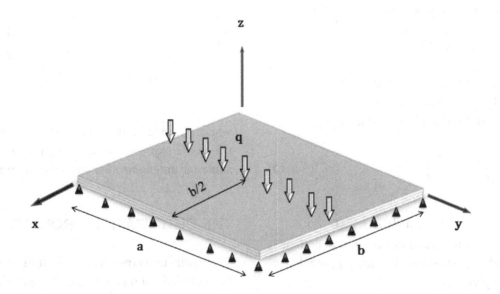

Figure 3. Composite Laminated Plate with hydrostatic load q(x,y)

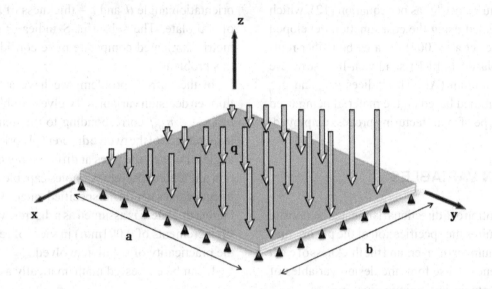

THE WEIGHT FUNCTION

The optimum weight of the composite plate is obtained for a given loading condition as follows. Initially the reduced stiffness and compliance matrices are obtained for the given CFRP material properties. For the applied load, the strains, curvatures, global stresses and material ply stresses are calculated. Finally, the stresses in each ply are checked for failure using various failure criteria. Total height h of the composite plate with $n_{\theta i}$ representing the number of layers at orientation angle θ_i with thickness of $t\ mm$ is given by;

$$h = [\{\Sigma_{i=1}^{12}(n_{\theta i})\} * t] \qquad (10)$$

The optimum weight of the plate is calculated using equation (11) such that all plies satisfy the failure conditions.

$$Weight,\ W_t = \rho * h * a * b \qquad (11)$$

where, ρ – density of the material of the composite plate.

THE COST FUNCTION

Due to the high cost of the composite materials, it is vital to consider the economical aspect of composite structure as a design objective. The main contribution to the material cost arises from the raw material for the composite plates. The manufacturing cost is a direct function of time (in minutes) associated with manufacturing of the composite plate, which includes the time lost in press form preparation, layer cutting, layer sequencing and final working. Hence the total cost of the composite plate is based on the material cost and manufacturing cost calculated using the relation

Total cost = Material Cost + Manufacturing Cost $\qquad (12)$

$$g = g_{matl} + g_{manufact}$$

$$g(x) = [g_{matl}\{W_t\} + 4 + g_{maf}\{$$
$$(\sum_{i=1}^{12}(n_{\theta i}) * 14_{\min} + 110_{\min}\}]units$$

The composite plate is optimized with respect to minimum cost 'g' as per equation (12), which is formulated using the cost function developed by Kovacs et al. (2005) for a carbon-fiber-reinforced plastic (CFRP) sandwich-like structure with aluminium (Al). The indices g_{matl} and g_{maf} are determined based on the material being used and the type of manufacturing process employed.

DESIGN VARIABLES

The definition of a directional laminated composite plate requires the specification of the ply orientation, the number of layers and the thickness of each layer. Hence, these form the design variables of the laminate design optimization process.

The stacking sequence describes the orientation of the plies, i.e., the number of plies placed at different orientation angles. This stacking sequence has a pronounced effect on the properties of the composites and greatly affects the strength of the composite component. In the current work, ply orientation angles are considered within the range of {-90°, +90°} in steps of 15°, hence leading to 12 different possible values of orientation angle θ which are as follows;

$$\theta = \{-75°/-60°/-45°/-30°/-15°/0°/15°/30°/45°/60°/75°/90°\} \quad (13)$$

Here we have considered the number of layers present at each of the different ply orientation angles as the actual decision variables for the optimization process. The second decision variable - the thickness of the lamina, has been considered to be within the range of 0.05mm to 0.5mm. In this case, we have assumed each layer to be of the same thickness, i.e., $\{t_1/t_2/......./t_i\} = t$.

The minimum value of weight and total cost is achieved by determining the optimal configuration, given by,

$$\{n_{\theta1}/n_{\theta2}/......./n_{\theta i}\}_S \{t_1/t_2/......./t_i\} \quad (14)$$

where, $n_{\theta i}$ – is the number of layers at a fibre orientation angle θ_i and t_i – thickness of ith layer of the plate. The subscript S indicates a symmetric laminated composite plate considered in this problem.

In the current problem, we have a total of thirteen decision variables. Twelve variables $\{n_{\theta1}/n_{\theta2}/......, /n_{\theta12}\}$ corresponding to the number of layers at each of the twelve different ply orientation angles. Further, the plies at different ply orientation angles are discrete variables capable of only integer values within the specified range. Also, the lamina thickness is assumed as a discrete variable (in increments of 0.001mm) in view of retaining the practicality of solution evolved.

It can be expressed mathematically as:

- **Minimize Weight:** $Min\{w = f(x)\}, f(x)$ *weight objective function*
- **Minimize Cost:** $Min\{c = g(x)\}, g(x)$ *cost objective function*
- Such that, *strength > Minimum strength*: *generic failure criterion,*
- Strengths are calculated using the reference structural mathematical model.

A possible solution,
$$x = \{[n_{\theta1} / n_{\theta2} / .... / n_{\theta i}], t\} \quad (15)$$

- *Number of layers at each orientation angle,* $n_{\theta i} \forall n_{\theta i} \in Z^+ \ni Z^+ = [0,....,50]$
- *Also,* $\sum_{i=1}^{12}(n_{\theta i}) > 0$ $\forall \theta_i \in \theta \ni \theta = [-90 + (15 * i)]° $ *for* $i = 1...12$.
- *Thickness,* $t \in S; S = [0.05mm, 0.5mm]$

DESIGN CONSTRAINTS: FAILURE CRITERIA

The strength of the lamina in different directions is considered as the design constraints. The stress in each ply should satisfy the design constraints, which are given by

(1) $X_C \leq \sigma_{LL} \leq X_T$

(2) $Y_C \leq \sigma_{TT} \leq Y_T$ and $\quad\quad$ (16)

(3) $-S \leq \sigma_{LT} \leq +S$

X_C, X_T, Y_C and Y_T are the laminate compressive and tensile strengths in X, Y directions respectively. S is the shear strength of the component in the X-Y plane and σ_{LL}, σ_{TT} and σ_{LT} are the stresses induced in the principal direction.

The stresses in an individual lamina are fundamental to control the failure initiation and progression in the laminate. The strength of each individual lamina is assessed separately by considering the stresses acting on it along material axes. Determination of failure load is very essential in understanding the failure process as well as the reliability of structures.

Failure Condition as per equation (16) in each lamina is verified using different failure criteria. This study considers some of the failure theories such as Maximum Stress, Failure Mechanism based, Tsai-Wu and Tsai-Hill failure criteria. The details about these criteria are explained in the following sections.

MAXIMUM STRESS FAILURE CRITERION

The composite plate ply fails when one of the following conditions is violated (Omkar et al., 2005).

$$X_c \geq \sigma_{LL} \leq X_t, Y_c \geq \sigma_{TT} \leq Y_t,$$
$$-S \geq \sigma_{LT} \leq S \quad\quad (17)$$

where, σ_{LL}, σ_{TT} and σ_{LT} are the stresses induced in the principal direction. The above equation stipulates the condition for non-failure for any particular ply of composite laminated plate.

FAILURE MECHANISM BASED FAILURE CRITERION (FMBFC)

This theory is based on micromechanics approach, which takes into consideration the initiating failure mechanisms to predict the strength parameters under the given loading conditions. In this theory it can be stated that the failure occurs when the failure load fraction is less than one. This failure criterion takes into account of initiating failure mechanisms such as fiber break, fiber cutting, matrix cracks, matrix meshing etc. Failure Envelopes developed by most of the failure criteria are based on the method of 'curve fits' and they do not consider the initiating failure mechanisms. Hence, this criterion will give better predictions of failure conditions. Refer Naik et al. (2005) for details about this criterion.

TSAI-WU FAILURE CRITERION

This failure theory is based on the total strain energy failure theory. This is a generic model proposed by Tsai-Wu et al. (2006). The failure theory states that the lamina failure occurs when the following condition is satisfied.

$$F_{LL}\sigma_L^2 + F_{TT}\sigma_T^2 + 2F_{LT}\sigma_L\sigma_T + F_L\sigma_L + F_T\sigma_L + F_{SS}\sigma_{LT}^2 > 1 \quad\quad (18)$$

$F_{LL} = \frac{1}{X_T X_C}$; $F_{TT} = \frac{1}{Y_T Y_C}$; $F_L = \frac{1}{(X_T - X_C)}$; $F_T = \frac{1}{(Y_T - Y_C)}$; $F_{SS} = \frac{1}{S^2}$; $F_{LT} = -0.5(F_{LL}F_{TT})^{0.5}$

where F_{LL}, F_{TT}, F_{LP}, F_L, F_T, F_{SS} are the strength parameters. This criterion predicts the immanency of failure but not the failure modes.

TSAI-HILL FAILURE CRITERION

Tsai-Hill theory is based on Von-Mises distortional energy yield criterion. Failure in the material takes place only when the distortion energy is greater than the failure distortion energy of the material. Failure occurs if the following condition is satisfied (Kaw, 2006).

$$\frac{\sigma_L^2}{X_T^2} + \frac{\sigma_T^2}{Y_T^2} - \frac{\sigma_L \sigma_T}{X_T^2} + \frac{\sigma_{LT}^2}{S^2} - 1 \leq 0 \qquad (19)$$

where σ_L, σ_T and σ_{LT} are the stress components referred to the material axes and X_T, Y_T and S are the corresponding allowable strengths.

ARTIFICIAL BEE COLONY (ABC)

ABC is a population based stochastic optimization technique which incorporates a flexible and well-balanced mechanism to accommodate the global and local exploration and exploitation abilities within a short computation time. In social insect colonies, each individual seems to have its own agenda; and yet the group as a whole appears to be highly systematized. Apparently, algorithms based on swarm intelligence and social insects begin to show their effectiveness and efficiency in solving difficult problems. A swarm is a group of multi-agent system such as bees, in which simple agents coordinate their activities to solve the complex problem of the allocation of labour to multiple forage sites in dynamic environments. An important and interesting behaviour of bee colonies is their foraging behaviour, and in particular, how bees find a food source based on the amount of nectar and successfully come back with information to the hive. In a real bee colony, there are some tasks performed by specialized individuals. These specialized bees try to maximize the nectar amount stored in the hive by performing efficient division of labour and self-organization. The minimal model of swarm-intelligent forage selection in a honey bee colony, that ABC algorithm adopts, consists of three kinds of bees: employed bees, onlooker bees, and scout bees. Half of the colony comprises employed bees and the other half includes the onlooker bees. Employed bees are responsible for exploiting the nectar sources explored before and giving information to the other waiting bees (onlooker bees) in the hive about the quality of the food source site which they are exploiting. Onlooker bees wait in the hive and decide a food source to exploit depending on the information shared by the employed bees. Scouts randomly search the environment in order to find a new food source depending on an internal motivation or possible external clues or randomly. The ABC parameters used in the current case are listed in Table 2. A detailed description about the artificial bee colony algorithm is found in Karaboga (2005).

Table 2. VEABC parameters

Convergence rate	β = [1,…,0.4], *adaptively allocated* (decreasing from 1 to 0.4 with each iteration)
Learning rate	γ = [1,…,0.4], *adaptively allocated* (decreasing from 1 to 0.4 with each iteration)
Randomness Amplitude of bee	α = [4,…,0], *adaptively allocated* (decreasing from 4 to 0 with each iteration)
Step size	S = 0.75
Number of Swarm Particles	N = 130
Maximum number of iterations	**Max_it = 1000 iterations**
End Condition (number of iterations without update in the best values)	**500 iterations**

VECTOR EVALUATED ARTIFICIAL BEE COLONY (VEABC) FOR WEIGHT AND COST MINIMIZATION OF THE LAMINATED COMPOSITE PLATE

The VEABC is a multi-objective ABC method inspired by the concept and main ideas of VEGA algorithm (Schaffer, 1984) and VEPSO algorithm (Omkar et al., 2008). The VEABC algorithm is conceptually simple. It is similar to two single objective functions being separately evaluated by separate bee particles. The multiple objectives being considered here are disparate in nature and hence this renders the separate/exclusive evaluation of the multiple objectives to be more appropriate. VEABC is very well suited for the current challenge, as it is capable of searching for multiple optimal solutions in a very vast solution space, in a single run.

VEABC Algorithm

1. Initialize both swarms (X1, X2) randomly within the feasible solution space.
   - All artificial bees are repeatedly initialized until it satisfies all the constraints.
2. While the end condition is false.
3. For both the swarms exclusively evaluate each of the objectives.
   - X1: Evaluates the weight-objective function.
   - X2: Evaluates the cost-objective function.
4. If the fitness value is better for the current solution than the previous best solution, then assign the current solution as the best solution—for both the swarms.
5. Adaptively generate the Randomness Amplitude of bee, the Convergence rate and the Learning rate.

6. For both the swarms update the bee positions towards the most favourable position (solution) from the other swarm.
7. After each update check whether the variables of each artificial bee of both the swarms satisfy the constraints.
8. IF the number of iterations without updating the best value of both Swarms >maximum number of iterations THEN condition = true, End. Else the whole process is repeated (Back to 2).

VEABC employs two or more swarms to probe the search space and information is exchanged among them. Each swarm is exclusively evaluated with one of the objective functions, but, information coming from other swarm(s) is used to influence its motion in the solution space. Thus, exchanging this information among swarms leads to Pareto optimal points. Specifically, in this case since there are two objective functions, two swarms (X_1, X_2) of N particles each are used. X_1 evaluates the weight objective function and X_2 evaluates the cost objective function. There is no necessity for a complicated information migration scheme between the swarms as only two swarms are employed. Each swarm is exclusively evaluated according the respective objective function. The best particle of the second swarm (X_2) is used for calculation of the new velocities of the first swarm's (X_1) particles and accordingly the best particle of the first swarm (X_1) is used for calculation of the new velocities of the second swarm (X_2).

The particle's position updates equations for the first swarm - X_1

$$^{[X_1]}D_{(i+1)} = \alpha * (r - S) + (1 - \beta) * {}^{[X_1]}D_i + \gamma * {}^{[X_2]}Dbest_i \tag{20}$$

The particle's position updates equations for the second swarm – X_2

$$^{[X_2]}D_{(i+1)} = \alpha * (r - S) + \tag{21}$$
$$(1 - \beta) * {}^{[X_2]}D_i + \gamma * {}^{[X_1]}Dbest_i$$

The particles of both the swarms (X_1, X_2) move in solution space according to the above mentioned equations, successively aligning themselves with respect to both the objective functions in each iteration and finally converging on the global optimum solution.

During initialization, it is ensured that all the particles are within the feasible solution space constrained by the failure criteria, since randomly initialized particles are not always confined to the feasible solution space. So initialization itself may take a longer time if the population size is too large. The VEABC parameters are the same for each swarm and for all simulation runs as shown in Table 2.

$$\beta = \beta_{max} - [\{(\beta_{max} - \beta_{min}) / it_{max}\} * it] \tag{22}$$

The Convergence rate β is adaptively allocated as per the equation (22). Where β_{max} is the initial Convergence rate value, β_{min} is the final Convergence rate value, it is the current iteration number and it_{max} is the maximum number of iterations. The Randomness Amplitude of bee α and the Learning rate γ are also adapted similar to β. A starting value of both β & γ of 1 are used initially to accommodate a more global search and dynamically are reduced to 0.4. For α, a starting value of 4 is used and dynamically reduced to $\alpha = 0$. The initial higher value may result in greater population diversity in the beginning of the optimization, whereas at a later stage lower values are favoured, causing a more focused exploration of the search space.

OBJECTIVE SWITCHING ARTIFICIAL BEE COLONY (OSABC) FOR WEIGHT AND COST MINIMIZATION OF LAMINATED COMPOSITE PLATE

In the current work we employ the principle of objective switching (Deepti Chafekar et al., 2003) for solving the given multi-objective optimization problem. The method of objective switching has a characteristic feature of each candidate solution being evaluated with respect to only one of the objectives at a time and therefore posing less risk of converging at individual objective optima (Omkar et al., 2008). This method is very well suited for the current challenge, as it is capable of fine searching for optimal solutions in a very vast solution space (Deepti Chafekar et al., 2003).

Objective Switching Artificial Bee Colony Algorithm (OSABC) implies the method of objective switching allows for alternative fitness evaluation of the objectives keeping the population of bees unchanged. The generic composite design optimization framework being presented employs the above-mentioned method and will henceforth be referred as Objective Switching Artificial Bee Colony (OSABC). In the given framework of optimization, we have to minimize both the objectives namely, weight and cost.

OSABC Algorithm

1. Initialize both swarms (X1, X2) randomly within the feasible solution space.
    ○ All artificial bees are repeatedly initialized until it satisfies all the constraints.
2. While the maximum number of iterations is not reached follow steps 3–8,
3. For odd numbered iterations,
    ○ Evaluate the weight-objective function. and for even numbered iterations,
    ○ Evaluate the cost-objective function

Table 3. Optimum design configurations evolved for variable stacking sequence - line loading

Loading (N/mm)	Failure Criteria Number of Ply	Ply-orientation	Thickness (mm)	Weight (Kgs)	Cost	(units)
q						
300	Tsai-Hill Failure criterion	79	$[0_{11}/15_7/-15_6/30_4/-30_9/45_{10}/-45_8/60_2/-60_6/75_7/-75_3/90_6]$	0.197	37.874	**188.191**
		76	$[0_{11}/15_8/-15_6/30_{10}/-30_9/45_9/-45_6/-60_4/75_7/-75_1/90_5]$	0.199	36.781	**182.874**
		80	$[0_9/15_3/-15_9/30_{11}/-30_9/45_{10}/-45_4/60_3/-60_6/75_6/-75_5/90_5]$	0.186	43.672	**236.57**
	Tsai – Wu Failure criterion	78	$[0_8/15_3/-15_7/30_{11}/-30_{12}/45_8/-45_8/60_2/-60_6/75_3/-75_5/90_5]$	0.197	37.370	**185.810**
		91	$[0_8/15_6/-15_7/30_{10}/-30_{10}/45_7/-45_5/60_9/-60_5/75_6/-75_9/90_9]$	0.183	40.463	**201.454**
		80	$[0_9/15_3/-15_8/30_{11}/-30_{10}/45_8/-45_6/60_6/-60_4/75_4/-75_5/90_6]$	0.199	38.717	**192.494**
	Maximum Stress based Failure criterion	66	$[0_{11}/15_4/-15_5/30_6/-30_4/45_{10}/-45_6/60_2/-60_7/75_3/-75_7/90_1]$	0.198	31.781	**158.033**
		61	$[0_{11}/-15_{10}/30_6/-30_9/45_4/-45_9/-60_4/75_6/-75_1/90_1]$	0.199	29.522	**146.802**
		67	$[0_8/15_4/-15_6/30_5/30_9/45_7/-45_8/60_1/-60_7/75_2/-75_4/90_6]$	0.199	32.425	**161.231**
	Failure Mechanism Based Failure Criterion	73	$[0_8/-15_7/30_{11}/-30_5/45_{19}/-45_7/-60_4/-75_5/90_7]$	0.2	30.642	**152.369**
		66	$[0_8/15_5/-15_2/30_8/-30_{11}/45_4/-45_4/60_7/-60_7/75_1/-75_5/90_4]$	0.2	32.102	**152.619**
		67	$[0_{10}/15_5/-15_8/-30_6/45_{11}/-45_7/60_1/-60_2/75_5/-75_6/90_6]$	0.198	32.262	**160.426**

4. If the fitness value is better for the current solution than the previous best solution, then assign the current solution as the best solution—for both the swarms.

5. Adaptively generate the Randomness Amplitude of bee, the Convergence rate and the Learning rate.

Table 4. The optimal configurations evolved for fixed standard stacking schemes of {0 / 90} and {0 / ±45 / 90} - line loading

Loading (N/mm)	Failure Criteria Number of Ply	Stacking Sequence	Thickness (mm)	Weight (Kgs)	Cost	(units)
q						
300	Tsai –Hill Failure criterion	85	$[0_{36}/45_{11}/-45_{11}/90_{20}]$	0.24	40.10	**242.421**
		82	$[0_{63}/90_{19}]$	0.25	41.29	**246.273**
	Tsai – Wu Failure criterion	86	$[0_{30}/45_{16}/-45_{16}/90_{24}]$	0.25	41.52	**232.604**
		81	$[0_{74}/90_7]$	0.26	42.16	**235.341**
	Maximum Stress based Failure criterion	78	$[0_{29}/45_{14}/-45_{14}/90_{21}]$	0.25	38.64	**212.96**
		75	$[0_{49}/90_{26}]$	0.245	40.12	**220.654**
	Failure Mechanism Based Failure Criterion	79	$[0_{54}/45_7/-45_7/90_{11}]$	0.25	36.04	**209.346**
		75	$[0_{60}/90_{15}]$	0.25	38.25	**211.148**

Figure 4. Minimum weights obtained for different failure theories with standard and non-standard stacking sequences for transverse line load

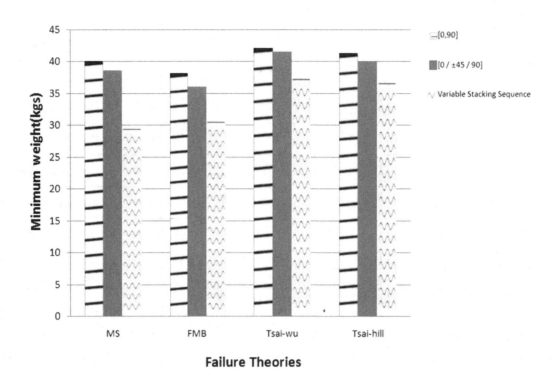

6. For both the swarms update the bee positions towards the most favourable position (solution) from the other swarm.

7. After each update check whether the variables of each artificial bee of both the swarms satisfy the constraints.

8. IF the number of iterations without updating the best value of both Swarms > maximum number of iterations THEN condition = true, End. Else the whole process is repeated (Back to 2).

RESULTS AND DISCUSSION

A laminated composite plate with dimensions of 800mm * 800mm is considered for the study.

The material properties considered for the study are given in Table 1. The Classical Laminate Plate Theory (CLPT) has been used for obtaining stresses in the plate and the failure criteria are used for the prediction of failure in a lamina. Different loading conditions are considered for the study. The results of the design optimization process, i.e., the evolved optimal designs, for each loading condition are presented and discussed in the coming sections. The optimum design of the composite plate is obtained in terms of thickness of ply, stacking sequence, number of plies at each orientation which does not fail under the considered failure criteria for each of the loading configuration. Optimal design evolved out of two of the most popularly used standard stacking schemes: *[0 / 90]$_S$* referred to as *SS$_1$* and *[0 /*

Figure 5. Minimum cost obtained for different failure theories with standard and non-standard stacking sequences for transverse line load

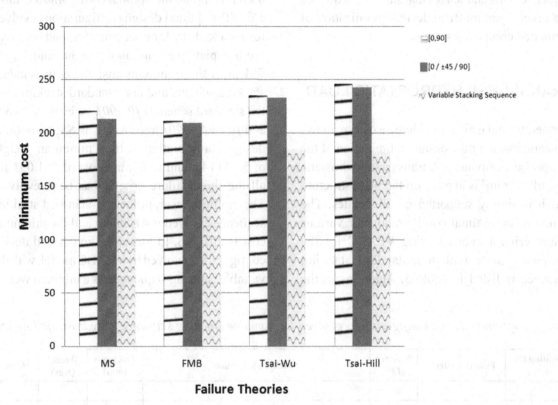

$\pm45 / 90]_S$ referred to as SS_2 are compared with the optimal design configurations evolved from variable stacking sequence. These comparisons help in understanding the influence of laminated stacking sequence on design optimization of the composite component.

TRANSVERSE LINE LOAD

This is a simple configuration of loading with the forces being applied on the plate structure in transverse direction. The plate is simply supported on all 4 edges. For the design optimization of composite plate, we have considered 300 N/mm transverse line load acting on the plate at the mid section (i.e. $x=a/2$ and $y=b/2$). The results of an optimal configuration for various failure criteria

evolving non-standard stacking sequence is listed in Table 3 and the optimal configuration evolved for the two standard sequences considered *[0 / 90]_S –SS_1* and *[0 / ±45 / 90]_S – SS_2* are listed in the Table 4. For non-standard stacking sequence, the minimum weight and the minimum cost are 29.522 kg and 146.802 respectively. The minimum weight and the minimum cost for optimal standard stacking schemes referred to as SS_1 and SS_2 are 36.04 kg and 209.34 respectively. *From this it is clearly seen that an optimal design evolved for non-standard stacking sequence has 18.09% savings in the weight and 18.07% savings in cost.* The Figures 4 and 5 show the comparison of weight and cost for the standard stacking schemes and non-standard stacking schemes for different failure criteria. From these figures it is found that a significant savings in weight and cost is obtained for optimal design evolved with variable stacking sequence. This enunciates that stacking sequence is a crucial parameter in design optimization of laminated composite structures.

TRANSVERSE HYDROSTATIC LOAD

Further, we have also considered a case with hydrostatic loads for the design optimization of the composite component. A transverse hydrostatic load of $1 N/mm^2$ is applied on the plate structure which is simply supported on all 4 sides. The results of an optimal configuration for various failure criteria evolved using VEABC for the composite plate with non-standard stacking sequence is listed in Table 5. Again, as in the

previous case the standard stacking schemes are used for finding the optimal configuration. Table 6 lists the optimal design configurations evolved for the fixed stacking schemes SS_1 and SS_2. Figure 6 depicts the minimum weights and Figure 7 depicts the minimum cost for both standard stacking scheme and non-standard stacking. For the standard schemes $[0 / 90]_S$ referred to as SS_1 and $[0 / \pm45 / 90]_S$ referred to as SS_2 an optimal design configuration with a minimum weight of 50.834 kg and minimum cost of 251.031 for all the three failure criteria has been evolved. The minimum weight for non-standard stacking sequence evolved is 41.893kg and the minimum cost is 207.596. In this case the optimal design configuration evolved by VEABC model, with the variable stacking sequence has a minimal weight

Table 5. Optimum design configurations evolved for variable stacking sequence - Hydrostatic loading

Loading (N/mm²)	Failure Criteria	Number of Ply	Stacking Sequence	Thickness (mm)	Weight (Kgs)	Cost
q						
1.0	Tsi-Hill Failure criterion	91	$[0_{10}/154/ -15_8/3012/- 3012/ 45_{13}/-458/607/ -6012 /754/-758/ 905]$	0.196	49.126	**244.092**
		102	$[0_{13}/158/ -15_9/307/- 3010/ 45_{13} /-457/605/-607/757/-759/ 907]$	0.199	49.364	**245.399**
		101	$[0_{12}/1511/ -15_{10}/3011/- 3010/ 45_8 /-459/609 / -605 /753/-757/ 906]$	0.2	49.126	**244.208**
	Tsai – Wu Failure criterion	103	$[0_{11}/1513/ -15_{14}/3015/- 305/ 45_{12}/-458/604/ -608/-758/ 905]$	0.199	49.848	**247.804**
		107	$[0_{11}/1510/ -15_{12}/3011/- 3012/ 45_{11}/-456/606/ -607 /758/-757/ 906]$	0.195	50.785	**252.281**
		103	$[0_{14}/1511/ -15_9/3013/- 3010/ 45_{12}/-456/605/ -6012 /751/-752/ 908]$	0.199	49.848	**247.804**
	Maximum Stress based Failure criterion / Failure Mechanism Based Failure Criterion	87	$[0_9 /158/ -15_5/3010/- 3012/ 45_{10} /-457/605 / -605 /755/-758/ 903]$	0.2	42.316	**210.372**
		87	$[0_{10} /154/ -15_7/309/- 307/ 45_{12} /-456/605 / -605 /759/-755/ 908]$	0.198	41.893	**208.282**
		86	$[0_7 /155/ -15_3/3010/- 3011/ 45_{10} /-458/608 / -607 /754/-756/ 907]$	0.2	41.896	**207.596**
		88	$[0_{12} /157/ -15_{10}/308/- 309/ 45_7 /-453/6010/ -608 /755/-758/ 901]$	0.198	42.423	**210.675**
		88	$[0_9 /156/ -15_6/3013/- 308/ 45_8 /-459/605 / -608 /756/-757/ 903]$	0.197	42.161	**209.618**
		86	$[0_{10} /158/ -15_8/3010/- 3012/ 45_4 /-454/605 / -604 /756/-759/ 906]$	0.199	48.589	**211.732**

Table 6. The optimal configurations evolved for fixed standard stacking schemes of {0 / 90} and {0 / ±45 / 90} - hydrostatic loading

Loading (N/ mm²)	Failure Criteria	Number of Ply	Ply-orientation	Thickness (mm)	Weight (Kgs)	Cost
q						
1.0	Tsai-Hill Failure criterion	85	[034/±4519/9032]	0.26	54.982	**254.531**
		81	[069/9012]	0.254	54.181	**253.13**
	Tsai – Wu Failure criterion Maximum Stress based Failure criterion	82	[058/±4511/9013]	0.251	53.21	**253.125**
		81	[071//9010]	0.254	52.126	**251.65**
		83	[066/9017]	0.245	51.693	**251.067**
		85	[045/±4521/9019]	0.262	51.346	**251.351**
	Failure Mechanism Based Failure Criterion	88	[049/±4524/9015]	0.251	50.834	**251.031**
		82	[055/9027]	0.25	53.247	**251.937**

compared to values obtained for standard stacking schemes. In this case it is clearly seen that an optimal design evolved for non-standard stacking sequence has a 25.5% savings in the weight and 25% savings in cost.

From the study, it is observed that the objectives more or less reinforce each other rather than

Figure 6. Minimum weights obtained for different failure theories with standard and non-standard stacking sequences for hydrostatic load

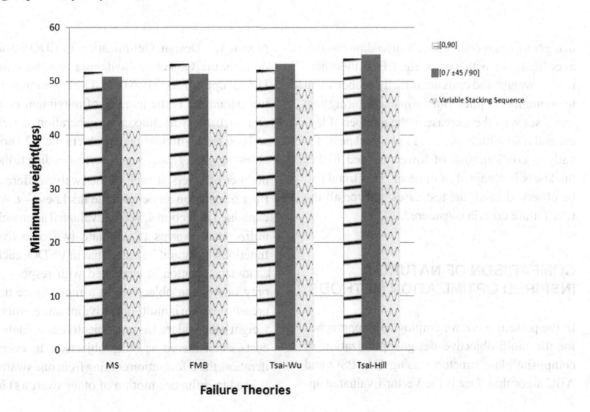

Figure 7. Minimum cost obtained for different failure theories with standard and non-standard stacking sequences for hydrostatic load

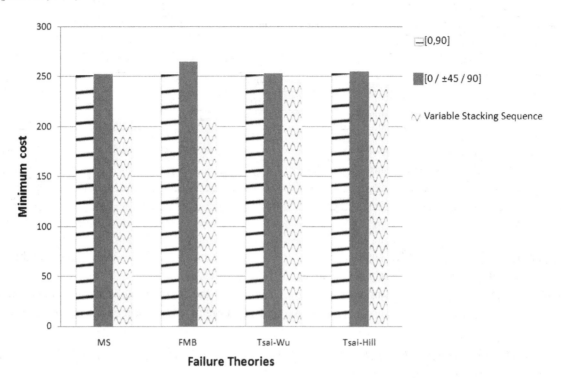

being completely conflicting, both tending towards a configuration with lower weight. Both the objectives—weight and costs are directly proportional to the number of plies. But the manufacturing cost increases with the increase in the number of layers and also with reduction in ply thickness. Finally, a configuration of lower ply and higher thickness is obtained as optimal. This trend can be observed in all the test cases and for all the four failure criteria considered.

COMPARISON OF NATURE-INSPIRED OPTIMIZATION METHODS

In the present work we employ two approaches for the multi-objective design optimization of composite plate structures using GA, PSO and ABC algorithm. One is the Vector Evaluated approach for Design Optimization (VEDO), and other is the Objective Switching approach for Design optimization (OSDO). The common control parameter is the number of maximum generation. In the work, maximum generation is fixed as 1000 for 13-dimentional problem. i.e. 12 variables $[n_{\theta 1} / n_{\theta 2} /, ... / n_{\theta 12}]$ corresponding to the number of layers at each of the twelve different fiber orientation angles and the thickness – t. As explained in section 4, vector evaluated approach utilize two swarms to evaluate two objective functions. It should be noted that in VEDO, each candidate solution is evaluated with respect to only one of the objectives at a time. Since the present problem is multi-objective in nature, where weight is considered as one objective and related cost is considered as other objective. In every iteration, the information coming from one swarm is used to influence motion of other swarm(s) in

Table 7. Comparison of results obtained from VEGA, VEPSO and VEABC algorithms

Failure Theory	Number of iterations	VEGA		VEPSO		VEABC	
		Minimum Weight (kg)	Minimum Cost	Minimum Weight (kg)	Minimum Cost	Minimum Weight (kg)	Minimum Cost
Line loading	1	43. 78	216.587	35.62	171.093	32.68	162.923
	2	42.56	211.624	34.31	169.546	33.19	165.376
	3	43. 72	216.587	36.38	165.992	31.19	165.376
Maximum Stress Failure criterion	4	45.75	228.243	33.85	168.234	33.27	166.058
	5	41.34	209.123	34.44	171.590	31.02	164.619
	6	43.86	217.891	35.81	172.002	32.03	164.619
	7	44.67	223.453	33.75	169.012	32.83	163.560
Hydrostatic Loading	1	46.51	216.312	42.53	211.201	42.68	212.595
	2	46.56	216.312	41.83	207.771	42.90	213.802
	3	45.43	212.657	43.85	217.842	42.99	213.974
Maximum Stress Failure criterion	4	46.72	219.431	43.71	217.047	42.80	213.097
	5	47.79	224.624	44.61	219.680	43.19	215.034
	6	47.16	223.512	45.29	221.024	43.19	215.034
	7	46.98	218.874	44.87	219.990	42.80	213.097

the solution space. The vector evaluated genetic algorithm (VEGA) and vector evaluated particle swarm optimization (VEPSO) are used for solving the given multi-objective optimization problem. In VEGA the population is divided into m different parts for n different objectives; the selection operation is done for each objective separately, filling equal portions of mating pool (Schaffer, 1984). Afterwards, the mating pool is shuffled, and crossover and mutation are performed as usual.

The VEPSO parameters are set similar to the parameters listed in the reference (Omkar et al., 2008) and VEGA parameters are as listed in the reference (Schaffer, 1984).

The generic composite design optimization framework being presented employs other method referred as Objective Switching approach for Design Optimization (OSDO). The main idea of OSDO is to use a single algorithm that optimizes multiple objectives in a sequential order. Every objective is optimized for a certain number of evaluations, then a switch occurs and the next objective is optimized. The population is not changed when objectives are switched. This continues till the maximum number of evaluations is complete. This method of objective switching allows for alternative evaluation of the objectives which proves to be very appropriate for the current problem. Objective switching technique with GA has been used for solving benchmark and engineering design domain problems by Deepti Chafekar et al. (2003). A co-variant of clonal selection principle called as Objective switching clonal selection algorithm has been implemented to multi-objective design optimization of composites by Omkar et al. (2008).

Table 8. Comparison of results obtained from OSGA, OSPSO and OSABC algorithms

Failure Theory	Number of iterations	OSGA		OSPSO		OSABC	
		Minimum Weight (kg)	Minimum Cost	Minimum Weight (kg)	Minimum Cost	Minimum Weight (kg)	Minimum Cost
Line loading	1	43. 78	216.587	33.46	166.373	**33.55**	165.572
	2	44.52	219.651	29.40	146.108	**33.03**	164.619
	3	43. 62	216.587	30.33	150.777	**32.68**	162.923
Maximum Stress Failure criterion	4	44.10	219.120	33.02	164.042	**33.36**	166.315
	5	43.41	216.322	32.97	163.970	**32.67**	162.923
	6	44.93	221.411	32.88	163.338	**30.19**	161.376
	7	45.23	222.571	33.55	166.582	**33.37**	166.315
Hydrostatic Loading	1	46.51	216.312	44.40	220.485	**43.05**	214.411
	2	46.51	216.312	42.19	209.611	**43.19**	215.034
	3	45.63	214.235	30.91	153.505	**42.68**	212.595
Maximum Stress Failure criterion	4	46.95	217.212	43.39	215.453	**42.80**	213.097
	5	47.12	217.665	43.75	217.321	**42.80**	213.097
	6	47.84	217.995	43.55	216.292	**42.68**	212.595
	7	46.32	216.421	42.87	212.816	**43.22**	215.043

In the given framework of optimization, we have to minimize both the objectives that are described in the section 5. For a fair comparison between algorithms, multiple runs are carried out and the optimum values of cost and weight obtained for seven runs are listed.

The results obtained by VEGA, VEPSO and VEABC for both loading conditions are listed in Table 7. The results obtained from the OSGA, OSPSO and OSABC for both loading conditions are listed in Table 8. It can be seen from the Tables 3 and 4 that in case of transverse line loading condition, the results of multi-objective design optimization of composite laminated plates are better in ABC in comparison with multi-objective variants of GA and PSO for both the Design optimization approaches. This is true for the other loading condition as well.

INTERPRETING A BOX PLOT THROUGH ANOVA TEST

In this paper a statistical comparison study similar to the one way ANOVA method between GA, PSO and ABC algorithm has been done employing both Vector Evaluated approach for Design Optimization (VEDO), and Objective Switching approach for Design optimization (OSDO) for the multi-objective design optimization of composite plate structure. Box plots display batches of data. Five values from a set of data are conventionally used; the extremes, the upper and lower hinges1 (quartiles), and the median. The basic configuration of the display is shown in Figure 8 (McGill et al., 1978). Such plots are becoming a widely used tool in exploratory data analysis and in preparing visual summaries for statisticians and non statisti-

Figure 8. Configuration of a basic box plot

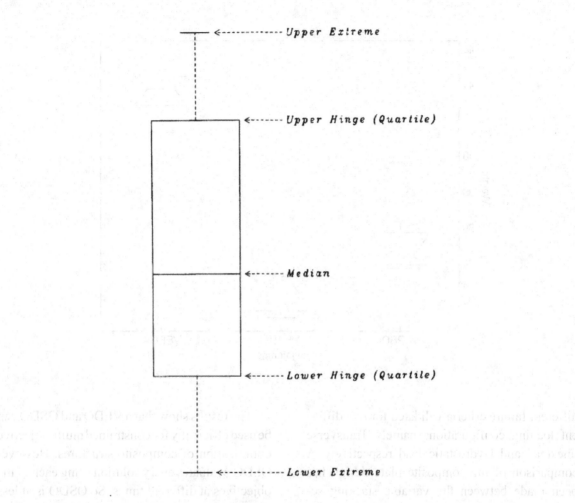

cians. Variants of the basic display in box plots are considered. The first visually incorporates a measure of group size; the second incorporates an indication of rough significance of differences between medians; the third combines the features of the first two. Hence the third variant a variable width notched box plot is used in this case for improvement in the basic display of the data for the ANOVA test. Figures 9 and 10 show the comparison of results for minimum weight and minimum cost using VEDO for line load condition. Figures 11 and 12 show the comparison of results for minimum weight and minimum cost using OSDO for line load condition.

CONCLUSION

Nature inspired optimization algorithms such as VEGA, VEPSO, VEABC, OSGA, OSPSO, OSABC have been considered for the strength based optimization of laminated composite structures. Multi-objective design optimization studies have been carried out. Weight and cost functions are the two objectives for the present study. Different failure criteria are considered for the optimum design of the composite plate. CLPT has been employed for obtaining stresses in the laminated composite plate and the stresses obtained has to satisfy the failure conditions of

Figure 9. Comparison of cost for VEGA, VEPSO and VEABC algorithms for line load with maximum stress failure criterion

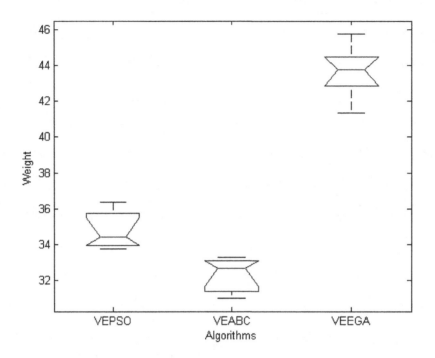

different failure criteria validated for two different loading configurations namely Transverse line load and Hydrostatic load respectively. A comparison of the composite plate design has been made between the variable stacking sequence and standard stacking sequence. These comparisons bring out weight and cost savings obtained by treating the stacking sequence also as a design variable instead of fixing it any of the commonly used standard stacking schemes like $[0/90]_S$ and $[0/\pm45/90]_S$. Regarding the different failure criteria, it was found that each criterion yields a different optimal design. It is seen from the comparison results that maximum stress criterion (MS) and Failure mechanism based failure criteria yielded fair and nearly same results for present problem. Hence, when optimizing laminated composite structures, the choice of a failure criterion corresponding to the real behavior of the structure is crucial for both economy and safety.

The results show that a VEDO and OSDO can be used efficiently for constrained multi-objective optimization of composite structures. However OSDO evaluates every solution using each of the objectives at different times. So OSDO is at less risk of converging at individual objective optima. The VEABC and OSABC based optimization models developed in this paper allows for easy incorporation of any changes in design parameters. Inclusion of further constraints and objectives can be brought into effect without necessitating any major changes to the current framework making it generic and robust and providing a great deal of flexibility to extend it to any number of different composite configurations.

Results in the form of Box plots give statistical evidence indicating ABC algorithm outperforming other meta-heuristic algorithms in terms of picking optimal solutions through the fair comparative ANOVA test. The ABC based optimization

Figure 10. Comparison of weight for OSGA, OSPSO and OSABC algorithms for line load with maximum stress failure criterion

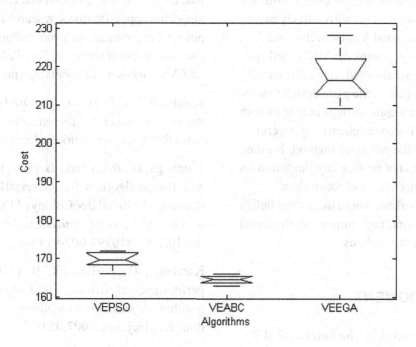

Figure 11. Comparison of cost for OSGA, OSPSO and OSABC algorithms for line load with maximum stress failure criterion

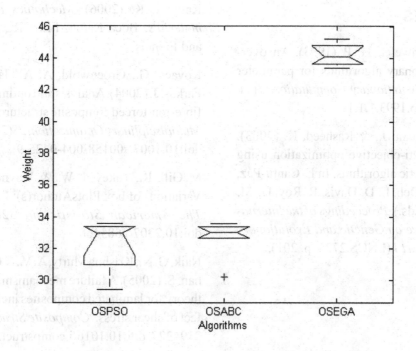

model has performed quite satisfactorily, evolving superior composite plate design that results in significant weight savings in case of both the approaches. ABC algorithms are relatively easy to implement and has good local search capability. For the considered problems the ABC based optimization model has performed quite satisfactorily, evolving superior composite plate design that results in significant weight savings in case of both the approaches. This comprehensively ascertains the robustness of the proposed method. Further, this approach does not impose any limitation on the number of objectives and constraints.

In our future work we would like to parallelize the algorithms in handling complicated structural design optimization problems.

ACKNOWLEDGMENT

This work is supported by the Aeronautical Research and Development Board, Defence Research and Development Organisation, New Delhi, India, Grant ARDB-204.

REFERENCES

Back, T., & Schwefel, H. P. (1993). An overview of evolutionary algorithms for parameter optimization. *Evolutionary Computation, 1*(1). doi:10.1162/evco.1993.1.1.1

Chafekar, D., Xuan, J., & Rasheed, K. (2003). Constrained multi-objective optimization using steady state genetic algorithms. In E. Cantú-Paz, J. A. Foster, K. Deb, L. D. Davis, R. Roy, U.-M. O'Reilly et al. (Eds.), *Proceedings of the International Conference on Genetic and Evolutionary Computation, Part I* (LNCS 2723, p. 201).

Gómez-Iglesias, A., Vega-Rodríguez, M. A., Castejon, F., Cardenas-Montes, M., & Morales-Ramos, E. (2010). Artificial bee colony inspired algorithm applied to fusion research in a grid computing environment. In *Proceedings of the 18th Euromicro Conference on Parallel, Distributed and Network-based Processing* (pp. 508-512).

Karaboga, D. (2005). *An idea based on honey bee swarm for numerical optimization* (Tech. Rep. No. TR06). Kayseri, Turkey: Erciyes University.

Karaboga, D., & Basturk, B. (2007). A powerful and efficient algorithm for numerical function optimization: Artificial Bee Colony (ABC) algorithm. *Journal of Global Optimization, 39*(3), 459–171. doi:10.1007/s10898-007-9149-x

Karaboga, D., & Basturk, B. (2008). On the performance of Artificial Bee Colony (ABC) algorithm. *Applied Soft Computing, 8*(1), 687–697. doi:10.1016/j.asoc.2007.05.007

Karaboga, D., & Ozturk, C. (2011). A novel clustering approach: Artificial Bee Colony (ABC) algorithm. *Applied Soft Computing, 11*(1), 652–657. doi:10.1016/j.asoc.2009.12.025

Kaw, A. K. (2006). *Mechanics of composite materials*. Boca Raton, FL: CRC Press/Taylor and Francis.

Kovacs, G., Groenwold, A. A., Jármai, K., & Farkas, J. (2004). Analysis and optimum design of fibre-reinforced composite structures. *Structural Multidisciplinary Optimization, 28*(2-3), 170–179. doi:10.1007/s00158-004-0425-9

McGill, R., Tukey, J. W., & Wayne, A. (1978). Variations of box PlotsAuthor(s): LarsenSource. *The American Statistician, 32*(1), 12–16. doi:10.2307/2683468

Naik, G. N., Krishna Murty, A. V., & Gopalakrishnan, S. (2005). A failure mechanism based failure theory for laminated composites including the effect of shear stress. *Composite Structures, 69*(2), 219–227. doi:10.1016/j.compstruct.2004.06.014

Omkar, S. N., Khandelwal, R., Yathindra, S., Naik, G. N., & Gopalakrishnan, S. (2008). Artificial immune system for multi-objective design optimization of composite structures. *Engineering Applications of Artificial Intelligence, 21*(8), 1416–1429. doi:10.1016/j.engappai.2008.01.002

Omkar, S. N., Mudigere, D., Naik, G. N., & Gopalakrishnan, S. (2008). Vector evaluated particle swarm optimization (VEPSO) for multi-objective design optimization of composite structures. *Computers & Structures, 86*(1-2), 1–14. doi:10.1016/j.compstruc.2007.06.004

Omkar, S. N., & Senthilnath, J. (2009). Artificial bee colony for classification of acoustic emission signal. *International Journal of Aerospace Innovations, 1*(3), 129–143. doi:10.1260/175722509789685865

Omkar, S. N., Senthilnath, J., Khandelwal, R., Narayana Naik, G., & Gopalakrishnan, S. (2011). Artificial Bee Colony (ABC) for multi-objective design optimization of composite structures. *Applied Soft Computing, 11*(1), 489–499. doi:10.1016/j.asoc.2009.12.008

Reddy, J. N. (1997). *Mechanics of laminated composite plates*. Boca Raton, FL: CRC Press/Taylor and Francis.

Schaffer, J. D. (1984). *Multi objective optimization with vector evaluated genetic algorithms*. Unpublished doctoral dissertation, Vanderbilt University, Nashville, TN.

Srinivasa Rao, R., Narasimham, S. V. L., & Ramalingaraju, M. (2008). Optimization of distribution network configuration for loss reduction using artificial bee colony algorithm. *International Journal of Electrical Power and Energy Systems Engineering, 1*(2).

Venkata Rao, R., & Pawar, P. J. (2010). Parameter optimization of a multi-pass milling process using non-traditional optimization algorithms. *Applied Soft Computing, 10*, 445–456. doi:10.1016/j.asoc.2009.08.007

Yao, X. (1999). *Evolutionary computation: Theory and applications*. Singapore: World Scientific.

APPENDIX

Nomenclature

a, b, h: Length, width, thickness of the plate.

E_{xx}, E_{yy}, E_{zz}: Longitudinal, Transverse, Normal Elastic Moduli

G_{xy}, G_{xz}, G_{yz}: Shear Moduli in Longitudinal, Lateral and traverse directions

S: Shear strength in the *X-Y* plane

X_t, X_C, Y_t, Y_C: Tensile and compressive strengths in *X-Y* directions

x, n, X: Vectors, components, domain are used to describe the design variables

t_k, $\theta\, \theta$, N : ply thickness, stacking sequence, number of plies

v_{xy}, v_{xz}, v_{yz}: Lamina Poisson's Ratio's in Longitudinal, Lateral and traverse directions

σ_{xx}, σ_{yy}, σ_{xy}: Longitudinal, Transverse, Shear Stress along material axes.

$\bar{Q}_{ij}$ Transformed reduced stiffness of the kth ply.

Z_k Thickness of respective kth ply.

Qmn: Transverse load coefficient.

s: Plate aspect ratio

This work was previously published in the International Journal of Applied Metaheuristic Computing, Volume 2, Issue 3, edited by Peng-Yeng Yin, pp. 1-26, copyright 2011 by IGI Publishing (an imprint of IGI Global).

Chapter 6
An Ant Colony System Algorithm for the Hybrid Flow-Shop Scheduling Problem

Safa Khalouli
University of Reims Champagne-Ardenne, France

Fatima Ghedjati
University of Reims Champagne-Ardenne, France

Abdelaziz Hamzaoui
University of Reims Champagne-Ardenne, France

ABSTRACT

An integrated ant colony optimization algorithm (IACS-HFS) is proposed for a multistage hybrid flow-shop scheduling problem. The objective of scheduling is the minimization of the makespan. To solve this NP-hard problem, the IACS-HFS considers the assignment and sequencing sub-problems simultaneously in the construction procedures. The performance of the algorithm is evaluated by numerical experiments on benchmark problems taken from the literature. The results show that the proposed ant colony optimization algorithm gives promising and good results and outperforms some current approaches in the quality of schedules.

INTRODUCTION

The hybrid flow shop (HFS) consists of a set of two or more processing stages (or centers) with at least one stage having two or more parallel machines. The hybrid characteristic of a flow-shop is ubiquitously found in various industries.

The duplication of the number of machines in some stages can introduce additional flexibility, increase the overall capacities, and avoid bottlenecks if some operations are too long. Scheduling in a HFS has a great importance from practical point of view.

DOI: 10.4018/978-1-4666-2145-9.ch006

In order to solve HFS problems we have to define both an assignment and a sequencing sub-problems. The former concerns the assignment of each operation to a machine. The latter orders the operations on the machines. In this paper, our objective is to find a schedule that minimizes the makespan, i.e., the time needed to complete all the jobs. This problem is *NP*-hard for the simplest case of two stages and at least two machines available in one of the stages (Gupta, 1988). The HFS models differ in the type of machines at the stages. The machines may be identical, uniform or unrelated (Blazewicz, Ecker, Pesch, & Schmidt, 1996).

The purpose of this paper is to present an integrated *ant colony optimization* algorithm. This proposed approach considers simultaneously the assignment and the sequencing sub-problems in order to solve an HFS problem with identical machines at each stage.

The remainder of this paper is organized as follows. In section 2, we review the literature for the different methods proposed to solve the HFS problem. Section 3 is dedicated to the description of the considered scheduling problem. In section 4, we introduce our proposed meta-heuristic algorithm. Computational results are provided in section 5. Finally, conclusions and further research directions are given.

LITERATURE REVIEW

In order to solve HFS problems, many solution methodologies are proposed in the literature (Linn & Zhang, 1999) and (Ruiz & Vázquez-Rodríguez, 2010). Exact approaches are suitable for small-sized problems but get very time consuming when the size is large. For practical purposes, it is more appropriate to use approximate methods. In fact, they are able to achieve good solutions (or eventually optimal) for scheduling problems in an acceptable time.

We quote hereafter, some recent studies carried out on the HFS problem minimizing the makespan. Exact techniques included branch and bound algorithms (B&B) (Vignier, Commandeur, & Proust, 1997; Néron, Baptiste, & Gupta, 2001; Haouari, Hidri, & Gharbi, 2006), mixed integer programming (Guinet, Solomon, Kedia, & Dussa, 1996), or dynamic programming-based heuristic (Riane, Artiba, & Elmaghraby, 1998). For solving the two-stage HFS problem, some heuristic methods based on Johnson's algorithm (Johnson, 1954) are developed (Guinet, Solomon, Kedia, & Dussa, 1996; Lee, Cheng, & Lin, 1993). Santos et al. (1996) adapted some pure flow-shop heuristics in the HFS environment. Various intelligent heuristics and meta-heuristics have become popular such as simulated annealing (SA) (Gourgand, Grangeon, & Norre, 1999; Jin, Yang, & Ito, 2006), tabu search (TS) (Nowicki & Smutnicki, 1998), genetic algorithms (GA) (Portmann, Vignier, Dardilhac, & Dezalay, 1998; Jin, Ohno, Ito, & Elmaghraby, 2002; Besbes, Loukil, & Teghem, 2006), approaches based on artificial immune systems (AIS) (Engin & Döyen, 2004).

Recently, ant colony optimization (ACO) approaches are increasingly used to solve combinatorial optimization problems (Yagmahan & Yenisey, 2010). For solving the HFS scheduling problem, Alaykýran et al. (Alaykýran, Engin, & Döyen, 2007) presented an improved ACO by adapting the classical ant system (AS) (Colorni, Dorigo, & Maniezzo, 1991). Likewise, in (Khalouli, Ghedjati, & Hamzaoui, 2008) another approach based on an ant colony system algorithm (ACS) (Dorigo & Gambardella, 1997) is proposed. It consists on solving the HFS problem by two phases: the first deals with the assignment by using some static heuristics and the second used the ACS approach to determine the processing sequences on the machines. In this paper, we propose a new ACS method different from existing ones. The particularity of this approach is the integration of the assignment and the sequencing decisions on the search process. In fact, the decisions, concerning assignment and sequencing, are usually treated separately.

PROBLEM NOTATION

The manufacturing environment of the HFS is an extension of the classical flow-shop. It presents a multistage production process with the property that n jobs need to be processed at all the stages in the same order, starting at stage 1 until finishing in stage S. Each stage i consists of a given number m_i, $(m_i \geq 1)$, of identical parallel machines available from time zero. They are denoted $M_s = \left(M_{s1}, M_{s2}, ..., M_{s,m_s} \right)$. So, each job j needs several operations $\left(O_{1j}, O_{2j}, ..., O_{Sj} \right)$, where an operation O_{ij} has to be processed by one machine of a set of given machines at the ith stage during uninterrupted p_{ij} time units (preemption is not allowed). An operation O_{ij} can start only after the completion of the $(i-1)$ previous operations. The starting time and the completion time of an operation are denoted by t_{ij} and C_{ij}, respectively. In this paper, we assume that:

- Setup times of machines and move times between operations are negligible.
- Machines are independent of each other.
- Jobs are independent of each other.
- At a given time, a machine can only execute one operation.

The objective is to find an optimal ordering through the S stages for the n jobs, by taking advantage of the multiple machines in stages, to minimize the makespan: $C_{\max} = \max_{1 \leq j \leq n}\{C_{Sj}\}$. Solving such a problem consists of assigning operations to machines on each stage (the routing problem) and sequencing the operations assigned to the same machine.

THE PROPOSED ANT COLONY SYSTEM ALGORITHM

The ACO approaches imitate the behavior of real ants when searching for food. They have been applied to a diverse set of combinatorial optimization problems. In ACO, a number of artificial ants build solutions to an optimization problem by using:

1. Some additional problem specific heuristic information.
2. The deposited quantity of pheromones according to the quality of the solution; moreover it is possible to have various types of pheromone.
3. Some data structures that contain the memory of their previous action.

Each ant uses the collective experience to find a solution to the problem. The construction step of a solution depends on the quantity of pheromone and the heuristic information. The first ACO algorithm, called ant system, was proposed applying it to the traveling salesman problem (TSP). Then, AS is improved and extended. The improved versions include ant colony system algorithm (ACS) (Dorigo & Gambardella, 1997) and MAX-MIN ant system (Stützle & Hoos, 1997).

In order to solve the considered HFS scheduling problem, the solution construction has to consider two decisions: the operations assignment to machines and the operations sequencing on machines. In this paper, we propose an ACO algorithm based on the framework of the ACS technique (Dorigo & Gambardella, 1997). To deal with the considered HFS scheduling problem, we have to take into consideration a number of design decisions (Blum & Sampels, 2002). Thus, we have to define:

1. An adequate pheromone model representation.
2. The mechanism of updating the quantity of pheromone.

3. A heuristic value, which can find information about the specific problem.

In fact, these components are used to guide moves selection and, consequently, they influence system performance. In order to achieve feasible solutions for the HFS, we use a restricted list (called candidate list) of all the operations that can be selected at each time. Algorithm 1 displays the steps of the proposed algorithm.

Constructive Procedure

While constructing a feasible solution, each ant k builds independently a sequence of operations by performing the construction step and selecting a machine to process each operation. Thus, an ant has to make two decisions representing a solution component: the selection of a machine and then the selection of an operation from the set of unsequenced operations. To respect the precedence constraints of the problem at each construction step, we propose to solve the problem stage by stage. So, the procedure is repeated, for each stage i, until all the operations are selected and assigned to an available machine $M_{i\ell} \in M_i$, $\ell = 1, \ldots, m_i$. The selected operations are successively stored in a list, which is used to record the visited operation of the current schedule.

Machine Selection

To select the next machine at the current position and according to a stage i, the ants employ the following transition rule in Equation 1:

$$m = \begin{cases} \arg\max_{m \in M_i}\{\eta_m\}, & if \quad q_m \leq q_{m0} \\ M, & otherwise \end{cases} \quad (1)$$

where η_m is the heuristic information, q_m is a random number uniformly distributed in $[0, 1]$ and q_{m_0} $\left(0 \leq q_{m_0} \leq 1\right)$ is a parameter that determines

the relative importance of intensification and diversification. Indeed, if $q_m \leq q_{m_0}$ the system tends towards intensification and consequently the algorithm exploits more information, collected by the system. Otherwise, the system tends to carry out a diversification and the next machine M is randomly determined according to the probability distribution defined by Equation 2:

$$P_k\left(m\right) = \frac{\eta_m}{\sum_{y \in M_i} \eta_y}, \quad if \quad m \in M_i \quad (2)$$

In this part, the heuristic information η_m (see Equation 3) indicates the desirability of choosing machine m based on the earliest available machine heuristic.

$$\eta_m = \frac{1}{t_m + 1} \quad (3)$$

where t_m is the sum of completion times of all the operations already assigned to machine m.

Operation Selection

After the selection of a machine, an ant has to choose an operation. The selection of an operation x from candidate list L_k depends on the pheromone value $\tau_m(x, p)$ and an heuristic information η_x related to the candidate operation x. So, the state transition rule is defined by the following Equation 4:

$$x = \begin{cases} \arg\max_{s \in L_k}\left\{[\tau_m(s, p)] \cdot [\eta_s]^\beta\right\}, & if \quad q \leq q_0 \\ X, & otherwise \end{cases}$$
$$(4)$$

L_k represents the list of operations that remain to be visited by ant k. The parameter β determines the relative importance between the pheromone and the heuristic information. q is a random number uniformly distributed in $[0,1]$ and q_0 $(0 \leq q_0 \leq 1)$ a parameter that determines the relative importance of intensification and diversification. X is a random variable selected according to the probability distribution defined by Equation 5:

$$P_k(x) = \frac{[\tau_m(x,p)] \cdot [\eta_x]^\beta}{\sum_{r \in L_k} [\tau_m(r,p)] \cdot [\eta_r]^\beta} \qquad (5)$$

Pheromone Model: The proposed pheromone model $\tau_m(x,p)$ indicates the desirability of placing operation x at the pth position of a machine m. So, to each machine m, we define a matrix of pheromone representing the favorability of placing the operation x at position p. Thus, pheromone influences the relative order of operations requiring the same machine. The initial pheromone trail is set equal to τ_0.

Heuristic Information: The heuristic information η finds information about the specific prob-

Algorithm 1. The proposed ACS algorithm for the HFS problem

```
Initialize: Set ACS parameters;
  while stopping criterion not satisfied do
    repeat
    for each ant do
  Select next machine by applying the state transition rule (1)
  and (2);
    Select next operation by applying the state transition rule (4)
  and (5);
  Apply local pheromone update to decrease the pheromone on
  the selected operation;
    end for
    until every ant has built a solution
    Evaluate current solutions;
    Apply local search to the best obtained solution;
  Apply global pheromone update to increase pheromone on the
  current best solutions;
  end while
```

lem. It is used to estimate the desirability of placing operation x on machine m. In case of the considered problem, the heuristic information represents information related to the partial sequencing. In order to bias the transitions probabilities, many dispatching rules and constructive heuristics have been used in the literature (Ying

Table 1. Values of the used parameters for the proposed algorithms

Parameter	Range	Best Value HACS-HFS	IACS-HFS
$nbANT$	—	5	5
t_{max}	—	2000	2000
τ_0	—	$(n \times LB)^{-1}$	$(n \times LB)^{-1}$
β	$0 \leq \beta \leq 10$	2	2
q_0	$0 \leq q_0 \leq 1$	0.8	$\log(t)/\log(t_{max})$
q_{m_0}	$0 \leq q_{m_0} \leq 1$	—	t/t_{max}
ρ_ℓ	$0 \leq \rho_\ell \leq 1$	0.5	0.1
ρ_g	$0 \leq \rho_g \leq 1$	0.5	0.1

Table 2. Performances solutions of the test problems

Problem	LB of C_{max}	$BestC_{max}$					
		HACS-HFS	IACS-HFS	B&B	Improved-AS	AIS	GA
j10s5a2	88	88	88	88	88	88	88
j10s5a3	117	117	117	117	117	117	117
j10s5a4	121	121	121	121	121	121	121
j10s5a5	122	122	122	122	124	122	122
j10s5a6	110	110	110	110	110	110	110
j10s5b1	130	130	130	130	131	130	130
j10s5b2	107	107	107	107	107	107	107
j10s5b3	109	109	109	109	109	109	109
j10s5b4	122	122	122	122	124	122	122
j10s5b5	153	153	153	153	153	153	153
j10s5b6	115	115	115	115	115	115	115
j10s5c1	**68**	**68**	**68**	**68**	**68**	**68**	**68**
j10s5c2	**74**	**74**	**74**	**74**	**76**	**74**	**74**
j10s5c3	**71**	**72**	**72**	**71**	**72**	**72**	**72**
j10s5c4	**66**	**66**	**66**	**66**	**66**	**66**	**66**
j10s5c5	**78**	**78**	**78**	**78**	**78**	**78**	**78**
j10s5c6	**69**	**69**	**69**	**69**	**69**	**69**	**69**
j10s5d1	**66**	**66**	**66**	**66**	**#**	**66**	**66**
j10s5d2	**73**	**73**	**73**	**73**	**#**	**73**	**74**
j10s5d3	**64**	**64**	**64**	**64**	**#**	**64**	**64**
j10s5d4	**70**	**70**	**70**	**70**	**#**	**70**	**70**
j10s5d5	**66**	**66**	**66**	**66**	**#**	**66**	**66**
j10s5d6	**62**	**62**	**62**	**62**	**#**	**62**	**62**
j10s10a1	139	139	139	139	#	139	139
j10s10a2	158	158	158	158	#	158	158
j10s10a3	148	148	148	148	#	148	148
j10s10a4	149	149	149	149	#	149	149
j10s10a5	148	148	148	148	#	148	148
j10s10a6	146	146	146	146	#	146	146
j10s10b1	163	163	163	163	163	163	163
j10s10b2	157	157	157	157	157	157	157
j10s10b3	169	169	169	169	169	169	169
j10s10b4	159	159	159	159	159	159	159
j10s10b5	165	165	165	165	165	165	165
j10s10b6	165	165	165	165	165	165	165
j10s10c1	113	115	115	127	118	115	115
j10s10c2	116	119	119	116	117	119	119
j10s10c3	98	116	116	133	108	116	116

continued on following page

Table 2. Continued

Problem	LB of C_{max}	$BestC_{max}$					
		HACS-HFS	IACS-HFS	B&B	Improved-AS	AIS	GA
j10s10c4	103	120	120	135	112	120	120
j10s10c5	121	126	126	145	126	126	126
j10s10c6	97	106	106	112	102	106	106
j15s5a1	178	178	178	178	178	178	178
j15s5a2	165	165	165	165	165	165	165
j15s5a3	130	130	130	130	132	130	130
j15s5a4	156	156	156	156	156	156	156
j15s5a5	164	164	164	164	166	164	164
j15s5a6	178	178	178	178	178	178	178
j15s5b1	170	170	170	170	170	170	170
j15s5b2	152	152	152	152	152	152	152
j15s5b3	157	157	157	157	157	157	157
j15s5b4	147	147	147	147	149	147	147
j15s5b5	166	166	166	166	166	166	166
j15s5b6	175	175	175	175	176	175	175
j15s5c1	**85**	**85**	**85**	**85**	**85**	**85**	**85**
j15s5c2	**90**	**90**	**90**	**90**	**90**	**91**	**91**
j15s5c3	**87**	**87**	**87**	**87**	**87**	**87**	**87**
j15s5c4	**89**	**90**	**89**	**90**	**89**	**89**	**89**
j15s5c5	**73**	**77**	**74**	**84**	**73**	**74**	**74**
j15s5c6	**91**	**91**	**91**	**91**	**91**	**91**	**91**
j15s5d1	**167**	**167**	**167**	**167**	**167**	**167**	**167**
j15s5d2	**82**	**85**	**85**	**85**	**86**	**84**	**84**
j15s5d3	**77**	**83**	**82**	**96**	**83**	**83**	**83**
j15s5d4	**61**	**86**	**85**	**101**	**84**	**84**	**84**
j15s5d5	**67**	**81**	**80**	**97**	**80**	**80**	**79**
j15s5d6	**79**	**83**	**82**	**87**	**79**	**82**	**81**
j15s10a1	236	236	236	236	236	236	236
j15s10a2	200	200	200	200	200	200	200
j15s10a3	198	198	198	198	198	198	198
j15s10a4	225	225	225	225	228	225	225
j15s10a5	182	182	182	183	182	182	182
j15s10a6	200	200	200	200	200	200	200
j15s10b1	222	222	222	222	222	222	222
j15s10b2	187	187	187	187	188	187	187
j15s10b3	222	222	222	222	224	222	222
j15s10b4	221	221	221	221	221	221	221
j15s10b5	200	200	200	200	#	200	200
j15s10b6	219	219	219	219	#	219	219

bold problems are hard problems

#: indicates that the solution is not found by the method

& Lin, 2007). In this paper, we make use of some dispatching rules as heuristic information:

- **SPT:** Select the job with the shortest processing time.
- **LPT:** Select the job with the longest processing time.
- **LWKR:** Select the job with the least work remaining.
- **MWKR:** Select the job with the most work remaining.
- **SRT:** Select the job with the shortest remaining work, excluding the operation under consideration.
- **LRT:** Select the job with the longest remaining work, excluding the operation under consideration.
- **FIFO:** First in first out.

Likewise, some flow-shop constructive heuristics have been also adapted as heuristic information (Nawaz, Enscore, & Ham, 1983; Santos, Hunsucker, & Deal, 1996).

Pheromone Updating Mechanism

The mechanism of updating pheromone simulates the changes of the amount of pheromone. Two kinds of pheromone updating strategies are proposed.

Local Updating Rule

The local updating rule reduces the pheromone level of the selected operation by an ant k according to Equation 6. Thus, the selection of the operation r in position y of the selected machine m becomes less attractive for the other ants. Indeed, this rule is used to shuffle the tour of the other ants and to avoid local optima.

$$\tau_m\left(r,y\right) = \left(1-\rho_\ell\right)\cdot\tau_m(r,y) + \rho_\ell\cdot\tau_0 \qquad (6)$$

where $\rho_\ell\left(0\leq\rho_\ell\leq 1\right)$ is the pheromone evaporation parameter.

Global Updating Rule

We update the pheromone trail at the end of the iteration. Only the best solution is globally updating. We define the global updating rule by Equation 7:

$$\tau_m\left(r,y\right) = \left(1-\rho_g\right)\cdot\tau_m\left(r,y\right) + \rho_g\cdot\Delta\tau_m\left(r,y\right) \qquad (7)$$

where

$$\Delta\tau(r,y) = \begin{cases} 1/C_{gb}, & if\ (r,y)\in\text{best solution} \\ 0, & otherwise \end{cases}$$

In the above equation, C_{gb} is the best solution of the schedules and $\rho_g\left(0\leq\rho_g\leq 1\right)$ is the pheromone evaporation parameter of the global updating rule. According to these equations, we apply the global updating rule to intensify the pheromone levels on the operations belonging to the best solution.

Local Search Approach

To improve the obtained solution, we include a local search strategy to our approach. We use the neighborhood structure proposed by (Nowicki & Smutnicki, 1998) as a local search strategy. This procedure can enable the reassignment of operations to other machines and/or the re-sequencing of the operations on machines.

COMPUTATIONAL EXPERIMENTS

In this section, we present the results of a series of computational experiments conducted to test the effectiveness of the proposed approach. Henceforth, we call our proposed algorithm IACS-HFS. The platform of our experiments is a personal

Table 3. Percentage deviations of the test problems

Problem	LB of C_{max}	%Deviation					
		HACS-HFS	IACS-HFS	B&B	Improved-AS	AIS	GA
j10s5a2	88	0	0	0	0	0	0
j10s5a3	117	0	0	0	0	0	0
j10s5a4	121	0	0	0	0	0	0
j10s5a5	122	0	0	0	1.64	0	0
j10s5a6	110	0	0	0	0	0	0
j10s5b1	130	0	0	0	0.77	0	0
j10s5b2	107	0	0	0	0	0	0
j10s5b3	109	0	0	0	0	0	0
j10s5b4	122	0	0	0	1.64	0	0
j10s5b5	153	0	0	0	0	0	0
j10s5b6	115	0	0	0	0	0	0
j10s5c1	**68**	**0**	**0**	**0**	**0**	**0**	**0**
j10s5c2	**74**	**0**	**0**	**0**	**2.7**	**0**	**0**
j10s5c3	**71**	**1.41**	**1.41**	**0**	**1.41**	**1.41**	**1.41**
j10s5c4	**66**	**0**	**0**	**0**	**0**	**0**	**0**
j10s5c5	**78**	**0**	**0**	**0**	**0**	**0**	**0**
j10s5c6	**69**	**0**	**0**	**0**	**0**	**0**	**0**
j10s5d1	**66**	**0**	**0**	**0**	**#**	**0**	**0**
j10s5d2	**73**	**0**	**0**	**0**	**#**	**0**	**1.37**
j10s5d3	**64**	**0**	**0**	**0**	**#**	**0**	**0**
j10s5d4	**70**	**0**	**0**	**0**	**#**	**0**	**0**
j10s5d5	**66**	**0**	**0**	**0**	**#**	**0**	**0**
j10s5d6	**62**	**0**	**0**	**0**	**#**	**0**	**0**
j10s10a1	139	0	0	0	#	0	0
j10s10a2	158	0	0	0	#	0	0
j10s10a3	148	0	0	0	#	0	0
j10s10a4	149	0	0	0	#	0	0
j10s10a5	148	0	0	0	#	0	0
j10s10a6	146	0	0	0	#	0	0
j10s5d5	**66**	**0**	**0**	**0**	**#**	**0**	**0**
j10s5d6	**62**	**0**	**0**	**0**	**#**	**0**	**0**
j10s10b1	163	0	0	0	0	0	0

continued on following page

computer with a Pentium (R), 2.27 GHz, and 3 Go of RAM. We code our algorithms in Java.

We performed the computational tests on 77 benchmark problems taken from (Carlier & Néron, 2000). The problem sizes are respectively:

- 10jobs × 5stages.
- 10jobs × 10stages.
- 15jobs × 5stages.
- 15jobs × 10stages.

The number of machines per stage varies from 1 to 3 machine(s). The notation of each instance is represented by three letters corresponding re-

Table 3. Continued

Problem	LB of C_{max}	%Deviation					
		HACS-HFS	IACS-HFS	B&B	Improved-AS	AIS	GA
j10s10b2	157	0	0	0	0	0	0
j10s10b3	169	0	0	0	0	0	0
j10s10b4	159	0	0	0	0	0	0
j10s10b5	165	0	0	0	0	0	0
j10s10b6	165	0	0	0	0	0	0
j10s10c1	113	1.77	1.77	12.39	4.42	1.77	1.77
j10s10c2	116	2.59	2.59	0	0.86	2.59	2.59
j10s10c3	98	18.37	18.37	35.71	10.2	18.37	18.37
j10s10c4	103	16.5	16.5	31.07	8.74	16.5	16.5
j10s10c5	121	4.13	4.13	19.83	4.13	4.13	4.13
j10s10c6	97	9.28	9.28	15.46	5.15	9.28	9.28
j15s5a1	178	0	0	0	0	0	0
j15s5a2	165	0	0	0	0	0	0
j15s5a3	130	0	0	0	1.54	0	0
j15s5a4	156	0	0	0	0	0	0
j15s5a5	164	0	0	0	1.22	0	0
j15s5a6	178	0	0	0	0	0	0
j15s5b1	170	0	0	0	0	0	0
j15s5b2	152	0	0	0	0	0	0
j15s5b3	157	0	0	0	0	0	0
j15s5b4	147	0	0	0	1.36	0	0
j15s5b5	166	0	0	0	0	0	0
j15s5b6	175	0	0	0	0.57	0	0
j15s5c1	85	0	0	0	0	0	0
j15s5c2	90	0	0	0	0	1.11	1.11
j15s5c3	87	0	0	0	0	0	0
j15s5c4	89	1.12	0	1.12	0	0	0
j15s5c5	73	5.48	1.37	15.07	0	1.37	1.37
j15s5c6	91	0	0	0	0	0	0
j15s5d1	167	0	0	0	0	0	0
j15s5d2	82	3.66	3.66	3.66	4.88	2.44	2.44
j15s5d3	77	7.79	6.49	24.68	7.79	7.79	7.79

continued on following page

spectively: (j) to job, (s) to stage and (a, b, c or d) to the machine configuration of each stage. For example, a 10-jobs, 5-stages with the machine structure d is denoted "j10s5d2". The processing times are integer generated from the uniform distribution $[3, 25]$. We have four types of machine configurations:

- Structure a is composed of one machine at the middle stage and three machines at the other stages.
- Structure b is composed of one machine at the first stage and three machines at the other stages.

Table 3. Continued

Problem	LB of C_{max}	%Deviation					
		HACS-HFS	IACS-HFS	B&B	Improved-AS	AIS	GA
j15s5d4	61	40.98	39.34	65.57	37.7	37.7	37.7
j15s5d5	67	20.9	19.4	44.78	19.4	19.4	17.91
j15s5d6	79	5.06	3.8	10.13	0	3.8	2.53
j15s10a1	236	0	0	0	0	0	0
j15s10a2	200	0	0	0	0	0	0
j15s10a3	198	0	0	0	0	0	0
j15s10a4	225	0	0	0	1.33	0	0
j15s10a5	182	0	0	0.55	0	0	0
j15s10a6	200	0	0	0	0	0	0
j15s10b1	222	0	0	0	0	0	0
j15s10a6	200	0	0	0	0	0	0
j15s10b1	222	0	0	0	0	0	0
j15s10b2	187	0	0	0	0.53	0	0
j15s10b3	222	0	0	0	0.9	0	0
j15s10b4	221	0	0	0	0	0	0
j15s10b5	200	0	0	0	#	0	0
j15s10b6	219	0	0	0	#	0	0

bold problems are hard problems

#: indicates that the solution is not found by the method

- Structure c is composed of two machines at the middle stage and three machines at the other stages.
- Structure d is composed of three machines at all the stages.

All the a, b and j10s10c structures are identified as easy problems. The rest of the problems are hard (Carlier & Néron, 2000; Néron, Baptiste, & Gupta, 2001). The authors calculated the lower bound (*LB*) of all these problems. They limited the computation time (CPU) to 1600 seconds (s) (Engin & Döyen, 2004).

In order to test these problems with our proposed IACS-HFS algorithm, we have to determine all the parameters of this latter. Indeed, these parameters have a great impact in the quality of the solution. These latter are $q_{m_0}, q_0, \beta, \tau_0, \rho_\ell, \rho_g$ the number of the ants *nbANT* and the number of the iterations t_{max} (*t* is the number of the current

iteration). Thus, we conducted some preliminary tests to find the best parameter settings (see Table 1). The algorithms terminates when either *LB* or the t_{max} iterations are reached. We use the percentage deviations (%Deviation) to evaluate our proposed approach. We compute it according to Equation 8:

$$\%\text{Deviation} = \frac{Best\,C_{max} - LB}{LB} \times 100 \qquad (8)$$

The different results obtained by our approach is presented and compared with a B&B procedure (Néron et al., 2001), two ACO algorithms (HACS-HFS) (Khalouli, Ghedjati, & Hamzaoui, 2008) and Improved-AS (Alaykýran, Engin, & Döyen, 2007), a GA (Besbes, Loukil, & Teghem, 2006) and an AIS (Engin & Döyen, 2004).

According to Table 2 and Table 3, better results have been obtained for a and b type problems than

Table 4. Performance of the compared approaches

Method	% Solved Problems	%Average Deviation
HACS-HFS	81.82	1.81
IACS-HFS	83.12	1.66
B&B	83.12	3.64
AIS	81.82	1.66
GA	80.52	1.64

c and d type problems. These results confirm that the machine configurations have an important effect on the complexity of problems. In fact, they affect the solution quality.

The IACS-HFS algorithm found the optimal solutions for all a and b type problems (47 problems) like HACS- HFS, AIS and GA. Concerning the j10s10c problems, the lower bounds were not reached by all considered approaches (except the j10s10c2 instance, which has been solved by the B&B). For the hard instances, we note that our proposed IACS-HFS algorithm found the LB value for 17 of the 24 problems, although B&B reached the LB values in 18 problems. For the HACS- HFS and AIS algorithms16 problems reached the LB and 15 problems for GA.

In Table 4, we present the percentages of solved problems and the average deviation values for all the considered approaches (except Improved-AS). Our IACS-HFS approach outperformed both of the HACS-HFS and the B&B methods in terms of %Solved problems (except the B&B). In terms of %Average Deviation, our proposed method is

better than B&B and HACS-HFS, like AIS but slightly higher than GA.

In Table 5, the percentage of reached to LBs solved problems (% solved) and the average deviation (% deviation) values for easy and hard problems are reported. The results show that:

- For easy problems, all the methods found the LB for **88.68%** of the considered problems but the B&B found the highest percentage deviation.
- For hard problems, both our proposed method IACS-HFS and B&B found the best percentage of solved problems (70.83%) than HACS-HFS, AIS and GA. In terms of %deviation, the IACS-HFS is better than both B&B and HACS-HFS. We note that the %deviation obtained by our proposed IACS-HFS method, the GA and the AIS are almost equal.

We notice that our IACS-HFS approach provides solutions with a reasonable time (less than 1 second for the majority of the easy problems). For hard problems, the CPU time is less than 1000s. In this case the smallest CPU time is obtained for the j15s5d1 problem (0,007s).

CONCLUSION

In this paper, we present an ant colony optimization algorithm based on an integrated approach

Table 5. Relative performance of the compared approaches

Method	Easy problems		Hard problems	
	% Solved	%Deviation	% Solved	%Deviation
HACS-HFS	88.68	0.99	66.67	3.60
IACS-HFS	88.68	0.99	70.83	3.13
B&B	88.68	2.17	70.83	6.88
AIS	88.68	0.99	66.67	3.13
GA	88.68	0.99	62.50	3.08

for solving the HFS scheduling problem (IACS-HFS). The proposed new approach is tested on a set of 77 instances taken from the literature. The computational results are compared with the two other ACO algorithms, a B&B procedure, a GA and an AIS approach. Extensive computational experiments suggest that IACS-HFS yields promising and good results.

The extension of the proposed ACS approach to multi-criteria HFS scheduling problems, particularly the earliness and tardiness criteria, will be investigated in the near future.

REFERENCES

Alaykýran, K., Engin, O., & Döyen, A. (2007). Using ant colony optimization to solve the hybrid flow shop scheduling problems. *International Journal of Advanced Manufacturing Technology, 35*(5-6), 541–550. doi:10.1007/s00170-007-1048-2

Besbes, W., Loukil, T., & Teghem, J. (2006). Using genetic algorithm in the multiprocessor flow shop to minimize the makespan. In . *Proceedings of the International Conference on Service Systems and Service Management, 2*, 1228–1233. doi:10.1109/ICSSSM.2006.320684

Blazewicz, J., Ecker, K. H., Pesch, E., & Schmidt, G. (1996). *Scheduling Computer and Manufacturing Processes*. Berlin, Germany: Springer.

Blum, C., & Sampels, M. (2002). Ant colony optimization for FOP shop scheduling: a case study on different pheromone representations. In *Proceedings of the 2002 Congress on Evolutionary Computation (CEC '02)* (Vol. 2, pp. 1558-1563). Los Alamitos, CA: IEEE Computer Society Press.

Carlier, J., & Néron, E. (2000). An exact method for solving the multi-processor flow-shop. *RAIRO: Recherche Operationnelle, 34*(1), 1–25. doi:10.1051/ro:2000103

Colorni, A., Dorigo, M., & Maniezzo, V. (1991). Distributed optimisation by ant colonies. In *Proceedings of ECAL91: First European Conference on Artificial Life* (pp. 134-142).

Dorigo, M., & Gambardella, L. (1997). Ant Colony System: A Cooperative Learning Approach to the Traveling Salesman Problem. *IEEE Transactions on Evolutionary Computation, 1*(1), 53–66. doi:10.1109/4235.585892

Engin, O., & Döyen, A. (2004). A new approach to solve hybrid flow shop scheduling problems by artificial immune system. *Future Generation Computer Systems, 20*(6), 1083–1095. doi:10.1016/j.future.2004.03.014

Gourgand, M., Grangeon, N., & Norre, S. (1999). Metaheuristics for the deterministic hybrid flow shop problem. In *Proceedings of the International Conference on Industrial and Production management (IEPM'99)* (pp. 136-145).

Guinet, A., Solomon, M., Kedia, P., & Dussa, A. (1996). A computational study of heuristics for two-stage flexible flowshops. *International Journal of Production Research, 34*, 1399–1415. doi:10.1080/00207549608904972

Gupta, J. (1988). Two-stage hybrid flowshop scheduling problem. *The Journal of the Operational Research Society, 39*, 359–364.

Haouari, M., Hidri, L., & Gharbi, A. (2006). Optimal scheduling of a two-stage hybrid flow shop. *Mathematical Methods of Operations Research, 64*(1), 107–124. doi:10.1007/s00186-006-0066-4

Jin, Z., Ohno, K., Ito, T., & Elmaghraby, S. (2002). Scheduling hybrid flowshops in printed circuit board assembly lines. *Production and Operations Management, 11*(1), 216–230.

Jin, Z., Yang, Z., & Ito, T. (2006). Metaheuristic algorithms for multistage hybrid flowshop scheduling problem. *International Journal of Production Economics, 100*, 322–334. doi:10.1016/j.ijpe.2004.12.025

Johnson, S. (1954). Optimal two- and three-stage production schedules with setup times included. *Naval Research Logistics Quarterly, 1*, 61–68. doi:10.1002/nav.3800010110

Khalouli, S., Ghedjati, F., & Hamzaoui, A. (2008). Method based on ant colony system for solving the hybrid flow shop scheduling problem. In *Proceedings of the 7th International Conference on Modelling, Optimization and SIMulation Systems (MOSIM'08)* (Vol. 2, pp. 1407-1416).

Lee, C., Cheng, T. C., & Lin, B. (1993). Minimizing the makespan in the 3-machine assembly-type flowshop scheduling problem. *Management Science, 39*(5), 616–625. doi:10.1287/mnsc.39.5.616

Linn, R., & Zhang, W. (1999). Hybrid flow shop scheduling: a survey. *Computers & Industrial Engineering, 37*(1-2), 57–61. doi:10.1016/S0360-8352(99)00023-6

Nawaz, M., Enscore, E., & Ham, I. (1983). A heuristic algorithm for the m-machine, n-job flowshop sequencing problem. *Omega, 11*, 91–95. doi:10.1016/0305-0483(83)90088-9

Néron, E., Baptiste, P., & Gupta, J. (2001). Solving an hybrid flow shop problem using energetic reasoning and global operations. *Omega, 29*, 501–511. doi:10.1016/S0305-0483(01)00040-8

Nowicki, E., & Smutnicki, C. (1998). The flow shop with parallel machines: A tabu search approach. *European Journal of Operational Research, 106*(2-3), 226–253. doi:10.1016/S0377-2217(97)00260-9

Portmann, M., Vignier, A., Dardilhac, D., & Dezalay, D. (1998). Branch and bound crossed with GA to solve hybrid flowshops. *European Journal of Operational Research, 107*(2), 389–400. doi:10.1016/S0377-2217(97)00333-0

Riane, F., Artiba, A., & Elmaghraby, S. (1998). A hybrid three stage flow-shop problem: efficient heuristics to minimize makespan. *European Journal of Operational Research, 109*, 321–329. doi:10.1016/S0377-2217(98)00060-5

Ruiz, R., & Vázquez-Rodríguez, J. (2010). The hybrid flow shop scheduling problem. *European Journal of Operational Research, 205*(1), 1–18. doi:10.1016/j.ejor.2009.09.024

Santos, D., Hunsucker, J., & Deal, D. (1996). An evaluation of sequencing heuristics in flow shops with multiple processors. *Computers & Industrial Engineering, 30*, 681–691. doi:10.1016/0360-8352(95)00184-0

Stützle, T., & Hoos, H. (1997). Improvement in the ant system: introducing min-max ant system. In *Proceedings of the International Conference on Artificial Neuronal Networks and Genetic Algorithms* (pp. 266-274).

Vignier, A., Commandeur, C., & Proust, P. (1997). New lower bound for the hybrid flowshop scheduling problem. In *Proceedings of IEEE Sixth International Conference on Emerging Technologies and Factory Automation (ETFA 97)* (pp. 446-451).

Yagmahan, B., & Yenisey, M. (2010). multi-objective ant colony system algorithm for flow shop scheduling problem. *Expert Systems with Applications, 37*, 1361–1368. doi:10.1016/j.eswa.2009.06.105

This work was previously published in the International Journal of Applied Metaheuristic Computing, Volume 2, Issue 1, edited by Peng-Yeng Yin, pp. 29-43, copyright 2011 by IGI Publishing (an imprint of IGI Global).

Chapter 7
Discrete Particle Swarm Optimization for the Multi-Level Lot-Sizing Problem

Laurent Deroussi
Blaise Pascal University, Clermont-Ferrand II, France

David Lemoine
Ecole des Mines de Nantes, France

ABSTRACT

This paper presents a Discrete Particle Swarm Optimization (DPSO) approach for the Multi-Level Lot-Sizing Problem (MLLP), which is an uncapacitated lot sizing problem dedicated to materials requirements planning (MRP) systems. The proposed DPSO approach is based on cost modification and uses PSO in its original form with continuous velocity equations. Each particle of the swarm is represented by a matrix of logistic costs. A sequential approach heuristic, using Wagner-Whitin algorithm, is used to determine the associated production planning. The authors demonstrate that any solution of the MLLP can be reached by particles. The sequential heuristic is a subjective function from the particles space to the set of the production plans, which meet the customer's demand. The authors test the DPSO Scheme on benchmarks found in literature, more specifically the unique DPSO that has been developed to solve the MLLP.

INTRODUCTION

Tactical planning plays a major part within industrial planning and consists in the elaboration of production plans which minimize logistic costs (for instance production, holding or setup costs).

It means that the aim of tactical planning is to determine, for each item, quantities to be manufactured per period in order to meet the customer's request at a lower cost. Referring to hierarchical planning system proposed by (Vollmann, Berry & Whybark, 1997), tactical planning is composed of three planning levels:

DOI: 10.4018/978-1-4666-2145-9.ch007

- The Sales and Operation Planning (S&OP).
- The Master Planning Schedule (MPS).
- The Material Requirement Planning (MRP).

The S&OP consists in finding a balance between sales and production level on a midterm horizon (around 1 year). On this horizon, production capacity is not considered as fixed but can vary thanks to the use of overtime, sub-contracting, hiring, firing and so on.

The MPS is defined by (Genin, 2003) as a problem for which the first goal is to find an optimal production plan which meets the customers' requests and provides release dates and amounts for all these products. The objective function lies in minimizing production, holding and setup costs.

Elaborating internal component planning is the main goal of MRP. By using bills of materials (BOM), these plans are deduced from the MPS.

Usually, tactical production planning are modelled thanks to mathematical models called "Lot-Sizing problems". A huge number of models have been proposed in the literature (Rizk & Martel, 2001; Drexl & Kimms, 1997).

The large diversity of these models can be explained by the complexity of modeling the system: some planning objectives require detailed modeling (setup time constraints, sequencing, etc.) while other ones need only rough one. Readers can refer to (Comelli, Gourgand & Lemoine, 2008; Rizk & Martel, 2001) for a review of the main "Lot-Sizing" mathematical models found in literature. Nevertheless, important discriminating criteria can be highlighted. Indeed, according to the planning levels concerned, some important assumptions have been made. For instance, "Lot-Sizing" models can be mono or multi-level, capacitated or uncapacitated formalization. With regard of level criteria, mono-level models deal with S&OP and MPS elaboration (only end items are planned) whereas multi-level ones take into account components production planning. Relating to capacitated criteria, MRP systems do not initially consider manufacturing capacity. With MRP II, the capacity becomes an important factor which is checked by a capacity requirements planning module. Nevertheless, (Han, Tang, Kaku & Mu, 2009) point out that uncapacitated "Lot-Sizing" problems are still largely used (in ERP for example) "since the implementation of capacitated approaches requires much data which firms are often reluctant to collect or maintain".

In this paper, we deal with the Multi-Level Lot-Sizing Problem (MLLP) which is an uncapacitated multi-level Lot-Sizing problem, dedicated to MRP system. We propose an effective heuristic method to solve it, based on a Particle Swarm Optimization (PSO). Our contribution is thus organized as follows: in the second section, we give a brief review of solution approaches for the MLLP. In the third section, we present the Particle Swarm Optimization method. In Section 4, the proposed approach is given and the last section is dedicated the last section is dedicated to our experimental results.

MLLP: STATE OF THE ART

As described before, the MLLP is an uncapacitated multi-level "lot sizing problem" mathematical model. It can be formulated as a mixed-integer programming model, using the following notations:

- **Parameters:**
  - N: Number of items.
  - T: Number of periods.
  - D_i: Customer's demand for item i at period t.
  - a_{ij}: "Gozinto" factor. Its value is zero if item is not an immediate successor of item j.
- Otherwise, it is the quantity of item that is directly needed to produce one item i:
  - h_i: Non-negative holding cost for item i.
  - s_i: Non-negative setup cost for item i.

○ I_{io}: Initial inventory for item i.

○ M: An arbitrary very large number.

Decision Variables

- Q_{it}:
  - ○ Production quantity for item i at period t.I_{it}:Inventory for item i at the end of period t.X_{it}:
  - ○ Binary variable which indicates whether a setup for item i occurs at period $t(X_{it} = 1)$ or not ($X_{it} = 0$).
- Q_{it}:
  - ○ Production quantity for itemat period t.
- I_{it}:
  - ○ Inventory for itemat the end of period t.
- X_{it}:
- Binary variable which indicates whether a setup for item i occurs at period $t(X_{it} = 1)$ or not ($X_{it} = 0$).

$$Minimize \sum_{t=1}^{T} \sum_{i=1}^{N} (h_i I_{it} + s_i X_{it}) \qquad (1)$$

$$I_{it} = I_{it-1} + Q_{it} - D_{it} - \sum_{j=1}^{N} a_{ji} Q_{jt} \qquad (2)$$
$$\forall (i,t) \in [1,N] \times [1,T]$$

$$Q_{it} \leq M X_{it} \quad \forall (i,t) \in [1,N] \times [1,T] \qquad (3)$$

$$(I_{it}, Q_{it}) \in \mathbb{N}^2, X_{it} \in \{0,1\} \qquad (4)$$
$$\forall (i,t) \in [1,N] \times [1,T]$$

The objective Function 1 is the sum of holding and setup costs for all items over the whole planning horizon. Equations 2 represent inventory balance. Constraints 3 determine if a production setup for item occurs at period. 4 are integrity and non negativity constraints.

Contrary to mono-level uncapacitated Lot-Sizing problem (Wagner & Whitin, 1958), MLLP is known to be NP-Hard (Arkin, Joneja & Roundy, 1989) in spite of the fact that there is at least one optimal solution which verifies the "zero-switch" property (Dellaert & Jeunet, 2000): $\forall (i,t) \in [1,N] \times [1,T], Q_{it} I_{it-1} = 0$. Thus, according to this property, there is at least one optimal solution in which a production can take place for an item during a period if, and only if, there is no stock for this item at the end of the period .

Few exact approaches have been developed for the MLLP. (Zangwill, 1966) has proposed a dynamic programming algorithm in the case of bill of material with serial structure. (Afentakis & Gavish, 1986) have used a branch and bound algorithm using the echelon stock notion introduced by (Clark & Scarf, 1960).

With regard to approximated methods, we can essentially distinguish two kinds of approaches:

- The sequential approaches, in which products are planned level by level thanks to the MRP concept.
- The simultaneous approaches, in which all levels are planned simultaneously.

The sequential approach is historically the first one. Indeed, this one is based on the MRP method: items are planned level by level, following the bill of material. The main idea of this kind of approach is to link an item and its components' planning by incorporating into item's logistic costs a part of those of its components. The generated costs are used to compute items' planning with mono-level algorithms (Wagner-Within dynamic programming algorithm, Silver-Meal heuristic, etc.). To do that, several modification costs procedures have been designed. For instance, readers can refer to (Blackburn & Millen, 1982; Bookbinder & Koch, 1990; Dellaert & Jeunet, 2002).

The simultaneous approaches have been developed much more recently and are essentially based on evolutionary computing methods. (Dellaert & Jeunet, 2000; Dellaert, Jeunet & Jonard, 2000) have designed a genetic algorithm for solving MLLP in case of bill of material with general structure. (Pitakaso, Almeder, Doener & Hartl, 2007) have presented a min-max ant system for this problem. Simulated annealing has also been developed by (Tang, 2004; Kuik & Salomon, 1990; Jeunet & Jonard, 2005). Particle Swarm Optimization (PSO) has already been used for this problem (Han, Tang, Kaku & Mu, 2009). Indeed, they have conceived such an algorithm in its discrete version, following the (Clerc, 2004) DPSO framework. Figure 1 illustrates this state of the art.

PARTICLE SWARM OPTIMIZATION

PSO is a nature inspired metaheuristic developed by (Eberhart & Kennedy, 1995) according to the observation of the social behaviour of bird flocking and fish schooling. This optimization method has met an increasing success over the past years. Despite PSO has been initially designed for continuous nonlinear optimization, some adaptations have been made by several authors in discrete way. In this section, we propose a general overview of PSO and its discrete variants.

Basic PSO

As Genetic Algorithm (GA), PSO is an evolutionary algorithm and many similarities can be noted between them: an initial population (resp. swarm in the PSO context) is generated, and each indi-

Figure 1. MLLP, state of the art

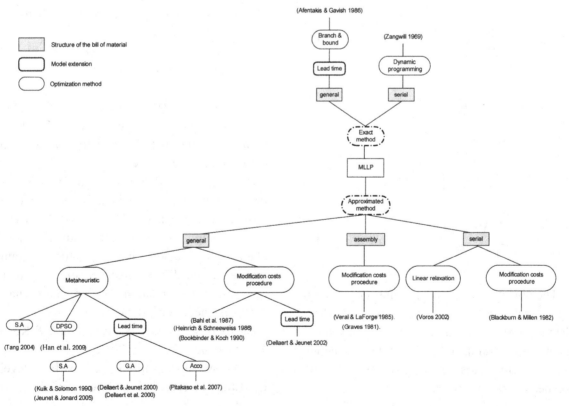

vidual (resp. particle) moves iteratively into the search space. Nevertheless, an essential difference between these two methods lies in the fact that there are no crossover and mutation operators in the PSO context. Indeed, each particle "flies" in the search space according to movement equations.

The framework of PSO is as follows (Hu, Shi & Eberhart, 2004):

- Each particle represents a solution of the optimization problem. All these solutions have a cost (fitness) value that is evaluated by the objective function.
- Each particle has a given velocity, which directs their trajectory in the search space.

PSO starts from a group of random particles. At each iteration, the current position of each particle is updated by using movement equations (see Equations (5) and (6)) which reflect a compromise between three criteria:

- The current velocity of the particle $\vec{v}_{ik}$.
- The best solution found by the particle (position $pbest_i$).

Figure 2. The affine representation of Equations 5 and 6.

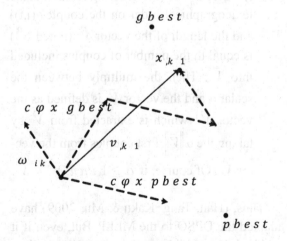

- The global best solution found by the particle swarm (position *gbest*) or a subset of this swarm which depends on the considered particle and which is called particle neighbourhood (position $nbest_i$).

We give the based movement equations given by (Clerc, 2004) and used in this paper:

$$\vec{v}_{i,k+1} = \omega\vec{v}_{i,k} + c_1\varphi_1 \overrightarrow{x_{lk}pbest_1} + c_2\varphi_2 \overrightarrow{x_{lk}gbest} \tag{5}$$

$$X_{i,k+1} = X_{ik} + \vec{v}_{i,k+1} \tag{6}$$

These equations make sense in an affine space (Audin, 2002), as shown in Figure 2.

Equation 5 calculates the new velocity for each particle according to the compromise described before. c_1 is the weight of the personal experience of the particle (individualistic behaviour), while c_2 quantifies the influence of the particle swarm (collective behaviour). These two coefficients are called confidence coefficients. φ_1 and φ_2 are random numbers uniformly chosen between 0 and 1. ω is called inertia weight (Shi & Eberhart, 1998). Its aim is to accelerate the convergence of PSO. Usually, particle's velocity is clamped to a maximum velocity called $\vec{v}_{max}$ in order to control the global exploration ability of particles swarm (Hu, Shi & Eberhart, 2004). Equation 6 determines the new position of the particle (for the next iteration) by using the new velocity (Equation 5).

A pseudo-code of a gbest model of PSO is given in Algorithm 1:

- **Algorithm 1:**
  ○ **Step 1:** Generate the initial swarm of n particles with given position and velocity.
  ○ **Step 2:** Evaluate the fitness of every particle. Initialize the local best position ($pbest_i$) of each particle. Compute

the global best position (gbest) of the swarm.

- Step 3: For each particle i:
  - Compute the new velocity of the particle using Equation 5.
  - Update the particle's position using Equation 6.
  - Evaluate the fitness of the particle.
  - If necessary, update the local best position () of the particle.
- Step 4: If necessary, update the global best position of the swarm ().
- Step 5: If terminal criteria are not met, go to Step 3.

Discrete PSO

Several attempts have been made in order to adapt PSO to discrete or binary variables. Indeed, we can distinguish three kinds of approaches, based on:

1. **A Stochastic Velocity Model:** (Kennedy & Eberhart, 1997) Propose the first binary PSO algorithm. The velocity is a vector of probabilities which determines if a dimension takes value 0 or 1.
2. **The Definition of Discrete Operators:** (Clerc, 2004) Proposes a general framework in which vector operators are redefined to work on a discrete space. As example, the velocity which is defined as the difference between two positions p_1 and p_2, is given by a list of exchanges which allows moving from p_1 to p_2. With this approach, PSO can be adapted to any combinatorial optimization problem.
3. **The Continuous PSO Algorithm:** Particles fly into a real multi-dimensional vector space. Each position of this search space is associated to a discrete solution of the studied problem. (Tasgetiren, Sevkli, Liang & Gencyilmaz, 2004) use this technique for the permutation flow shop problem. The authors

transform a real vector to a job scheduling using the Smallest Position Value (SPV) rule.

Reader can refer to (Anghinolfi & Paolucci, 2009) which propose a recent survey on DPSO algorithms.

(Han, Tang, Kaku & Mu, 2009) have already used DPSO for solving the MLLP. The authors based their optimization method on the Clerc's framework (Clerc, 2004):

- A position of a particle is represented by a binary matrix A with N rows (the number of products) and T columns (the number of periods) in which a value of 1 in entry $A_{i,j}$ means that the production of item i can take place at period t. Thus, the set of possible positions for each particle is a discrete space whose cardinal is equal to 2^{NT}.
- In order to adapt the movements equations of a particle (see (5) and (6)), a discretization of the vector space operations is proposed. A vector (such as velocity) is defined as the boolean difference between two positions. Thus a vector is a binary matrix. In order to define the product between a vector $\vec{V}$ and a scalar α, the vector is rewritten as the ordinate set of couples (i, t) where $V_{it} = 1$ (the ordering relation is the lexicographical order on the couples (i,t)) and the length of the vector $\vec{V}$ (noted $\|\vec{V}\|$) is equal to the number of couples included into $\vec{V}$. Then the multiply between the scalar α and the vector $\vec{V}$ is defined as the vector $\vec{V}'$ which is extracted from $\vec{V}$ by taking the $\alpha\|\vec{V}\|$ first couples from the vector $\vec{V}$. Of course, if $\alpha \geq 1$ *then* $\vec{V}' = \vec{V}$.

Thus, (Han, Tang, Kaku & Mu, 2009) have adapted the DPSO to the MLLP. But, even if it is effectively possible to apply the PSO paradigm to a combinatorial optimization problem by using the (Clerc, 2004) framework, the definition of such discrete operations does not really respect the swarm optimization paradigm: the definition of velocity and the product between a scalar and a vector cannot define a direction as in an affine space. Moreover, these kind of discrete operations are not always easily understandable. Therefore, it is for this reason that we decide to apply another DPSO scheme: the particles move freely into a continuous space (using classical continuous operators) as in the original particles swarm optimization concept and we use a rule in order to assign a solution of the MLLP to each position of the particles. This approach is motivated by the fact that we prove that all the solutions of the MLLP can be found according to this rule. Indeed, we are going to prove that all the solutions of the MLLP can be reached by particles of the swarm (see proposition 1).

A PSO APPROACH BASED ON COST MODIFICATION

Contrary to (Han, Tang, Kaku & Mu, 2009) which used the second approach for designing their DPSO, we propose to use the third one: the continuous PSO algorithm. The main idea of our approach is to combine PSO with sequential approach heuristic as shown in Figure 3, in the following way:

- For the PSO, we define a particle $x_{i,k}$ as a matrix of logistics' costs (holding and setup cost):

  ○ $x_{i,k} = (x_{i,k,1} \ x_{i,k,2} \ \ldots \ x_{i,k,T})$

  where $x_{i,k,t} = \begin{pmatrix} h_{1t}^{(i,k)} & h_{2t}^{(i,k)} \ldots h_{Nt}^{(i,k)} \\ s_{1t}^{(i,k)} & s_{2t}^{(i,k)} \ldots s_{Nt}^{(i,k)} \end{pmatrix}$ and $h_{jt}^{(i,k)}$ and $s_{jt}^{(i,k)}$ are respectively the holding and the setup costs for item j at period t for particle i at iteration k. Thus, $x_{i,k,t} \in M_{2 \times N}(\mathbb{R})$ vector space and $x_{i,k} \in M_{2 \times N \times T}(R)$ vector space.

- Each particle $x_{i,k}$ is linked to a production planning $Q_{i,k}$ which is determined thanks

Figure 3. Our DPSO Scheme

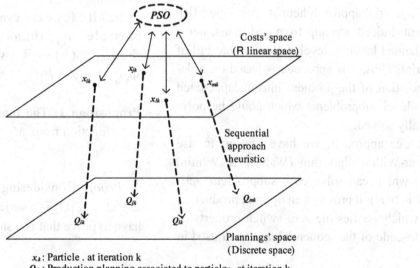

x_k: Particle . at iteration k
Q_k: Production planning associated to particle x_k at iteration k

to a sequential approach heuristic using $x_{i,k}$ as costs parameters. Then, the production planning $Q_{i,k}$ is evaluated thanks to real logistics' costs.

Particle's Move

Equations 5 and 6 suppose that we have defined, for each particle I and iteration k, the velocity $\vec{v}_{i,k+1}$ and the two vectors $\overrightarrow{x_{ik}pbest_i}$, $\overrightarrow{x_{ik}gbest}$. Such vectors are defined as a difference between two particles positions. Since each particle position is a vector of $M_{2 \times N \times T}(\mathbb{R})$ real vector space, difference between two particles is well defined and the result is also a vector of $M_{2 \times N \times T}(\mathbb{R})$. Thus, Equation 5 can be easily calculated thanks to usual operations defined in $M_{2 \times N \times T}(\mathbb{R})$ real vector space.

Now, if we consider $M_{2 \times N \times T}(\mathbb{R})$ underlying affine space, we can see that Equation 6 which can be seen as a translation in this affine space, makes sense. Thus, Equations 5 and 6 are well defined and can be easily calculated thanks to usual operations defined in $M_{2 \times N \times T}(\mathbb{R})$ vector space and in its underlying affine space.

Sequential Approach Heuristic

The sequential approach heuristic uses the MRP concept. Indeed, starting from end-items, items are planned level by level, following the bill of material. Thus, this approach is based on a decomposition of the problem into uncapacitated mono-level subproblems which could be polynomially solved.

In our approach, we have chosen to use Wagner-Within algorithm (Wagner & Whitin, 1958) which can solve each subproblem optimally in $0(T^2)$: it provides an optimal production plan which verifies the zero-switch property. A pseudo-code of the sequential approach used in

our method is proposed in Algorithm 2. N (resp N_e) is the number of products (resp. end-products). For each item j, the demand (internal or external) is denoted by $D(j),Q_{i,k}(j)$ points out the production planning for this item and α_{jl} the "Gozinto" factor. Wagner-Within $(D(j), x_{i,k})$ points out the Wagner-Within algorithm using $D(j)$ as demand and $x_{i,k}$ as logistics' costs. In our algorithm, we assumed that items are sorted according to the bill of material, i.e. the first items are the end-items and components follow according to their level in the bill of material (BOM).

- **Algorithm 2:**
  - For $j = 1$ to N_e do
  - $D(j)$customer's demand for item
  - End For
  - For j = 1 to N do
  - $Q_{i,k}(j)$= Wagner-Within $(D(j), x_{i,k})$
  - For l =1 to N do
  - $D(l) = D(l) + Q_{i,k}(j) \times \alpha_{jl}$
  - End For
  - End For

By using such a heuristic, we define a function $P : M_{2 \times N \times T}(\mathbb{R}) \to \Omega$ where Ω is the set of the production plans which respect the customers' demand and verify the zero-switch property. Thus, by defining $Z : \Omega \to \mathbb{R}^+$ as the objective function of the MLLP, we can evaluate the fitness of each particle $x_{i,k}$ thanks to the function $\Phi : M_{2 \times N \times T}(\mathbb{R}) \to \mathbb{R}^+$ where $\Phi(x_{i,k}) = (Z \circ P)(x_{i,k})$.

Proposition 1: The function Φ is a surjective function from $M_{2 \times N \times T}(\mathbb{R})$ to the set $Z[\Omega]$.

Proof: Considering that $\phi = Z \circ P$ and $Z : \Omega \to Z[\Omega]$ is a surjective function, we only have to prove that is a surjective function.

Figure 4. The scheme of the sequential heuristic

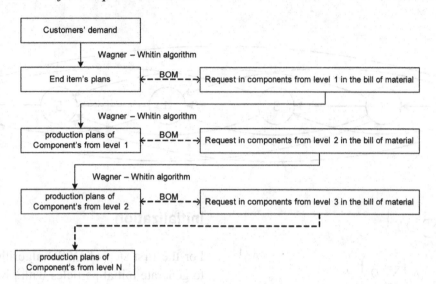

The function *P*, defined b Algorithm 2, is a sequential heuristic which uses Wagner-Whitin algorithm (Wagner & Whitin, 1958) following the scheme illustrated in Figure 4.

We can see that, thanks to Wagner-Within algorithm, production plans which verify the zero-switch property, are computed for each item and, by using the bill of material, the need in terms of components for each item is also defined. This means that from a production plan, internal demands are defined and Wagner-Whitin algorithm computes production plans which meet these demands. Then, we can easily deduce that, if Wagner-Whitin algorithm, tuned with appropriated parameters, can compute from a demand (external or internal), any production plans which meet this demand, then we will be able to generate all the production plans which verify the zero-switch property, for end items and all the components in order to meet the customers' request. As a result, *P* will be a surjective function.

Thus, we have to prove that, considering a demand, Wagner-Within algorithm can compute all the production plans which meet this demand, under the assumption of the use of appropriated parameters. To prove this, we can notice that the set of production plans which meet this demand and verify the zero-switch property, can be modelled according to a directed graph (Wagner & Whitin, 1958), as shown in Figure 5.

In this directed graph, each node represents the beginning of one period and each arc between each couple of nodes(t, t') represents the decision to produce, during the period t, the needed quantity in order to meet the demand between the period t and t' - 1. Thus, each path between the source and the sink represents a production plan which meets the whole demand and verifies the zero-switch property. Then, if each edge is valuated according to the logistic cost $C_{t,t'}$ generated by the production represented by the arc (t, t'), the optimal plan is given by the shortest path (In fact, Wagner-Whitin algorithm is a shortest path algorithm). $C_{t,t'}$ can be easily computed thanks the following formula:

Figure 5. Directed graph which represents the set of production plans for customer demand during 5 periods

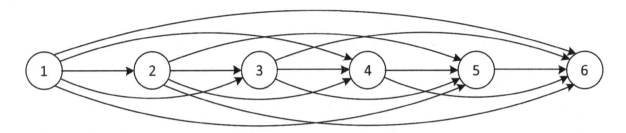

$$C_{t,t'} = \sum_{k=t}^{t'-2} \left(\sum_{l=k+1}^{t'-1} D_l h_k \right)$$
$$+ \left[1 - \delta_0 \left(\sum_{k=t}^{t'-1} D_k \right) \right] S_t \qquad (7)$$

where is the demand at period , and are respectively the non negative holding and setup cost and is equal to 1 if , 0 otherwise. Then, by tuning the holding and setup costs, each path between the source and the sink can be the shortest path. As a result, Wagner-Whitin algorithm can compute all the production plans which meet the demand, under the assumption of the use of appropriated parameters which are the logistic costs. Thus we have proved that ϕ is a surjective function and, therefore, Figure 6 is a commutative one:

Figure 6.

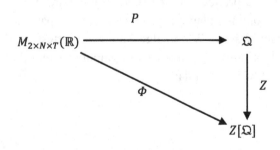

Initialization

For the first step of DPSO algorithm, we have to generate initial particles with given positions and velocities. Concerning initial position of each particle, we decide to use a cost modification procedure inspired from (Dellaert & Jeunet, 2002). For each itemand period,

- Modified setup cost can be seen as cumulated setup cost and is computed thanks to the Formula 7:

- $$s_{j,t}^{(i,0)} = s_j + \rho \sum_{\substack{l \in [1,N] \\ a_{jl} \neq 0}} \frac{s_{l,t}^{(i,0)}}{a_{jl}} \quad (8)$$

  where $\rho \sim U[0,10]$. In fact, uniformly distributed random variable ρ is motivated by the necessity to widely scatter the swarm.

- Modified holding cost is based on echelon cost and is computed according to Formulas 8 to 11:

- $h_{j,t}^{(i,0)} = Max[h_j, e_j + \rho(H_{j,t} - e_j)] \, (9)$

  where $e_j = h_j - \sum_{\substack{l \in [1,N] \\ a_{jl} \neq 0}} a_{jl} h_l \;(10)$

  $$H_{j,t} = e_j - \sum_{\substack{l \in [1,N] \\ a_{jl} \neq 0}} f_{jt} a_{jl} H_{l,t} \;(11)$$

  and $\rho \sim U[0,2]$;

  $$f_{jt} = max\left(1, \sqrt{\frac{h_j s_{l,t}^{(i,o)}}{s_j h_{l,t}^{(i,o)}}}\right) \;(12)$$

Our DPSO Scheme

The proposed DPSO scheme is given in Algorithm 3.

- **Algorithm 3:**
  - **Step 1:** Generate the initial swarm of n particles with given position and velocity (Formulas 8 to 12)
  - **Step 2:** Evaluate the fitness of each particle by using Algorithm 2. Initialize the local best position (pbest$_i$) of each particle i. Compute the global best position (gbest) of the swarm.
  - **Step 3:** For each particle:
    - Compute the new velocity of the particle (Equation 5).
    - Update the particle's position (Equation 6).
    - Evaluate the fitness of the particle by using Algorithm 2
    - If necessary, update the local best position (*pbest$_i$*) of the particle.
  - **Step 4:** If necessary, update the global best position of the swarm (*gbest*).
  - **Step 5:** If terminal criteria are not met, go to Step 3.

EXPERIMENTAL FRAMEWORK AND RESULTS

In this section, we use benchmark designed by (Han, Tang, Kaku & Mu, 2009) to test our DPSO method. We can find three kinds of instances: two small-sized problems and a medium-sized problem.

In order to compare with results given by (Han, Tang, Kaku & Mu, 2009), we have defined our DPSO parameters in the same way: we use a swarm of 60 particles, confidence coefficients $c_1 = c_2 = 2$. Concerning the matrix of the maximum velocity $\vec{v}_{\max}$, we have distinguished entries relating to holding and setup costs. Since setup costs are much higher than holding costs, we decide to set $\vec{v}_{\max}$ setup entries on 10 and 1 otherwise. We also choose that ω geometrically decreases at each iteration, from 2 to 0,64. We have also decided to restrict the solution space by forcing all the modified costs to be positive. Like (Han, Tang, Kaku & Mu, 2009), the max iteration for our DPSO is 500.

For each problem (number of items x number of periods), Han, Tang, Kaku & Mu, 2009), represented by H.T.K.M in Table 1, have tested three methods: the first one is their DPSO algorithm, the second one, a Genetic Algorithm (GA) and finally a Wagner Within sequential heuristic (WW). Their results as well as ours are summarized in Table 1.

For each instance and each method, 100 runs have been computed and we give:

- The best solution obtained.
- The worst solution obtained.
- The mean solution.
- The standard deviation.
- The relative gap between Best and Mean Cost.
- The mean computational times.

For their tests, (Han, Tang, Kaku & Mu, 2009) have used a PC with a 2.8 GHz CPU and 1G RAM and their DPSO has been coded in Visual Basic 6.0, whereas we have used a PC with 2.4 GHz CPU and 2G RAM and our DPSO has been coded in Visual Basic.Net)

In Table 1, the Best results are emphasized in bold. Even if Table 1 shows that our DPSO always finds the best solution, the most important result concerns the stability of our method. Indeed, contrary to (Han, Tang, Kaku & Mu, 2009), the relative gap between best cost and mean cost

Table 1. Computational results for instances of (Han, Tang, Kaku & Mu, 2009)

Problem	Author	Algorithm	Best	Worst	Mean cost	Std cost	GAP	Mean time(s)
6X10	H.T.K.M	DPSO	1493	1917	1560,7	84,4	4,53%	4.2
		GA	1493	2656	1904,89	248,63	27,59%	5.6
		WW	1707	n.a.	n.a.	n.a.	n.a.	<0.1
	Our approach	DPSO	**1493**	**1628**	**1499,38**	**26,54**	**0,43%**	**1.5**
6X12	H.T.K.M	DPSO	1895	5565	2143,93	578,4	13,14%	5.3
		GA	1895	6446	2526,9	601,79	33,35%	8.0
		WW	2123	n.a.	n.a.	n.a.	n.a.	<0.1
	Our approach	DPSO	**1895**	**2033**	**1901,88**	**20,7**	**0,36%**	**2.1**
6X15	H.T.K.M	DPSO	2546	8214	3664,73	1503,7	43,94%	6.3
		GA	2623	9170	3982,04	1305,2	51,81%	10.7
		WW	2909	n.a.	n.a.	n.a.	n.a.	<0.1
	Our approach	DPSO	**2546**	**2749**	**2580,34**	**48,25**	**1,35%**	**3.2**
9X10	H.T.K.M	DPSO	2043	2877	2158,9	136,2	5,67%	6.5
		GA	2043	5813	2581,79	767,47	26,37%	10.1
		WW	2807	n.a.	n.a.	n.a.	n.a.	<0.1
	Our approach	DPSO	**2043**	**2321**	**2058,06**	**52,58**	**0,74%**	**3.1**
9X12	H.T.K.M	DPSO	2522	9525	3057,8	935,7	21,25%	7.2
		GA	2522	9951	3887,2	1457,55	54,13%	12.9
		WW	3498	n.a.	n.a.	n.a.	n.a.	<0.1
	Our approach	DPSO	**2522**	**2764**	**2552,87**	**74,1**	**1,22%**	**4.6**
9X15	H.T.K.M	DPSO	3448	12457	5951,9	2771,6	72,62%	9.2
		GA	3714	12966	8302,23	2906,44	123,54%	16.3
		WW	4834	n.a.	n.a.	n.a.	n.a.	<0.1
	Our approach	DPSO	**3448**	**4056**	**3531,5**	**159,04**	**2,42%**	**5.4**
50X10	H.T.K.M	DPSO	4921	13339	5938,28	2106,2	20,67%	30.9
		GA	4921	25037	9241,86	5012,88	87,80%	33
		WW	13207	n.a.	n.a.	n.a.	n.a.	0.1
	Our approach	DPSO	**4921**	**4921**	**4921**	**0**	**0%**	**10.2**

shows their closeness. Furthermore, the standard deviations obtained after 100 replications by our DPSO are very slight compared to those obtained by (Han, Tang, Kaku & Mu, 2009). Thus, our method clearly outperforms methods proposed by (Han, Tang, Kaku & Mu, 2009) and seems to be more stable.

CONCLUSION

In this paper, we propose a DPSO optimization scheme which is relevant to solve MLLP Problem. Despite the fact that MLLP is a combinatorial optimization problem, each particle flies over a vector space such as in the initial Particle Swarm Optimization method. In order to associate each particle to a solution for the MLLP, we used a sequential heuristic which used the MRP concept and

Wagner-Whitin algorithm in order to determine a production plan which meets the demand and verifies the zero-switch property. We prove that this heuristic is a subjective function and, as a result, it proves that all the solutions for the MLLP can be reached by particles. Then, we have tested our optimization method on literature instances and we have obtained rather good results and, more specifically, our method clearly outperforms the unique DPSO which has been designed for the MLLP. In our future study, we want to extend this method to the Multi-Level Capacitated Lot-Sizing Problem. Nevertheless, in this case, the zero switch property does not remain valid and there is no algorithm, such as Wagner-Whitin algorithm, which can provide the optimal solution in a polynomial time for the mono-level capacitated lot sizing problem (CLSP). Indeed, CLSP is an NP-Hard problem (Bitran & Yanasse, 1982).

REFERENCES

Afentakis, P., & Gavish, B. (1986). Optimal Lot-Sizing for complex product structures. *Operations Research*, *34*, 237–249. doi:10.1287/opre.34.2.237

Anghinolfi, D., & Paolucci, M. (2009). A new discrete particle swarm optimization approach for the single-machine total weighted tardiness scheduling problem with sequence-dependent setup times. *European Journal of Operational Research*, *193*(1), 73–85. doi:10.1016/j.ejor.2007.10.044

Arkin, E., Joneja, D., & Roundy, R. (1989). Computational complexity of uncapacitated multi-echelon, production planning problems. *Operations Research Letters*, *8*, 61–66. doi:10.1016/0167-6377(89)90001-1

Audin, M. (2002). *Geometry*. Berlin, Germany: Springer.

Bahl, H., Ritzman, L., & Gupta, J. (1987). Determining lot sizes and ressource requirements: a review. *Operations Research*, *35*(3), 329–345. doi:10.1287/opre.35.3.329

Bitran, G., & Yanasse, H. (1982). Computational complexity of the capacitated lot size problem. *Management Science*, *46*(5), 724–738.

Blackburn, J., & Millen, R. (1982). Improved heuristics for multi-stage requirements planning systems. *Management Science*, *28*(1), 44–56. doi:10.1287/mnsc.28.1.44

Bookbinder, J., & Koch, L. (1990). Production planning for mixed assembly/arborescent systems. *Journal of Operations Management*, *9*, 7–23. doi:10.1016/0272-6963(90)90143-2

Clark, A., & Scarf, H. (1960). Optimal policies for a multi-echelon inventory problem. *Management Science*, *6*, 475–490. doi:10.1287/mnsc.6.4.475

Clerc, M. (2004). Discrete Particle Swarm Optimization, illustrated by the Traveling Salesman Problem . In *New Optimization Techniques in Engineering*. Berlin, Germany: Springer-Verlag.

Comelli, M., Gourgand, M., & Lemoine, D. (2008). A review of tactical planning models. *Journal of Systems Science and Systems Engineering*, *18*(2), 204–229. doi:10.1007/s11518-008-5076-8

Dellaert, N., & Jeunet, J. (2000). Solving large unconstrained multilevel Lot-Sizing problems using a hybrid genetic algorithm. *International Journal of Production Research*, *38*(5), 1083–1099. doi:10.1080/002075400189031

Dellaert, N., & Jeunet, J. (2002). Randomized multi-level Lot-Sizing heuristics for general product structures. *European Journal of Operational Research*, *148*(1), 211–228. doi:10.1016/S0377-2217(02)00403-4

Dellaert, N., Jeunet, J., & Jonard, N. (2000). A genetic algorithm to solve the general multi-level lotsizing problem with time-varying costs. *International Journal of Production Economics, 68,* 241–257. doi:10.1016/S0925-5273(00)00084-0

Drexl, A., & Kimms, A. (1997). Lot sizing and scheduling - Survey and extensions. *European Journal of Operational Research, 99,* 221–235. doi:10.1016/S0377-2217(97)00030-1

Eberhart, R. C., & Kennedy, J. (1995). *A new optimizer using particle swarm theory.* Paper presented at the Sixth International Symposium on Micro Machine and Human Science.

Genin, P. (2003). *Planification tactique robuste avec usage d'un A.P.S. Proposition d'un mode de gestion par plan de référence.* Unpublished doctoral dissertation, Ecole supérieure des mines de Paris, France.

Graves, S. (1981). In multi-stage production / inventory control systems: Theory and practice. *TIMS Studies in the Management Science, 16,* 95–110.

Han, Y., Tang, J., Kaku, I., & Mu, L. (2009). Solving uncapacitated multilevel Lot-Sizing problems using a particle swarm optimization with flexible inertial weight. *Computers & Mathematics with Applications (Oxford, England), 57*(11-12), 1748–1755. doi:10.1016/j.camwa.2008.10.024

Heinrich, C., & Schneeweiss, C. (1986). Multi-stage Lot-Sizing for general production systems . In Axsater, S., Schneeweiss, C., & Silver, E. (Eds.), *Multi-Stage Production Planning and Inventory Control.* Berlin, Germany: Springer Verlag.

Hu, X., Shi, Y., & Eberhart, R. (2004). Recent advances in particle swarm Evolutionary Computation. In *Proceedings of the 2004 Congress on Evolutionary Computation* (Vol. 1, pp. 90-97).

Jeunet, J., & Jonard, N. (2005). Single point stochastic search algorithms for the multi-level Lot-Sizing problem. *International Journal of Production Economics, 32,* 985–1006.

Kennedy, J., & Eberhart, R. (1997). A discrete binary version of the particle swarm algorithm. In *Proceedings of the International Conference on Systems, Man and Cybernetics* (Vol. 5, pp. 4104-4108).

Kuik, R., & Solomon, M. (1990). Multi-level Lot-Sizing problem: Evaluation of a simulated-annealing heuristic. *European Journal of Operational Research, 45*(1), 25–37. doi:10.1016/0377-2217(90)90153-3

Pitakaso, R., Almeder, C., Doener, K. F., & Hartl, R. (2007). A max-min ant system for unconstrained multi-level Lot-Sizing problems. *Computers & Operations Research, 34,* 2533–2552. doi:10.1016/j.cor.2005.09.022

Rizk, N., & Martel, A. (2001). *Supply chain flow planning methods: a review of the Lot-Sizing literature.* Quebec City, QC, Canada: Université Laval.

Shi, Y., & Eberhart, R. (1998). A modified particle swarm optimizer. In *Proceedings of the Evolutionary Computation Conference* (pp. 69-73).

Tang, O. (2004). Simulated annealing in lot sizing problem. *International Journal of Production Economics, 88*(2), 173–181. doi:10.1016/j.ijpe.2003.11.006

Tasgetiren, M. F., Sevkli, M., Liang, Y., & Gencyilmaz, G. (2004). Particle Swarm Optimization Algorithm for Permutation Flowshop Sequencing Problem. In M. Dorigo et al. (Eds.), *Proceedings of ANTS 2004* (LNCS 3172, pp. 382-389).

Veral, E., & LaForge, R. (1985). The performance of a simple incremental Lot-Sizing rule in a multilevel inventory environment. *Decision Sciences, 16,* 57–72. doi:10.1111/j.1540-5915.1985.tb01475.x

Vollmann, T. E., Berry, D. W., & Whybark, D. C. (1997). *Manufacturing planning and control Systems* (4th ed.). New York, NY: McGraw-Hill.

Voros, J. (2002). On the relaxation of multi-level dynamic Lot-Sizing models. *International Journal of Production Economics*, 77(1), 53–61. doi:10.1016/S0925-5273(01)00202-X

Wagner, H., & Whitin, T. (1958). Dynamic version of the economic lot size model. *Management Science*, 5, 89–96. doi:10.1287/mnsc.5.1.89

Zangwill, W. (1966). A deterministic multiproduct, multifacility production and inventory model. *Operations Research*, 486–507. doi:10.1287/opre.14.3.486

This work was previously published in the International Journal of Applied Metaheuristic Computing, Volume 2, Issue 1, edited by Peng-Yeng Yin, pp. 44-57, copyright 2011 by IGI Publishing (an imprint of IGI Global).

Chapter 8

Adaptive Non–Uniform Particle Swarm Application to Plasmonic Design

Sameh Kessentini
University of Technology of Troyes, France

Thomas Grosges
University of Technology of Troyes, France

Dominique Barchiesi
University of Technology of Troyes, France

Laurence Giraud-Moreau
University of Technology of Troyes, France

Marc Lamy de la Chapelle
University of Paris XIII, France

ABSTRACT

The metaheuristic approach has become an important tool for the optimization of design in engineering. In that way, its application to the development of the plasmonic based biosensor is apparent. Plasmonics represents a rapidly expanding interdisciplinary field with numerous transducers for physical, biological and medicine applications. Specific problems are related to this domain. The plasmonic structures design depends on a large number of parameters. Second, the way of their fabrication is complex and industrial aspects are in their infancy. In this study, the authors propose a non-uniform adapted Particle Swarm Optimization (PSO) for rapid resolution of plasmonic problem. The method is tested and compared to the standard PSO, the meta-PSO (Veenhuis, 2006) and the ANUHEM (Barchiesi, 2009).These approaches are applied to the specific problem of the optimization of Surface Plasmon Resonance (SPR) Biosensors design. Results show great efficiency of the introduced method.

INTRODUCTION

In plasmonics, the Surface Plasmon Resonance (SPR) based sensors have an increasing expansion since the success in the engineering control of nanofabrication (Hoaa, Kirk & Tabrizian, 2007).

They became one of the most successful label-free and commercially developed optical sensors. This technique is currently employed in biomolecular engineering, drug design, monoclonal antibody characterization, virus-protein interaction, en-

DOI: 10.4018/978-1-4666-2145-9.ch008

vironmental pollutants detection, among other interesting fields.

The fabrication process of the SPR biosensors-nanometer scale- is complex and costly. Moreover, specific problems are related to this domain. Therefore, the design optimization of SPR biosensors is of great interest although it has been rarely addressed (Barchiesi, 2009; Barchiesi, Kremer, Mai & Grosges, 2008a; Lecaruyer, Canva & Rolland 2006; Ekgasit, Thammacharoen & Knoll, 2005; Kolomenskii, Gershon & Schuessler, 1997). Therefore appropriate adjustments to the most successful optimization technique must be introduced to solve these problems.

The Particle Swarm Optimization (PSO) was first introduced by Kennedy and Eberhart in 1995 and imitates the swarm behaviour to search the globally best solution. In this method, particle moves using its own experience and collaboration with neighbor swarm particles. This technique attracted a high level of interest because of its encouraging results and was the subject of many improvements. A first set of improvements is related to the PSO parameters. We will be basically interested with this issue to introduce a new optimization algorithm as it will be fully reported in the section 2.3. A second set of improvements is related to testing different topologies for different problems. The main discussed issues (Kennedy & Mendes, 2002; Hu & Eberhart, 2002; Mendes, Kennedy, & Neves, 2004; Liang, Qin, Suganthan & Baskar, 2006) are:

- Whether to adopt GBEST (a global search technique for best position) or one of the LBEST (local) configurations.
- If the PSO with small or high neighborhood might perform better.
- The weight/influence of each particle on its neighbours.
- How to move if the collaboration and self experience part dictates displacements in opposite sides.

In this paper, we will introduce a new PSO algorithm – basically using adaptive PSO parameters with non-uniform law- and compare its efficiency with some other techniques while applied to the plasmonic problem. The second section will introduce SPR biosensor optimization problem and give the related fitness function and decision variables. The proposed optimization method as well as other methods used for the benchmarking will be presented in section 2. In the third section, numerical results will be reported and discussed. Finally, in the last section, some conclusions and perspectives will be drawn.

THE OPTIMIZATION PROBLEM OF SPR BIOSENSORS

The operating principle of the SPR biosensors is based on the shift of the position of mathematical poles. This mathematical issue is in fact related to a critical change of the interaction between light and matter/substrate that happens in presence of slight environment changes (especially the presence of substances to be detected by the sensor) (Barchiesi et al., 2008b; Kretschman & Raether, 1968). Basically, a sudden absorption of light by metal layer of the biosensor occurs, for a given incidence angle of the illumination, leading to a device with high sensitivity to any change in the surrounding biological environment (Leracuyer et al., 2006; Kolomenskii et al., 1997).

The problems encountered in plasmonics depend on complex parameters (characteristics of materials, n_i $(i=1,...,4)$ in Figure 1) varying with the illumination condition (Homola, 1997). Moreover, in general case, they could be a function of more than ten parameters (Davy, Barchiesi, Spajer, & Courjon, 1999; Fikri, Grosges, & Barchiesi, 2003; Fikri, Grosges, & Barchiesi, 2004, Pagnot, Barchiesi, Labeke & Pieralli, 1997). Therefore, an appropriate optimization method should be introduced.

Figure 1. Schematic of the biosensor. The decision variables are the thickness e1 and e2 and the illumination angle θ. The optical characteristics of each medium (n$_i$) -complex parameters-are supposed to be known in this study. The objective function is given by the minimization of the reflected light R(e1,e2,θ).

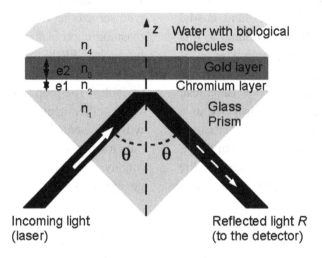

The plasmonic structure, considered in this paper, is illuminated by light with incidence angle θ and made of a glass prism and two metallic layers: a stick Chromium layer and a Gold one (Figure 1). The performance of the structure corresponds to a maximum of energy transfer or, in other words, a vanishing reflection $R(x)$ (the light reflected by the biosensor), where x is the vector of decision variables. The resulting objective function is then the minimization of $R(x)$ which calculation has been fully described by Barchiesi (2009) and Li (1994). The vector of decision variables x is a vector of three decision variables: $e1$ the thickness of a stick Chromium layer (nm), $e2$ the thickness of the Gold layer (nm) and θ (degree). Two plots of $R(x)$ are shown for illustration in Figure 2 (log scale). They exhibit a high sensitivity of the fitness function $R(x)$ in the search space S, governed by physical constraints (Barchiesi et al., 2008b; Neff, Zong, Lima, Borre, & Holzhuter, 2006), $(e1,e2,\theta)$ in $S=[1; 5]\times[30, 70]\times[58, 89.5]$.

The specificity of plasmonic models (and optimization of biosensor based above) can be summarized by the following mathematical properties, x being a vector of variables:

- $R(x)$ is multimodal, non convex, non-uniform or inhomogeneous (Michalewicz, 1992), continuous problem with poles, which is formulated to associate minimums (near 0) to the mathematical poles, with $0\leq R(x)\leq 1$. R depends on real variables (x is in R^3 here, but two other parameters are complex: the optical constants of each layer). Of course, the derivative of $R(x)$ is hard to be computed and could not be used to find optimums.

- Bounded search area (x in $[xmin; xmax]$) and single-objective: $min(R)<10^{-4}$ typically.

Two plots of $R(x)$ are shown for illustration in Figure 2 (log scale). They exhibit a high sensitivity of the fitness function $R(x)$ to the three parameters in the domain search area: $x=(e1, e2, \theta)$ in $[1; 5]\times[30, 70]\times[58, 89.5]$. The search area is governed by physical constraints (Neff, 2006; Barchiesi, 2009). The optical characteristics of the biosensor correspond to a monochromatic red light (670nm, $n_1=1.5$, $n_2= 3.08 +3.35j$, $n_3= 0.163+3.46j$, $n_4=1.33$ where $j^2=-1$).

The quick detection of the wells with sharp shapes (Figure 2), corresponding to poles within

Figure 2. Two maps of R(e1,e2,θ) in log scale showing the sensitivity of the multimodal fitness function R(x)

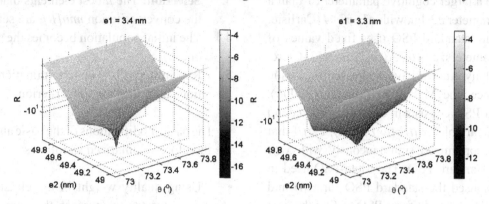

a multidimensional objective function is a specific problem of meta-heuristic optimization, requires preserving diversity of research as well as keeping the information of the potential global optimum. The aim is to find a compromise between the success rate of the optimization and the computational time or convergence speed-up, that maybe heavy in case of the recent nanostructured plasmonic devices (Işıl, Güllü, & Kürşat, 2009).

For the above reasons, the Adaptive Non-Uniform PSO is developed and its efficiency is compared with that of some other methods.

THE ADAPTIVE NON-UNIFORM OPTIMIZATION METHOD FOR PLASMONICS

The PSO (Kennedy & Eberhart, 1995) as well as Evolutionary Methods (Schwefel, 1995) are direct search, stochastic methods, used to find an optimal solution. Despite the interest and effort achieved in the last fifteen years to improve these methods, no generic method is able to deal with all models. In this section brief overviews of the standard PSO, the Meta PSO (Veenhuis, 2006) and the ANUHEM (Barchiesi, 2009) to be used for the benchmark of plasmonic problem- are given. Then the proposed Adaptive Non-Uniform PSO is fully described.

Overview of PSO

The PSO is basically a cooperative method where the vector of variables $x(t)$ is considered as a particle, at the current iteration step t. Members of a swarm communicate good positions to each other and adjust their own position and velocity based on these good positions following:

$$V(t+1) = \omega(t)\, V(t) + c1(t)\,(p(t) - x(t)) + c2(t)\,(g(t) - x(t)) \quad (1)$$

$$x(t+1) = x(t) + V(t+1) \quad (2)$$

where $p(t)$ is the particle best position, $g(t)$ is the global best, $\omega(t)$ is the inertial weight linearly decreasing from 0.9 to 0.4 and $ci(t)(i=1,2)$ are the accelerations, with $ci(t)=U(0,1).ci$, where $U(0,1)$ is the uniform law in *[0,1]*. Equation 1 is used to calculate the particle new velocity according to its previous velocity and the distances between its current position and its own best found position in history i.e. its own best experience $(p(t))$ and the group best experience $(g(t))$. Then the particle moves toward a new position following the Equation 2.

The success of PSO strongly depends on values taken by $c1$ and $c2$. Initially, a lot of PSO models have used equal values for these parameters but recent work reports that it might be even better to choose a larger cognitive parameter $c1$ than a social parameter $c2$, but with $c1+c2<4$ (Carlisle, 2001). The standard PSO uses fixed values of these parameters: $c1=0.738$ and $c2=1.51$ (Clerc, 2009). Many studies have addressed this issue for instance, the Meta-PSO (Veenhuis, 2006). The Meta-PSO consists in considering a second vector of variables $x'(t)=(\omega(t), c1(t), c2(t))$ that has to be optimized simultaneously with $x(t)$, by using standard PSO. The Meta-PSO used in this paper used the standard PSO for $x'(t)$ and the Adaptive Non-Uniform PSO to find the best decision parameters $x(t)$.

Overview of ANUHEM

To improve the classical evolutionary method, Barchiesi (2009) proposed the ANUHEM. The algorithm of the evolutionary loop involves the following steps:

1. **Recombination (or Crossover):** ρ randomly extracted elements of the initial population are combined together to lead to a secondary population of λ elements. The quality of each element (inverse value of the fitness function) is used to weight each element. This is a hyper-ellitist crossover: $x(t+1) = \Sigma(x(t)/R(x(t)) / \Sigma(1/R(x(t))$, Σ being the sum over ρ.
2. **Mutation:** These elements are randomly mutated through a non-uniform law and evaluated. The non-uniform law is $NU(T,b,t)=1-(U(0,1)^{(1-t/T)})^{b(t)}$, where $U(0,1)$ is the uniform law in $[0,1]$, t is the generation (iteration step), T is the maximal generation number, $b(t)$ is the adaptive term determining the degree of non-uniformity of the fitness function and given by the ratio of the standard deviation of the population to the standard deviation (*std*) of the fitness function, and this, for each element of $x(t)$:
    $b(1) = [1,..., 1]$ (3)
    $b(t+1) = std(R(t))/std(x(t))$ (4)

3. **Selection:** The μ best elements that satisfy the convex criterion *min[R]*, are selected.
4. The initial population becomes the selected one.
5. Repeat steps (1-4) up to termination conditions until the stopping criterion.

The basic modifications of the basic algorithm are therefore summarized as follows:

*   Using quality weight in the elitist breeding operator to increase the convergence speed-up.
*   Using the non-uniform law in the mutation operator proposed by Michalewicz (1992) to maintain the diversity with an adaptive power $b(t)$.

The application of ANUHEM to plasmonics revealed a strong decrease of the number of generations to find the global optimum, compared to the classical evolutionary scheme (Schwefel, 1995).

The Adaptive Non-Uniform PSO

As dressed by Barchiesi (2009), for the plasmonic problems, a compromise between diversity search and convergence speed-up to the global optimum must be found. Consequently, the proposed adaptive PSO algorithm, will exploit topological information gathered about fitness function $R(t)$ at each iteration. The initialization of variables and speed for each particle is carried following the scheme proposed by Clerc (2009):

$V(1) =0.5 (U(xmin,xmax)- U(xmin,xmax))$ (5)

$x(1) =U(xmin,xmax)$ (6)

Then, the non-uniform law $b(t)$ used in ANUHEM (Equation 4) is used to compute the parameters:

- The inertia weight is set to: $\omega(t) = max(b(t))$ at each iteration t.
- The acceleration coefficients are set to: $c_i = NU_i(T,b,t) = 4 (1-(U_i(0,1)^{(1-t/T)})^{b(t)})$, $(i=1,2)$, updated at each iteration t. The domain of variation of c_i includes the classical values used for PSO that are mentioned above.

All the adaptive parameters are therefore monitored by the topology of the fitness function. Moreover, $b(t)$ can be seen as the "slope" of the solutions, i.e. the spreading/ dispersion of the population with regards to their quality. $b(t)$ can differ for each component of $x(t)$ and therefore takes into account the "sensitivity" of $R(x)$ to the variations of all decision variables. Zhao, Cao and Hu (2007, pp. 9-10) explain how $b(t)$ determines the degree of non-uniformity: the diversity of parameters increases instead the average speed of the diversity decreases, when $b(t)$ increases.

The proposed adaptive PSO can be compared to the method proposed by Zhan, Zhang, Li, and Chung (2009) in which the authors outlined the necessity of updating the acceleration coefficients following the phase (exploration, exploitation or convergence phase). Three phases are expected during the current PSO loop. The first one is the exploration phase: no adequate set of particles is reached. The second one is the exploitation phase where some particles with "good" quality are found (for them $R(x)$ is typically less than 0.1). In the third one, $R(x)$ decreases to fit with the objective value (10^{-4} in this application).

Through the value of $b(t)$ (an example is given in Figure 3):

- The inertia weight $\omega = max(b(t))$ is:
  - Close to one in the exploration phase and then decreasing.
  - Increasing in the exploitation state. This preserves the diversity of solutions and prevents the convergence to a local minimum.
  - Decreasing toward 0 in the convergence state (when many particles are close to those for the global minimum).
- The acceleration coefficients $c_1(t)$ and $c_2(t)$ are obtained from non-uniform law and therefore tends toward 0 if t tends to-

Figure 3. An example of b(t) evolution as a function of generation t showing the three phases: exploration (generations 1-2), exploitation (generations 2-5) and convergence (generation 4,6-7)

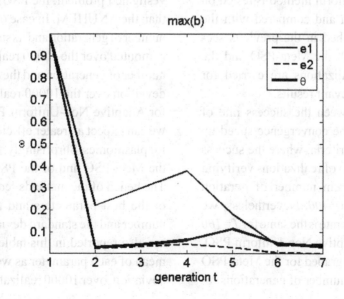

ward T (Zhao, 2008). Therefore T can be considered as a breaking parameter and remains the only non-adaptive parameter of the method. The values of $b(t)$ govern the convergence speed-up as follows:

- In the exploration state, $b(t)$ is close to one as well as $c1(t)$ and $c2(t)$. The value of $b(t)$ gives equal weight to the different contributions of the velocity. If the speed of the particle is too high, the particle (obtained at the generation t) leaves the search space and is therefore replaced by a particle generated randomly in the domain, through a uniform law, as in the initialization step. This contributes to the diversity at this stage.
- In the exploitation state, $b(t)$ increases and also the convergence speed-up (Zhao, 2008).
- In the final state of convergence, b(t) decreases again and the wells of the objective function are more carefully exploited.

NUMERICAL BENCHMARKING

The Adaptive Non-Uniform method is tested on the plasmonic problem and compared with the methods briefly described in the previous section: the standard PSO, the Meta-PSO and the ANUHEM. 10000 realizations are carried for each method to get relevant results.

A compromise between the success rate of the optimization and the convergence speed-up is taken as efficiency criteria; where the success rate is given by the ratio of realizations verifying $R(x) < 10^{-4}$. The maximum number of iteration is fixed arbitrarily to $T=400$. Nevertheless, we checked that the success rate is the same for $T=100$ and $T=400$ for the Adaptive Non-Uniform PSO. Instead, it is slightly degraded for the Meta-PSO (89%). Given that the number of generations re-

quired for convergence of the optimization methods is smaller than 50, the influence of T is low.

The three varieties of PSO tested have a population of 20 particles, as in many numerical studies in the litterature. However, we do not observe strong modification of the behaviour of the PSO methods if the number of particles is in the interval [20;40], but the number of evaluations of $R(x)$ increases with the number of particles. For each iteration, the number of evaluations of the fitness function $R(x)$ has been chosen to be the same for all the methods. Therefore, the selection pressure (ratio of the sizes of the final population (λ) to the initial population (μ) in the evolutionary scheme) for ANUHEM has been fixed to 4 with $\lambda=20$ and $\mu=5$, with elitist scheme). It should be mentioned that the selection pressure is not optimal for ANUHEM and deteriorates the success rate of the evolutionary algorithm. Barchiesi (2009) found an optimum for the selection pressure equal to 10 for the ANUHEM.

Table 1 summarizes the obtained results: success rate, mean number of generation and its standard deviation (std). If the optimization code converges before reaching the maximum number of allowed iterations (T=400), with a value of the fitness function lower than 0.0001, we consider that the optimization was successful. In the investigated problem, the PSOs are more efficient than the ANUHEM. In case of success, the mean number of generation and its standard deviation are computed over the 10000 realizations. The mean number of generation and the associated standard deviation over the 10000 realizations is minimal for Adaptive Non-Uniform PSO, and therefore, we can expect a greater efficiency of this method for plasmonics. Surprisingly, the performances of the Meta-PSO and of the PSO are very similar. The bench of the methods requires an inspection of the best parameters and therefore the mean number and the standard deviation of θ, $e1$ and $e2$ are also reported in this table. The values of the mean of each parameter as well as their standard deviation, over 10000 realizations, reveal that the stop criterion related to the fitness value defines a narrow well around the real minimum of the fitness function. Consequently, all the described methods could benefit from a end-descent method to reach the minimum of the fitness function if required (for example a simulated annealing).

To summarize, the best success rate in Table 1 is for adaptive PSO than for ANUHEM, but the mean number of generation required to converge is divided by four. The Meta-PSO and PSO are intermediate score methods, with twice mean number of generations and lower success rate. All PSO methods have about the same success rate. The Adaptive Non-Uniform PSO requires approximately half evaluation of the fitness function than PSO and meta-PSO and the associated standard deviation shows more stability over realizations. These remarks give advantage to the Adaptive Non-Uniform PSO. Hence, the adaptive PSO may be considered as more efficient for the investigated plasmonic problem. Let us note that the mean values and standard deviations of decisions parameters *x*, are very close together. The decision parameters cannot be compared directly to those obtained by Barchiesi (2009), the numerical values of n_2 and n_3 as well as the search area being different. Otherwise, the results are in good agreement with the experimental results

obtained by Neff et al. (2006). In their study, Neff et al. consider a Chromium layer with fixed thickness (2.5 nm) and find a pronounced minimum reflectance for illumination incidence θ=68° and for a Gold layer thickness in the range of 50-60 nm. Therefore, we can rely on the mathematical model and the introduced optimization algorithm to get the optimal performance of plasmonic structure without carrying costly experiences.

CONCLUSION

In this paper, we have proposed the Adaptive Non-Uniform PSO. The main idea is to preserve the diversity of the population and the convergence speed-up using adaptive exogenous parameters *(ω, c1, c2)* and a non-uniform law. Results show that the method is more efficient than ANUHEM and standard PSO (success rate and number of evaluations needed to get the objective value) for the plasmonic problem. Further improvements of the method could be introduced. First, a standard simulated annealing method could be used at the end of the adaptive PSO to get a lower value of the fitness function, if required. One could also consider improving the meta-PSO especially in the determination of the exogenous parameters,

Table 1. Performance over 10000 realizations of Adaptive Non-Uniform Hyper Elitist Method (ANUHEM), meta-PSO, standard PSO and the Adaptive Non-Uniform PSO, for T=400

	ANUHEM (λ=20, μ=5) (Barchiesi, 2009)	Meta-PSO Mean(c1)=mean(c2)=2.11. std(c1)=0.57, std(c2)=0.56.	PSO c1=0.738; c2=1.51; ω is linearly decreasing from 0.9 to 0.4.	Adaptive Non-Uniform PSO
Success rate	48%	99.64%	99.98%	100%
Mean number of generations (std)	84 (182)	41 (42)	41 (47)	18 (7)
Mean value of θ (std)	73.423 (0.06)	73.426 (0.06)	73.427 (0.07)	73.427 (0.06)
Mean value of e1 (std)	3.181 (1.01)	3.239 (1.18)	3.243 (1.28)	3.248 (1.18)
Mean value of e2 (std)	49.439 (0.67)	49.396 (0.79)	49.386 (0.85)	49.386 (0.79)

and to study the stability of optimization method. Nevertheless, the results obtained by the Adaptive Non-Uniform PSO method are already encouraging and offer scope for effective application in plasmonics. Regarding the plasmonic biosensors, the optimization of the sensitivity with the concentration of the biological material is in process. A further purpose is to apply the optimization to the last generation of biosensors having complex geometry with structured metallic nano-patterns. For this class of biosensors, the model depends on more than six parameters and requires Finite Element computation, with auto-adaptive mesh of the volume of computation.

ACKNOWLEDGMENT

This project is supported by the Nanoantenna European Project (FP7 Health-F5-2009-241818), the Conseil Régional de Champagne-Ardennes, and the Conseil général de l'Aube. Authors thank them for financial support.

REFERENCES

Barchiesi, D. (2009). Adaptive non-uniforme, hyperellitist evolutionary method for the optimization of plasmonic biosensors. In *Proceedings of the International Conference on Computers & Industrial Engineering* (pp. 542–547).

Barchiesi, D., Kremer, E., Mai, V., & Grosges, T. (2008a). A Poincare's approach for plasmonics: The plasmon localization. *Journal of Microscopy, 229*(3), 525–532. doi:10.1111/j.1365-2818.2008.01938.x

Barchiesi, D., Macias, D., Belmar-Letellier, L., Labeke, D. V., de la Chapelle, M. L., Toury, T., Kremer, E., et al. (2008b). Plasmonics: Influence of the intermediate (or stick) layer on the efficiency of sensors. Applied Physics. B, Lasers and Optics, 93(1), 177–181

Carlisle, C., & Dozier, G. (2001). An off-the-shelf PSO. In *Proceedings of the Particle Swarm Optimization Workshop* (pp. 1-6).

Clerc, M. (2009). *A method to improve standard PSO*. Retrieved January 29, 2010, from http://clerc.maurice.free.fr/pso/Design_efficient_PSO.pdf

Davy, S., Barchiesi, D., Spajer, M., & Courjon, D. (1999). Spectroscopic study of resonant dielectric structures in near–field. *The European Physical Journal Applied Physics, 5*, 277–281. doi:10.1051/epjap:1999140

Ekgasit, S., Thammacharoen, C., Yu, F., & Knoll, W. (2005). Influence of the metal film thickness on the sensitivity of surface plasmon resonance biosensors. *Applied Spectroscopy, 59*, 661–667. doi:10.1366/0003702053945994

Fikri, R., Grosges, T., & Barchiesi, D. (2003). Apertureless scanning near-field optical microscopy: The need of the tip vibration modelling. *Optics Letters, 28*(22), 2147–2149. doi:10.1364/OL.28.002147

Fikri, R., Grosges, T., & Barchiesi, D. (2004). Apertureless scanning near-field optical microscopy: Numerical Modeling of the lock-in detection. *Optics Communications, 238*(1-6), 15–23.

Hoaa, X., Kirk, A., & Tabrizian, M. (2007). Towards integrated and sensitive surface plasmon resonance biosensors: A review of recent progress. *Biosensors & Bioelectronics, 23*, 151–160. doi:10.1016/j.bios.2007.07.001

Homola, J. (1997). On the sensitivity of surface Plasmon resonance sensors with spectral interrogation. *Sensors and Actuators. B, Chemical, 41*, 207–211. doi:10.1016/S0925-4005(97)80297-3

Hu, X., & Eberhart, R. C. (2002). Multiobjective optimization using dynamic neighborhood particle swarm optimization. In *Proceedings of the IEEE Congress Evolutionary Computation,* Honolulu, HI (pp. 1677–1681).

Işıl, B., Güllü, K., & Kürşat, S. (2009). A multiobjective optimization framework for nano-antennas via normal boundary intersection (NBI) method. In *Proceedings of the IEEE AP-S International Symposium on Antennas and Propagation and USNC/URSI National Radio Science Meeting,* Charleston, SC.

Kennedy, J., & Eberhart, R. (1995). Particle swarm optimization. In *Proceedings of the IEEE International Conference on Neural Networks,* Path, Australia (Vol. 4, pp. 1942–1948).

Kennedy, J., & Mendes, R. (2002). Population structure and particle swarm performance. In *Proceedings of the IEEE Congress Evolutionary Computation,* Honolulu, HI (pp. 1671–1676).

Kolomenskii, A., Gershon, P., & Schuessler, H. (1997). Sensitivity and detection limit of concentration and absorption measurements by laser-induced surface-plasmon resonance. *Applied Optics,* 36, 6539–6547. doi:10.1364/AO.36.006539

Kretschman, E., & Raether, H. (1968). Radiative decay of nonradiative surface plasmons excited by light. *Zeitung Naturforschung A,* 23, 2135–2136.

Lecaruyer, P., Canva, M., & Rolland, J. (2006). Metallic film optimization in a surface plasmon resonance biosensor by the extended Rouard method. *Applied Optics,* 46(12), 2361–2369. doi:10.1364/AO.46.002361

Lee, C. Y., & Yao, X. (2004). Evolutionary programming using mutations based on the Levy probability distribution. *IEEE Transactions on Evolutionary Computation,* 8, 1–13. doi:10.1109/TEVC.2003.816583

Li, L. (1994). Multilayer-coated diffraction gratings: Differential method of Chandezon et al. revisited. *Journal of the Optical Society of America. A, Optics, Image Science, and Vision,* 11, 2816–2828. doi:10.1364/JOSAA.11.002816

Liang, J. J., Qin, A. K., Suganthan, P. N., & Baskar, S. (2006). Comprehensive learning particle swarm optimizer for global optimization of multimodal functions. *IEEE Transactions on Evolutionary Computation,* 10(3), 281–295. doi:10.1109/TEVC.2005.857610

Mendes, R., Kennedy, J., & Neves, J. (2004). The fully informed particle swarm: Simpler maybe better. *IEEE Transactions on Evolutionary Computation,* 8, 204–210. doi:10.1109/TEVC.2004.826074

Michalewicz, Z. (1992). *Genetic Algorithms + Data Structure = Evolution Programs.* New York, NY: Springer-Verlag.

Neff, H., Zong, W., Lima, A., Borre, M., & Holzhuter, G. (2006). Optical properties and instrumental performance of thin gold films near the surface plasmon resonance. *Thin Solid Films,* 496, 688–697. doi:10.1016/j.tsf.2005.08.226

Pagnot, T., Barchiesi, D., Labeke, D. V., & Pieralli, C. (1997). Use of a SNOM architecture to study fluorescence and energy transfer near a metal. *Optics Letters,* 22, 120–122. doi:10.1364/OL.22.000120

Schwefel, H. (1995). *Evolution & Optimum Seeking.* New York, NY: John Wiley & Sons.

Veenhuis, C. (2006). Advanced Meta-PSO. In *Proceedings of the IEEE Sixth International Conference on Hybrid Intelligent Systems* (pp. 54–59).

Zhan, Z.-H., Zhang, J., Li, Y., & Chung, H.-S. (2009). Adaptive Particle Swarm Optimization. *IEEE Transactions on Systems, Man, and Cybernetics. Part B, Cybernetics*, 6(39), 1362–1381. doi:10.1109/TSMCB.2009.2015956

Zhao, X. (2008). Convergent analysis on evolutionary algorithm with non-uniform mutation. In *Evolutionary Computation* (pp. 940–944).

Zhao, X., Cao, X.-S., & Hu, Z.-C. (2007). Evolutionary programming based on non-uniform mutation. *Applied Mathematics and Computation*, *192*, 1–10. doi:10.1016/j.amc.2006.06.107

This work was previously published in the International Journal of Applied Metaheuristic Computing, Volume 2, Issue 1, edited by Peng-Yeng Yin, pp. 18-28, copyright 2011 by IGI Publishing (an imprint of IGI Global).

Chapter 9
New Evolutionary Algorithm Based on 2-Opt Local Search to Solve the Vehicle Routing Problem with Private Fleet and Common Carrier

Jalel Euchi
University of Sfax, Tunisia

Habib Chabchoub
University of Sfax, Tunisia

Adnan Yassine
University of Le Havre, France

ABSTRACT

Mismanagement of routing and deliveries between sites of the same company or toward external sites leads to consequences in the cost of transport. When shipping alternatives exist, the selection of the appropriate shipping alternative (mode) for each shipment may result in significant cost savings. In this paper, the authors examine a class of vehicle routing in which a fixed internal fleet is available at the warehouse in the presence of an external transporter. The authors describe hybrid Iterated Density Estimation Evolutionary Algorithm with 2-opt local search to determine the specific assignment of each tour to a private vehicle (internal fleet) or an outside carrier (external fleet). Experimental results show that this method is effective, allowing the discovery of new best solutions for well-known benchmarks.

DOI: 10.4018/978-1-4666-2145-9.ch009

INTRODUCTION

In the last years, the price of fuels increased in a dramatic and spectacular way. This price increase obliges the companies to define their own program of transport and an effective management of use of their vehicles fleet. Any modern method of management of the increase of cost of fuel is equivalent to reducing the cost of transport and the planning of vehicle routing.

Numerous organizations are involved in the production and the distribution of goods. The manager has to make a decision and to specify the expeditions assigned to every truck for the delivery. Many manufacturers and distributors use private fleets, or the public carrier, with the purpose to collect or deliver goods for their installations. In addition, to offer greater control over goods movement private fleets may reduce costs over common carrier prices. Whereas common carriers typically require that shipments be processed at consolidation terminals, private fleets can transport shipments directly from origin to destination via multiple stop routes.

It is necessary to cultivate the interest in the selection of service of trucks because it has a particular impact on the organization which uses private vehicles, especially that the decision-makers are responsible for the use of the fleet and the setting of strategies who determines the balance between the public carrier and carrier's private use. If private vehicles are used, routing the vehicles for best utilization, sizing the vehicles and determining the required number of common choice must be made. The rate negotiation, shipment consolidation and routing are important considerations to use a common carrier. Owing to his scale economies, the common carrier may be able to offer a lower price for small shipments in particular. As mentioned in the literature we show that the success stories of the operations research community is derived from the Vehicle Routing Problem (VRP). The interplay between

theory and practice is recognized as a major driving force for this success.

Despite its wealth and abundance of work that are devoted to him, the Vehicle Routing Problem with Private fleet and common Carrier (VRPPC) represent only a subset of a larger family known as VRP. It is one of the optimization problems most studied. This problem holds the attention of several researchers, and everywhere in the world, since many years. Several authors have made a literature review that deal with VRP these include those of Clarke and Wright (1964) which propose a saving heuristic to solve this variant of problem. A state of the art is proposed by Bodin et al. (1983) with purpose to give a good review for the Routing and scheduling of vehicles and crews.

Laporte (1992a, 1992b) studied the vehicle routing and proposed an overview of exact and approximate algorithms to solve the TSP and VRP. Toth and Vigo (2002) addressed the VRP, the authors in this book collected contributions from renowned experts on vehicle scheduling, who describe the main algorithms built for the VRP over the last 40 years. Like other authors, Golden et al. (1984) presented the VRP as problem easy to explain but difficult to solve.

More specifically, two types of problems have been addressed in literature: the VRP with limited fleet and the VRP with private fleet and common carrier. Several authors have studied the vehicle routing problem with limited fleet.

Osman and Salhi (1996) proposed local search strategies to solve the vehicle fleet mix problem. The Tabu Search (TS) heuristic is proposed to solve the heterogeneous fleet vehicle routing problem in the paper of Gendreau et al. (1999). Taillard (1999) implemented a heuristic column generation method (HCG) to solve the Heterogeneous fixed fleet vehicle routing problem (HFFVRP). Tarantilis et al. (2004) introduce a new metaheuristic called Back-tracking Adaptive Threshold Accepting (BATA) in order to solve the HFFVRP. A column generation method to solve the Heterogeneous fleet VRP is proposed by Choi and Tcha (2007). Li et al. (2007) developed a record-to-record travel algorithm for the HFFVRP. They have built an integer programming model and solved the linear relaxation by column generation. Recently, Euchi and Chabchoub (2010) implemented a hybrid tabu search to solve the heterogeneous fixed fleet vehicle routing problem, this metaheuristic produced very competitive results in the literature.

All these authors considered a mixed fleet limited but sufficient capacity to serve all customers. At the level of VRP with external carrier are Volgenant and Jonker (1987), which demonstrated that the problem involving a fleet of one single vehicle and external carriers can be rewritten as Travelling Salesman Problem (TSP). This problem also have been studied by Diaby and Ramesh (1995) whose objective was to decide that customers visited with external carrier and optimize the tour of remaining customers. Several approaches have been used to solve the classical VRP, exact methods, heuristics and metaheuristics solution principally are proposed as in Desrochers and Verhoog (1991) were developed a further extended savings-based approaches to solve the fleet size and mix vehicle-routing problem.

The VRPPC is more complex problem because it involves an internal fleet of several vehicles. To our knowledge, the VRPPC was introduced by Ball et al. (1983). They decide to determine the optimal homogeneous fleet size in the presence of an external carrier. The VRPPC has been constantly studied since the appearance of the second publication on it by Klincewicz et al. (1990). They have presented the problem in a context to divide the customers into sectors and the private fleet size and common carrier must be determined for each sector.

Chu (2005) put forward an interesting formulation for the VRPPC and solve it with a heuristic economies improved by inter and intra routes. Thereafter, Bolduc et al. (2007) have improved the results of Chu using more sophisticated customer exchanges using two different initial solutions.

Recently, to solve the VRPPC, Bolduc et al. (2008) they present a metaheuristic procedure based on different neighborhood structures which is enhanced by two diversification strategies, a randomized construction procedure and a perturbation mechanism.

Evolution computation (EC) motivated by evolution in the real world, has become one of the most widely used techniques, because of its effectiveness and versatility. It maintains a population of solution, which evolves subject to selection and genetic operators (such as recombination and mutation). Each individual in the population receives a measure of its fitness, which is used to guide selection. Iterated Density Estimation Evolutionary Algorithm (IDEA) family is a new type of metaheuristics which has attained interest during the last 5 years. IDEA is relatively recent type of optimization and learning techniques based on the concept of using a population of tentative solutions to iteratively approach the problem region where the optimum is located (Larrañaga, 2002; Mühlenbein et al., 1999; Bosman et al., 2002).

One of the main impulses that triggered the emergence of the IDEA field has been the research into the dynamics of discrete Genetic Algorithm (GA) (Goldberg, 1989). Because probability distributions are used to explicitly guide the search in IDEA, the probability distribution itself is an explicit model for the inductive search bias. Because a lot is known about how probability distributions can estimated from data the flexibility of the inductive search bias of IDEA is potentially large. In addition, the tuning of the inductive search bias in this fashion also has a rigorous foundation in the form of the well-established field of probability theory. The IDEA explores the search space by sampling a probability distribution (Pelikan et al., 2002) that is developed during the optimization. They work with a population of candidate solutions. In each generation, the fittest solutions are used for the model building or model updating and new solutions are generated by sampling this distribution. These new solutions are evaluated and incorporated into the original population, replacing the crossover and mutation operators of the genetic algorithm. This process is repeated until the termination criterion is met.

The IDEA is good at identifying promising areas in the search space, but lacks the ability of refining a single solution. A very successful way to improve the performance of IDEA is to hybridize it with local search techniques (Lozano et al., 2006). We propose 2-opt local search for the vehicle routing with private fleet and common carrier.

Under this paper we are studying the vehicle Routing Problem with Private fleet and Common carrier (VRPPC). The network studied consists of a depot and several customers. One or more products are distributed; each customer must be served one and only once by the private fleet either by a common carrier; the internal fleet is composed by a limited number of vehicles. The capacity of vehicles is determined in terms of units produced. The vehicles have both a fixed cost of use and a variable cost depending on the distance travelled. The costs of external transport are set for each customer depending on its geographical location and quantities required.

Hence, in that case we can see our contribution in two different manners. Our first contribution in this work concerns the presentation of a new variant of the vehicle routing problem and the proposition of a mathematical model which represents the problem.

The main and second contribution of this paper is to show that it is actually possible to make local decisions and a choice between common carrier and private carrier usage. In the presence of the limited fleet constraints and the external transporter, the problem become more complex, what deduce that the choice of a good meta-heuristics can provide a good result, so the main

strategy is intended to support efficient solution procedures based on evolutionary algorithm to solve the vehicle routing with private fleet and common carrier.

The rest of this paper is organized as follows. In Section 2 we describe a problem statement and the main definition of the VRPPC. In Section 3 we give the main paradigm of the IDEA meta-heuristic. In Section 4 we present our proposed approach to solve the VRPPC. In Section 5 we perform experiments results. We conclude this paper in Section 6.

PROBLEM STATEMENT: DESCRIPTION OF THE VRP WITH PRIVATE FLEET AND COMMON CARRIER

One of the most general versions of the vehicle routing problem is the vehicle routing problem with private fleet and common carrier (VRPPC), which can be formally described in the following way.

Let $G = (V, A)$ be a graph where $V = \{0, 1, \ldots, n\}$ is the vertex set and $A = \{(i, j) : i, j \in V, i \neq j\}$ is the arc set. Vertex 0 is a depot, while the remaining vertices represent customers. A private fleet of m vehicles is available at the depot. The fixed cost of vehicle k is denoted by f_k, its capacity by Q_k, and the demand of customer i is denoted by q_i. A travel cost matrix (c_{ij}) is defined on A. If travel costs are vehicle dependent, then c_{ij} can be replaced with c_{ijk}, where $k \in \{1, \ldots, m\}$. Each customer i can be served by a vehicle of the private fleet, in which case it is called an internal customer or by a common carrier at a cost equal to e_i, in which case it is called an external customer.

The VRPPC consists of serving all customers in such a way that:

Assumptions:

- Each customer is served exactly once by a private fleet vehicle either by a common carrier vehicle.
- All routes associated with the private fleet start and end at the depot.
- Each private fleet vehicle performs only one route.
- The total demand of any route does not exceed the capacity of the vehicle assigned to it.
- The total cost is minimized. In practice, several common carriers may be used to serve any of the customers unvisited by the private fleet.

Typically, the one selected is the lowest cost carrier. It is not necessary to specify the routes followed by the common carrier because it charges a fixed amount e_i for visiting customer i, irrespective of visit sequence.

Notation and Model Formulation

Index:

n : Number of customers
m : Number of vehicles in the internal fleet
$i \in \{0, ..., n\}$: Index of customer (depot = 0)
$j \in \{0, ..., n\}$: Index of customer (depot = 0)

Parameters:

F_k : Fixed cost of vehicle k
c_{ijk} : Variable cost for vehicle k travelling from customer i to customer j
L_i : Fixed cost to supply customer i with the external transporter
q_i : Demand of customer i
Q_k : Capacity of vehicle k

Decisions Variables:

The formulation uses the flowing variables:

$$x_{ijk} = \begin{cases} 1 \text{ if the vehicle } k \text{ traveling from} \\ \quad \text{customer } i \text{ to customer } j \\ 0 \text{ otherwise} \end{cases}$$

$$y_{ik} = \begin{cases} 1 \text{ if the demand of customer } i \text{ is} \\ \quad \text{supplied by the vehicle } k \\ 0 \text{ otherwise} \end{cases}$$

$$Z_i = \begin{cases} 1 \text{ if the demand of customer } i \text{ is} \\ \quad \text{supplied by the external transporter} \\ 0 \text{ otherwise} \end{cases}$$

The vehicle routing problem with private fleet and common carrier can be formulated as follows:
Minimize

$$\sum_{k=1}^{m} F_k y_{0k} + \sum_{i=0}^{n} \sum_{j=0}^{n} \sum_{k=1}^{m} c_{ijk} x_{ijk} + \sum_{i=1}^{n} L_i Z_i \quad (1)$$

subject to

$$\sum_{k=1}^{m} y_{0k} \leq m \qquad (k = 1, ..., m) \quad (2)$$

$$\sum_{\substack{j=0 \\ j\neq h}}^{n} x_{hjk} = \sum_{\substack{i=0 \\ i\neq h}}^{n} x_{ihk} = y_{hk} \quad (h \in \{0,...,n\}; k \in \{1,...,m\}$$

$$(3)$$

$$\sum_{k=1}^{m} y_{ik} + Z_i = 1 \qquad (i \in \{1,...,n\}) \quad (4)$$

$$\sum_{i=1}^{n} q_i y_{ik} \leq Q_k \qquad k \in \{1,...,m\} \quad (5)$$

$$\sum_{ij \in s} x_{ijk} \leq |S| - 1 \text{ for all } S \subseteq \{2,...,n\},$$

$$(6)$$

$$(k = 1,...,m) \qquad x_{ijk}, y_{ik}, Z_i \in \{0,1\}$$

The objective functions (1) minimize the transportation cost: fixed and variable cost of the internal fleet and fixed cost of the external transporter. Constraint (2) specify that at most m private fleets vehicles can be used in the solution, while constraint (3) indicate that the same vehicle k must enter and leave customer h, (4) apply that each customer is served either by the private truck or the external transporter. (5) Ensure that the vehicle capacity is never exceeded. (6) Are the sub tour-breaking constraints.

MAIN PARADIGM OF THE IDEA METAHEURISTIC

IDEA is listed as a type of evolutionary algorithms in which an initial population of individuals, each one encoding a possible solution to the problem, is iteratively improved by the application of stochastic operators. Every individual encodes a solution that is weighted with respect to the others by assigning a fitness value according to the objective function being optimized.

IDEA iterates the three steps listed below, until some termination criterion is satisfied:

1. Select good candidates (i.e., solutions) from a (initially randomly generated) population of solutions.
2. Estimate the probability distribution from the selected individuals.
3. Generate new candidates (i.e., offspring) from the estimated distribution.

SOLUTION METHODOLOGY

The need for optimization using methods of operations research in VRP has become more and more important in recent years. One way of tackling the VRPPC is by allowing the search space of the IDEA with 2-opt local search. Local search remains the main practical tool for finding

good solutions for large instances of the VRPPC. Figure 1 presents the scheme of the IDEA / 2-opt approach.

Hybrid IDEA to Solve the VRPPC (IDEA / 2-opt)

To solve the VRPPC, it is recommended to hybridize it with a local search. In this way, we propose to use a 2-opt local search (Lin, 1965) to improve the solution generated after the creation of the initial population and after the generation of new solution.

With each generation t, IDEA/2-opt local search maintains a population of solutions, which $Pop(t) = \{\pi^1, \pi^2, ..., \pi^N\}$ and the probability

matrix is $p(t) = \begin{pmatrix} p_{11}(t)....p_{1n}(t) \\ \vdots \qquad\qquad \vdots \\ p_{n1}(t)....p_{nn}(t) \end{pmatrix}$; Where $p(t)$

models the distribution of promising solutions in the search space. More precisely, $p_{kj}(t)$ is the probability that vehicle k is assigned to customer j in the assignment.

Below the implementation of each part of the IDEA / 2-opt local search to solve the VRPPC is described.

Vehicle Routing Representation

A suitable presentation of solution to VRP is i.e. a chromosome consisting of several routes, each containing a subset of customers that should be visited in the same order as they appear. Every customer has to be a member of exactly one route.

The solution is presented by one vector of dimension $n + k$ where n is the number of customers and k is the number of vehicles. It is assumed that every solution is started from the depot. Each vector has a combined value of 1 to n and k value $= 0$.

Each value $\in [1, n]$ indicates the customer and each value 0 indicates the return to the depot

Figure 1. General scheme of the IDEA / 2-opt local search

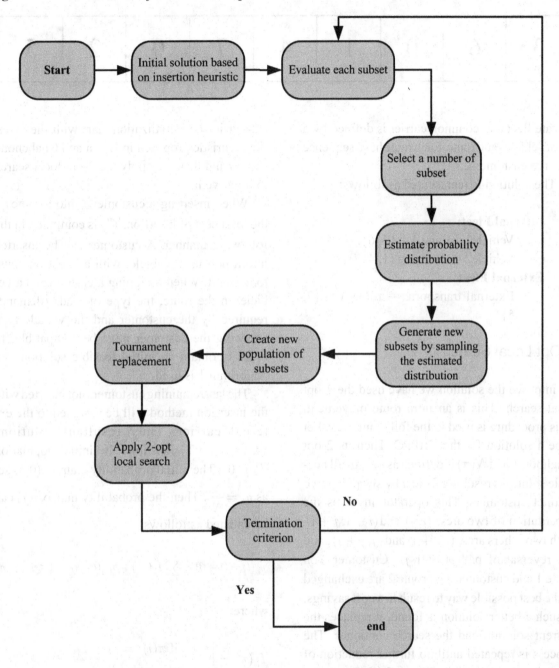

(depot for the new truck) in the $k-1$ value 0 in the sequence. For the last value 0 indicates the end of the cycle allocated to vehicles. The last remaining customers will be assigned to the external carrier. The solution is represented in Figure 2.

Figure 2 presents the solution with one vector. Each value between 1 and n represent the index of the customers when each customer is listed in the order in which they are visited. In our example, we have 7 customers and 2 private vehicles. The solution to the vehicle routing problem with

Figure 2. Vehicle routing representation

3	6	0	1	4	7	0	2	5

private fleet and common carrier is defined by a set of vehicles and route. Each route has a sequence to serve customers.

The solution is represented as follows:

- **Internal Fleet:**
  - Vehicle1 = route 1 = {3, 6}.
  - Vehicle2 = route 2 = {1, 4, 7}.
- **External Fleet:**
  - External transporter = route 3 = {2, 5}.

2-Opt Local Search

To improve the solution we have used the 2-opt local search. This is an intra route movement. This procedure is used in the following way. Let π be a solution for the VRPPC. Then its 2-opt neighbourhood $N(\pi)$ is defined as a set of all possible solutions resulting from π by swapping two distinct customers. This operator involves the substitution of two arcs, (i, j) and $(i + 1, j + 1)$ with two others arcs, $(i, i + 1)$ and $(j, j + 1)$, and the reversal of path $p(i + 1, j)$. Customer i on route 1 and customer j on route 2 are exchanged in the best possible way to result in a cost savings. If such a better solution is found, it replaces the current solution and the search continues. The process is repeated until no further reduction of route length is possible. Figure 3 illustrates the main 2-opt procedure used in this paper.

Initialization

In this section, Insertion Heuristic is developed and used for obtaining an initial solution. IDEA

/ 2-opt in the initialization start with the insertion heuristic proposed in Euchi and Chabchoub (2009) and then we apply the 2-opt local search to improve it.

When inserting a customer v_i into a route r_k, the least cost of insertion, C_{ik} is computed in the following manner. A customer can be inserted into a non used vehicle, with at least one customer in it. When assigning a customer to a vehicle on the route, the type of loads (demand) required by the customer and the vehicle type servicing the customer must be compatible. In order to obtain an initial feasible solution, the Algorithm 1 is used.

The last remaining customers not inserted with the insertion method will be assigned to the external carrier. The N resultant solutions $\{\pi^1, \pi^2, ..., \pi^N\}$ constitute the initial population $Pop(0)$. The initial probability matrix $p(0)$ is set as $p_{ij} = \dfrac{1}{N}$. Then the probability matrix $p(t)$ can be updated as follows:

$$p_{ij}(t) = (1 - \beta)\frac{1}{N}\sum_{k=1}^{N} I_{ij}(\pi^k) + \beta p_{ij}(t-1), \quad (1 \le i, j \le n)$$

where

$$I_{ij}(\pi) = \begin{cases} 1 & \text{if } \pi(i) = j \\ 0 & \text{otherwise} \end{cases}$$

$0 \le \beta \le 1$ is a learning rate. The bigger β is, the greater the contribution of the solutions in $Pop(t)$ is to the probability matrix $p(t)$.

Figure 3. 2-Opt Neighbourhood

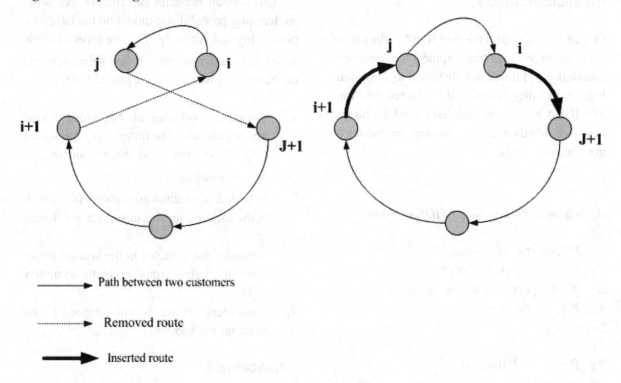

Path between two customers

Removed route

Inserted route

Selection Operators

Selection in evolutionary algorithms meant to select the better solutions of the population to perform variation with. In our IDEA/ 2-opt local search algorithm we use the selection Pressure

Towards Diversity of Bosman and Thierens (2002). The selection operator proposed are presented as follows: the selection operator selects $\lfloor \tau\, pop \rfloor$ solutions, where pop is the population size and $\tau \in \left[\dfrac{1}{pop}, 1 \right]$ is the selection percentile.

Algorithm 1. Initial solution algorithm

```
1: begin
2: sort all available vehicles in increasing order of capacity
3: for each v_i ∈ V(i = 1,..,n) loop
4: for each available vehicle k := 1 to m_max loop
5: if (v_i ∈ r_ki and q_i ≤ Q_k) then
6: Insert a customer v_i into route r_ki using the insertion method described up
7: end if
8: end for
9: customers not inserted will be assigned to the external transporter
10: Execute 2-Opt local search
11: end for
12: end
```

Probabilistic Model

One of the successful ways of IDEA is the use of a probabilistic model that captures the important correlations of the search distribution, assigning high probability values to the selected solution. The IDEA builds a probabilistic model with the best individuals and then samples the model to generate new ones.

Algorithm 2. Pseudo code of IDEA / 2-opt

```
1:  P :Population size
2:  π⁰ : initial solution
3:  π*, π** :intermediate solution
4:  F : fitness value
5:  begin
6:  t := 0;
7:  P = popoulation (t);
8:  // initialization using insertion
       heuristic
9:  π⁰ = π⁰ ∪ (Insertion heuristic)
10: π* = 2 - opt local search (π⁰)
11: //selection operator and
       probabilistic model
12: while (stopping condition is
       not met) do
13: S = selection (pop(t))
14: M = learn model (S)
15: Pop = S
16: for (i ≤ |τ pop| to n ≤ 1) do
17: Pop = Pop ∪ (Popₛ(z))
18: //improve solution with
       local search
19: π** = 2 - opt local search(π*)
20: if (F(π**) > F(π*)) then
21: π* = π**;
22: end if
23: end for
24: t = t + 1
25: end while
26: end
```

This section presents an efficient technique for learning probabilistic model on the base of a probability matrix $p = (p_{ij})_{n \times n}$, we generate a new solution. In order to create a new solution based on the probabilistic model we proceed as follows:

1. Divides the vehicles into two groups based on their capacity. The first group has assigned to $[\alpha n]$ customers and the second one to $n - [\alpha n]$ customers.
2. Vehicle k is assigned to location $\pi(k)$, which is the location for this customer in solution π.
3. Arranges the vehicles in the second group sequentially, based on the probability matrix $p(t)$.
4. Customers not served are assigned to the external transporter.

Replacement

The role of the replacement is keeping the population size constant. To do so, some individuals from the population have to be substituted by some of individuals created during the probabilistic model. This can be done using the tournament replacement.

A subset of α individuals is selected at random, and the worst one is selected for replacement (for $\alpha > 1$).

Stopping Criterion

The stopping criterion is inherent to the complexity of used probabilistic model. We use a maximum number of iteration and maximum of non improvement iteration in the solution as a stopping condition. We can give a pseudo-code of the proposed approach in Algorithm 2.

COMPUTATIONAL RESULTS

In this section we discuss the performance of our IDEA / 2-opt local search algorithm over the benchmark test problem presented in the paper of Bolduc et al. (2008).

Implementation and Benchmarks Description

We consider two sets of instances to evaluate the performance of IDEA / 2-opt local search algorithm. For the 34 instances of the set (Table 1), the fleet is composed of a limited number of homogeneous vehicles, divide into two subsets: the 14 instances subset of Christofides and Eilon (1969) and the 20 instances proposed by Golden et al. (1998). For the 44 instances of the second set (Table 2) is limited and heterogeneous where composed with the five small instances beginning with Chu-H used by Chu (2005), and the five instances B-H used by Bolduc et al. (2007, 2008)and the set generated by Christofides and Eilon (1969) and from those of Golden et al. (1998) (CE-H and G-H). Some details on the instances generated in this way are reported in Tables 1 and 2.

Instance description of the data set for the VRPPC, where n is the number of customers, $\bar{n}$ is the average number of customers per route, m is the number of vehicles, q_{min} and q_{max} are the lowest and highest demands, respectively, Q is

the capacity of vehicle, f is the fixed cost and c the vehicle variable cost is set equal to 1 or 1.5 per unit of distance.

The algorithm described here has been implemented in C++ using Visual Studio C++ 6.0. Experiments are performed on a PC Pentium 4, 3.2 GHz with 512MB of RAM.

Parameter Settings

To illustrate the effectiveness and performance of IDEA / 2-opt local search algorithm for the VRPPC, the algorithm will be compared to the results of the metaheuristic specified in Bolduc et al. (2008). In order to have a uniform comparison, the algorithm was evaluated with the parameters in Box 1.

We experimented with various factor. We run the algorithm to find the best results within a maximum number of 10^5 evaluations (After 10 runs).

Evaluation Method

In order to verify the effectiveness and efficiency of proposed IDEA / 2_opt local search, various computational experiments were conducted. In particular, computation results reported in this paper are compared with those proposed by Bolduc et al. (2008). The comparison results of all algorithms are carried out on an equal foot-

Box 1. Algorithm evaluation parameters

Parameters	Description	Value
$iter_{max}$	Maximal number of iteration	500
N_{noimp}	Maximum of non improvement iteration	200
τ	Selection percentage	100/5=20%
pop	Population	100
α	Factor	$\alpha \in \{1, 10\}$

ing and the experimentations is performed with respecting of the conversion factor.

Figures 4 and 5 present the comparison of results between the IDEA / 2-opt, SRI and RIP metaheuristic in the generation space of homogeneous and heterogeneous instances. According to the figures we notice that the average percentage deviation tends towards zero for the IDEA / 2_opt algorithm in both homogeneous and heterogeneous instances. In fact, the IDEA / 2_opt finds slightly fitter solutions than the previous approach using SRI and RIP metaheuristic. We notice that this comparison is around 1000 generation but when we increase the number of generation both algorithms has a percentage deviation nearly zero.

Therefore we demonstrate the effectiveness of the probabilistic model to diversify the solution and the local search to minimize the objective function and to increase the gap between our proposed approach and the others algorithms.

Figure 6 indicates how the proposed probabilistic model in the generation of new solution in the VRPPC. We measure the efficiency of the IDEA / 2-opt changes during the optimization. Finally, in Figure 7 we observe that with our proposed approach, we obtain quickly, a best fitness value in a number of reasonable generations.

The comparative result for the homogeneous and heterogeneous instances of the IDEA/2-opt based on 15000 and 30000 iterations are shown in Tables 3 and 4, respectively with the result reported in Bolduc et al. (2008). Table 3 gives the solution values for the instances of homogeneous fleet, while Table 4 contains results for the heterogeneous fleet instances. These tables compare the results obtained with IDEA / 2-opt to those obtained with the SRI and RIP metaheuristic of Bolduc et al. (2008) for each instances.

From these tables, we can see that the IDEA/2-opt with 15000 and 30000 iterations is on average 0.23% and 0.007% respectively over the best known solutions with the 0.63% for the RIP metaheuristic. Table 4 show that the IDEA/2-opt with

15000 and 30000 iterations is on average 0.99% and 0.24% with 1.2% for the RIP metaheuristic over the best known solutions.

The results in Tables 3 and 4 clearly show the strong performance of hybrid IDEA algorithm. Over the 34 homogeneous instances, 31 best solutions were produced with our algorithm. So, for the all heterogeneous instances, a new best solution was found. Although IDEA / 2-opt local search is better than the RIP metaheuristic with a good parameter setting. We want to turn to the evaluation of the solution quality obtained by hybridization between the IDEA and the local search 2-opt over time. To that end Tables 3 and 4 show for each instance the time needed to obtain solution. It can be seen from this table that IDEA / 2-opt algorithm finds reasonably good and best execution time in all instances.

For the homogeneous instances we save the CPU time on average 346.33 and 438.28 seconds for the IDEA/2-opt with 15000 and 30000 iterations respectively, then for the heterogeneous instances we report 326.805 and 418.92 seconds. We conclude that IDEA/2-opt algorithm is faster than the RIP metaheuristic with respect of the conversion ratio (conversion factor). In Tables 5 and 6 we give the best solution founded with the IDEA/2-opt and the others methods reported in Bolduc et al. (2008).

CONCLUSION

We have proposed an Iterated Density Estimation Evolutionary Algorithm with 2-opt local search to solve the vehicle routing problem with private fleet and common carrier. The main features of this metaheuristic are a simple and flexible local search as well as an acceptance criterion for the search space and the use of the probabilistic model. In the process of comparing the algorithm, the contributions of the local search operator and the use of probabilistic model to the performance of the IDEA are also displayed.

The combination of the IDEA / 2-opt local search to solve the VRPPC has shown its robustness and its place among the most effective algorithms of combinatorial optimization.

The results of the metaheuristic are compared to the results of SRI, RIP metaheuristic obtained in Bolduc et al. (2008) on each benchmark problems. The results demonstrated the competitiveness and accuracy of proposed IDEA algorithm. From the experiments carried out here we can conclude that IDEA / 2-opt local search obtained the best solution in the homogeneous instances, although it was the best performer on the heterogeneous data sets. Moreover, the IDEA runs quickly, even for problems with many variables. The results of this research show that the proposed metaheuristic method is a very effective tool for finding good solutions for the VRPPC.

REFERENCES

Ball, M. O., Golden, A., Assad, A., & Bodin, L. D. (1983). Planning for truck fleet size in the presence of a common-carrier option. *Decision Sciences, 14*, 103–120. doi:10.1111/j.1540-5915.1983.tb00172.x

Bodin, L. D., Golden, B. L., Assad, A. A., & Ball, M. O. (1983). Routing and scheduling of Vehicles and crews. The state of the Art. *Computers & Operations Research, 10*, 69–211.

Bolduc, M. C., Renaud, J., & Boctor, F. F. (2007). A heuristic for the routing and carrier selection problem. Short communication. *European Journal of Operational Research, 183*, 926–932. doi:10.1016/j.ejor.2006.10.013

Bolduc, M. C., Renaud, J., Boctor, F. F., & Laporte, G. (2008). A perturbation metaheuristic for the vehicle routing problem with private fleet and common carriers. *The Journal of the Operational Research Society, 59*, 776–787. doi:10.1057/palgrave.jors.2602390

Bosman, P. A. N., & Thierens, D. (2002). Multi-objective optimization with diversity preserving mixture-based iterated density estimation evolutionary algorithms. *International Journal of Approximate Reasoning, 31*, 259–289. doi:10.1016/S0888-613X(02)00090-7

Choi, E., & Tcha, D. W. (2007). A column generation approach to the heterogeneous fleet vehicle routing problem. *Computers & Operations Research, 34*, 2080–2095. doi:10.1016/j.cor.2005.08.002

Christofides, N., & Eilon, S. (1969). An algorithm for the vehicle dispatching problem. *Operations Research, 20*, 309–318. doi:10.1057/jors.1969.75

Chu, C. W. (2005). A heuristic algorithm for the truckload and less-than-truckload problem. *European Journal of Operational Research, 165*, 657–667. doi:10.1016/j.ejor.2003.08.067

Clarke, G., & Wright, J. W. (1964). Scheduling of vehicles from a central depot to a number of delivery points. *Operations Research, 12*, 568–581. doi:10.1287/opre.12.4.568

Desrochers, M., & Verhoog, T. W. (1991). A new heuristic for the fleet size and mix vehicle routing problem. *Computers & Operations Research, 18*(3), 263–274. doi:10.1016/0305-0548(91)90028-P

Diaby, M., & Ramesh, R. (1995). The Distribution Problem with Carrier Service: A Dual Based Penalty Approach. *ORSA Journal on Computing, 7*, 24–35.

Euchi, J., & Chabchoub, H. (2009, July 6-9). Iterated Density Estimation with 2-opt local search for the vehicle routing problem with private fleet and common carrier. In *Proceedings of the International Conference on Computers & Industrial Engineering (CIE 2009)* (pp.1058-1063).

Euchi, J., & Chabchoub, H. (2010). A Hybrid Tabu Search to Solve the Heterogeneous Fixed Fleet Vehicle Routing Problem. *Logistics Research, 2*(1), 3–11. doi:10.1007/s12159-010-0028-3

Gendreau, M., Laporte, G., Musaraganyi, C., & Taillard, E. D. (1999). A tabu search heuristic for the heterogeneous fleet vehicle routing problem. *Computers & Operations Research, 26,* 1153–1173. doi:10.1016/S0305-0548(98)00100-2

Goldberg, D. E. (1989). *Genetic Algorithms in Search, Optimization, and Machine Learning.* Reading, MA: Addison-Wesley.

Golden, B., Wasil, E., Kelly, J., & Chao, I.-M. (1998). The impact of metaheuristics on solving the vehicle routing problem: algorithms, problem sets, and computational results . In Crainic, T., & Laporte, G. (Eds.), *Fleet management and logistics* (pp. 33–56). Boston, MA: Kluwer.

Golden, B.-L., Assad, A.-A., Levy, L., & Gheysens, F.-G. (1984). The fleet size and mix vehicle routing problem. *Computers & Operations Research, 11*(1), 49–66. doi:10.1016/0305-0548(84)90007-8

Klincewicz, J. G., Luss, H., & Pilcher, M. G. (1990). Fleet size planning when outside carrier services are available. *Transportation Science, 24,* 169–182. doi:10.1287/trsc.24.3.169

Laporte, G. (1992a). The Vehicle Routing Problem: An overview of exact and approximate algorithms. *European Journal of Operational Research, 59*(3), 345–358. doi:10.1016/0377-2217(92)90192-C

Laporte, G. (1992b). The traveling salesman problem: An overview of exact and approximate algorithms. *European Journal of Operational Research, 59*(2), 291–247. doi:10.1016/0377-2217(92)90138-Y

Larrañaga, P. (2002). A review on estimation of distribution algorithms . In Larrañaga, P., & Lozano, J. A. (Eds.), *Estimation of Distribution Algorithms. A New Tool for Evolutionary Computation* (pp. 80–90). Boston, MA: Kluwer.

Li, F., Golden, B. L., & Wasil, E. A. (2007). A record-to-record travel algorithm for solving the heterogeneous fleet vehicle routing problem. *Computers & Operations Research, 34,* 2734–2742. doi:10.1016/j.cor.2005.10.015

Lin, S. (1965). Computer solutions of the traveling salesman problem. *The Bell System Technical Journal, 44,* 2245–2269.

Lozano, J. A., Larrãnaga, P., Inza, I., & Bengoetxea, E. (2006). *Towards a New Evolutionary Computation: Advances on Estimation of Distribution Algorithms (Studies in Fuzziness and Soft Computing).* New York, NY: Springer-Verlag.

Mühlenbein, H., & Mahnig, T. (1999). FDA – a scalable evolutionary algorithm for the optimization of additively decomposed functions. *Evolutionary Computation, 7*(4), 353–376. doi:10.1162/evco.1999.7.4.353

Osman, I. H., & Salhi, S. (1996). Local search strategies for the VFMP . In Rayward-Smith, V. J., Osman, I. H., Reeves, C. R., & Smith, G. D. (Eds.), *Modern Heuristic Search Methods* (pp. 131–153). New York, NY: Wiley.

Pelikan, M., Goldberg, D. E., & Lobo, F. G. (2002). A survey of optimization by building and using probabilistic models. *Computational Optimization and Applications, 21*(1), 5–20. doi:10.1023/A:1013500812258

Taillard, E. D. (1999). A heuristic column generation method for the heterogeneous fleet VRP. *RAIRO*, *33*(1), 1–14. doi:10.1051/ro:1999101

Tarantilis, C., Kiranoudis, C., & Vassiliadis, V. (2004). A threshold accepting metaheuristic for the heterogeneous fixed fleet vehicle routing problem. *European Journal of Operational Research*, *152*, 148–158. doi:10.1016/S0377-2217(02)00669-0

Toth, P., & Vigo, D. (2002). Models relaxations and exact approaches for the capacitated vehicle routing problem. *Discrete Applied Mathematics*, *123*, 487–512. doi:10.1016/S0166-218X(01)00351-1

Volgenant, T., & Jonker, R. (1987). On some generalizations of the traveling salesman problem. *The Journal of the Operational Research Society*, *38*, 1073–1079.

APPENDIX

Table 1. Characteristics of the instances with homogeneous limited fleet

Instances	n	$\overline{n}$	m	Q	qmin	qmax	f	c
CE-01	50	12,5	4	160	3	41	120	1
CE-02	75	8,3	9	140	1	37	100	1
CE-03	100	16,7	6	200	1	41	140	1
CE-04	150	16,7	9	200	1	41	120	1
CE-05	199	15,3	13	200	1	41	100	1
CE-06	50	12,5	4	160	3	41	140	1
CE-07	75	8,3	9	140	1	37	120	1
CE-08	100	16,7	6	200	1	41	160	1
CE-09	150	15	10	200	1	41	120	1
CE-10	199	15,3	13	200	1	41	120	1
CE-11	120	20	6	200	2	35	180	1
CE-12	100	12,5	8	200	10	50	120	1
CE-13	120	20	6	200	2	35	260	1
CE-14	100	14,3	7	200	10	50	140	1
G-01	240	34,3	7	550	10	30	820	1
G-02	320	40	8	700	10	30	1060	1
G-03	400	50	8	900	10	30	1380	1
G-04	480	60	8	1000	10	30	1720	1
G-05	200	50	4	900	10	30	1620	1
G-06	280	56	5	900	10	30	1700	1
G-07	360	51,4	7	900	10	30	1460	1
G-08	440	55	8	900	10	30	1480	1
G-09	255	23,2	11	1000	29	300	60	1
G-10	323	24,9	13	1000	25	300	60	1
G-11	399	28,5	14	1000	23	300	80	1
G-12	483	32,3	15	1000	20	300	80	1
G-13	252	12	21	1000	60	300	60	1
G-14	320	13,9	23	1000	52	300	60	1
G-15	396	15,2	26	1000	47	300	60	1
G-16	480	16,6	29	1000	42	300	60	1
G-17	240	13,3	18	200	10	40	40	1
G-18	300	13,6	22	200	10	40	60	1
G-19	360	13,9	26	200	10	40	60	1
G-20	420	13,6	31	200	10	40	60	1

Table 2. Characteristics of the instances with heterogeneous limited fleet

Instances	n	m_A	Q_A	f_A	c_A	m_B	Q_B	f_B	c_B	m_C	Q_C	f_C	c_C
		Vehicle A				**Vehicle B**				**Vehicle C**			
Chu-H-01	5	1	40	60	1.50	1	30	50	1.50				
Chu-H-02	10	1	75	120	1.50	1	65	100	1.50				
Chu-H-03	15	1	110	150	1.50	1	100	140	1.50	1	90	130	1.50
Chu-H-04	22	1	4500	250	1.50	1	4000	200	1.50				
Chu-H-05	29	1	4500	250	1.50	1	4000	200	1.50	1	3500	180	1.50
B-H-01	5	1	40	60	1.50	1	30	50	1.50				
B-H-02	10	1	75	120	1.50	1	65	100	1.50				
B-H-03	15	1	110	150	1.50	1	100	140	1.50	1	90	130	1.50
B-H-04	22	1	4500	250	1.50	1	4000	200	1.50				
B-H-05	29	1	4500	250	1.50	1	4000	200	1.50	1	3500	180	1.50
CE-H-01	50	2	160	140	1.00	2	192	168	1.00				
CE-H-02	75	4	112	80	1.00	5	168	120	1.00				
CE-H-03	100	2	160	112	1.00	2	200	140	1.00	2	240	168	1.00
CE-H-04	150	2	160	96	1.00	4	200	120	1.00	3	240	144	1.00
CE-H-05	199	7	160	80	1.00	5	200	100	1.00	2	240	120	1.00
CE-H-06	50	1	128	112	1.00	2	160	140	1.00	1	192	168	1.00
CE-H-07	75	4	112	96	1.00	3	140	120	1.00	2	168	144	1.00
CE-H-08	100	1	160	128	1.00	1	200	160	1.00	4	240	192	1.00
CE-H-09	150	4	160	96	1.00	3	200	120	1.00	3	240	144	1.00
CE-H-10	199	2	160	96	1.00	5	200	120	1.00	6	240	144	1.00
CE-H-11	120	2	160	144	1.00	2	200	180	1.00	2	240	216	1.00
CE-H-12	100	2	160	96	1.00	3	200	120	1.00	3	240	144	1.00
CE-H-13	120	1	160	208	1.00	4	200	260	1.00	1	240	312	1.00
CE-H-14	100	1	160	96	1.00	1	200	120	1.00	5	240	144	1.00
G-H-01	240	3	440	656	1.00	1	50	820	1.00	3	660	984	1.00
G-H-02	320	2	560	848	1.00	2	700	1060	1.00	4	840	1272	1.00
G-H-03	400	3	720	1104	1.00	3	900	1380	1.00	2	1080	1656	1.00
G-H-04	480	2	800	1376	1.00	4	1000	1720	1.00	2	1200	2064	1.00
G-H-05	200	2	720	1296	1.00	2	900	1620	1.00				
G-H-06	280	3	720	1360	1.00	2	900	1700	1.00	1	1080	2040	1.00
G-H-07	360	3	720	1168	1.00	1	900	1460	1.00	3	1080	1752	1.00
G-H-08	440	1	720	1184	1.00	2	900	1480	1.00	5	1080	1776	1.00
G-H-09	255	6	800	48	1.00	3	1000	60	1.00	3	1200	72	1.00
G-H-10	323	3	800	48	1.00	3	1000	60	1.00	6	1200	72	1.00
G-H-11	399	6	800	64	1.00	8	1000	80	1.00	1	1200	96	1.00
G-H-12	483	6	800	64	1.00	6	1000	80	1.00	4	1200	96	1.00

continued on following page

Table 2. Continued

Instances	n	Vehicle A				Vehicle B				Vehicle C			
		m_A	Q_A	f_A	c_A	m_B	Q_B	f_B	c_B	m_C	Q_C	f_C	c_C
G-H-13	252	6	800	48	1.00	4	1000	60	1.00	10	1200	72	1.00
G-H-14	320	11	800	48	1.00	2	1000	60	1.00	11	1200	72	1.00
G-H-15	396	7	800	48	1.00	9	1000	60	1.00	10	1200	72	1.00
G-H-16	480	12	800	48	1.00	6	1000	60	1.00	11	1200	72	1.00
G-H-17	240	4	160	32	1.00	7	200	40	1.00	6	240	48	1.00
G-H-18	300	7	160	48	1.00	9	200	60	1.00	6	240	72	1.00
G-H-19	360	9	160	48	1.00	7	200	60	1.00	10	240	72	1.00
G-H-20	420	16	160	48	1.00	6	200	60	1.00	10	240	72	1.00

Table 3. Comparative result for the homogeneous instances

Instances	IDEA/2-op with 15000 iteration			IDEA/2-op with 30000 iteration			RIP metaheuristic	
	Best solution	Average	Average CPU (s)	Best solution	Average	Average CPU (s)	Solution	Average CPU (s)
CE-01	1119.47	1119.47	4.21	1119.47	1119.47	7.49	1132.91	25.00
CE-02	1814.52	1818.12	2.55	1814.52	1814.52	3.73	1835.76	73.00
CE-03	1920.36	1946.97	15.71	1920.36	1920.36	22.98	1959.65	107.00
CE-04	2520.58	2550.74	35.64	2511.63	2521.50	37.72	2545.72	250.00
CE-05	3102.54	3173.81	110.04	3087.95	3099.81	172.31	3172.22	474.00
CE-06	1211.05	1217.51	110.98	1204.56	1206.99	131.21	1208.33	25.00
CE-07	2006.52	2010.17	114.53	2004.02	2008.37	159.14	2006.52	71.00
CE-08	2050.32	2071.33	28.56	2045.63	2051.55	48.09	2082.75	110.00
CE-09	2431.12	2473.72	50.67	2427.99	2431.12	53.26	2443.94	260.00
CE-10	3403.94	3413.81	110.29	3391.23	3404.98	112.86	3464.90	478.00
CE-11	2332.21	2332.21	23.15	23.29.01	2333.01	26.41	2333.03	195.00
CE-12	1953.55	1953.55	13.82	1950.64	1952.88	18.21	1953.55	128.00
CE-13	2861.39	2864.21	27.51	2857.03	2859.17	31.78	2864.21	188.00
CE-14	2214.14	2222.32	15.79	2214.14	2217.04	19.57	2224.63	110.00
G-01	14206.51	14206.51	81.12	14206.51	14206.51	81.12	14 388.58	651.00
G-02	19171.62	19235.64	201.63	19169.84	19234.69	220.98	19 505.00	1178.00
G-03	24925.28	24925.28	1071.46	24763.11	24956.83	1145.91	24 978.17	2061.00
G-04	34645.19	34645.19	936.52	34601.79	34618.23	1985.27	34 957.98	3027.00
G-05	14249.82	14325.42	602.53	14249.82	14292.57	654.09	14 683.03	589.00

continued on following page

Table 3. Continued

Instances	IDEA/2-op with 15000 iteration			IDEA/2-op with 30000 iteration			RIP metaheuristic	
	Best solution	Average	Average CPU (s)	Best solution	Average	Average CPU (s)	Solution	Average CPU (s)
G-06	21550.39	21838.81	559.73	21550.39	21550.39	612.99	22 260.19	1021.00
G-07	23525.15	23925.11	764.10	23525.15	23525.15	958.36	23 963.36	1628.00
G-08	30025.13	30123.48	875.13	30025.13	30179.06	1072.75	30 496.18	2419.00
G-09	1321.73	1326.38	93.41	1316.53	1316.53	103.91	1341.17	832.00
G-10	1599.01	1607.18	194.23	1583.10	1596.71	204.30	1612.09	1294.00
G-11	2229.56	2230.12	850.99	2123.98	3180.44	1020.36	2198.45	2004.00
G-12	2485.00	2496.12	765.91	2485.00	2485.01	825.66	2521.79	2900.00
G-13	2268.32	2288.33	72.18	2266.21	2266.21	78.56	2286.91	802.00
G-14	2746.06	2704.01	548.7	2688.31	2708.99	698.71	2750.75	1251.00
G-15	3141.72	3242.59	1079.08	3104.68	3142.98	1249.13	3216.99	1862.00
G-16	3601.23	3695.45	980.31	3595.22	3598.69	1180.49	3693.62	2778.00
G-17	1685.22	1710.05	57.36	1631.29	1631.29	65.92	1701.58	806.00
G-18	2701.05	2766.15	105.16	2686.54	2688.32	125.97	2765.92	1303.00
G-19	3513.16	3513.16	745.69	3413.56	3421.98	988.21	3576.92	1903.00
G-20	4361.74	4460.38	526.77	4268.12	4360.08	784.12	4378.13	2800.00
%deviation	0.23%	0.49%	346.33	0.007%	0.13%	438.28	0.63%	1047,14

Table 4. Comparative result for the heterogeneous instances

Instances	IDEA/2-op with 15000 iteration			IDEA/2-op with 30000 iteration			RIP metaheuristic	
	Best solution	Average	Average CPU (s)	Best solution	Average	Average CPU (s)	Solution	Average CPU (s)
Chu-H-01	387.50	387.50	0.13	387.50	387.50	0.02	387.50	0.35
Chu-H-02	586.00	586.00	0.18	586.00	586.00	0.08	586.00	1.90
Chu-H-03	826.50	826.50	1.05	826.50	826.50	1.02	826.50	3.50
Chu-H-04	1389.00	1389.00	3.20	1389.00	1389.00	3.00	1389.00	5.85
Chu-H-05	1444.50	1444.50	6.15	1441.50	1441.50	5.70	1444.50	10.40
B-H-01	423.50	423.50	2.20	423.50	423.50	1.30	423.50	1.85
B-H-02	476.50	476.50	2.00	476.50	476.50	2.00	476.50	3.50
B-H-03	778.50	778.50	3.40	777.00	777.00	2.80	778.50	4.75
B-H-04	1564.50	1564.50	8.50	1564.50	1564.50	10.23	1564.50	15.85
B-H-05	1609.50	1609.50	9.26	1609.50	1609.50	10.29	1609.50	12.90
CE-H-01	1183.91	1186.19	3.58	1183.91	1183.91	5.18	1192.72	26.00

continued on following page

Table 4. Continued

Instances	IDEA/2-op with 15000 iteration			IDEA/2-op with 30000 iteration			RIP metaheuristic	
	Best solution	Average	Average CPU (s)	Best solution	Average	Average CPU (s)	Solution	Average CPU (s)
CE-H-02	1781.59	1781.59	12.15	1781.59	1781.59	16.15	1798.26	72.00
CE-H-03	1921.77	1938.16	40.58	1918.29	1920.35	45.98	1934.85	105.00
CE-H-04	2446.95	2446.95	20.62	2428.22	2431.14	24.76	2493.93	251.00
CE-H-05	3142.81	3159.66	245.61	2059.86	2117.07	321.83	3195.66	490.00
CE-H-06	1204.36	1204.89	0.55	1202.99	1202.99	1.51	1210.23	25.00
CE-H-07	2036.81	2040.86	10.54	2014.65	2021.78	14.6	2042.79	74.00
CE-H-08	1984.49	1989.36	59.37	1981.17	1981.17	77.18	2015.72	112.00
CE-H-09	2433.52	2436.87	87.04	2409.46	2424.94	95.73	2445.88	267.00
CE-H-10	3225.41	3304.21	182.05	3207.20	3265.84	191.78	3304.69	482.00
CE-H-11	2315.56	2348.30	72.53	2246.87	2269.42	93.62	2308.76	188.00
CE-H-12	1902.05	1908.54	14.25	1902.05	1902.05	23.74	1908.74	130.00
CE-H-13	2836.11	2840.19	88.42	2814.10	2824.99	90.47	2842.18	195.00
CE-H-14	1913.87	1923.44	8.54	1903.01	1912.15	12.44	1920.36	114.00
G-H-01	14382.07	14405.41	410.78	14345.09	14395.22	489.39	14 408.31	647.00
G-H-02	18537.40	18557.24	769.16	18506.39	18557.82	852.33	18 663.15	1254.00
G-H-03	25482.0	25559.09	743.24	25309.61	25511.80	972.66	25 561.55	2053.00
G-H-04	34797.50	34960.39	678.43	34473.26	34886.27	952.57	35 495.66	2049.00
G-H-05	15685.69	15711.43	358.39	15609.30	15621.01	376.36	16 138.50	512.00
G-H-06	19987.69	20280.11	666.01	19960.21	20183.03	787.10	20 329.04	1005.00
G-H-07	23639.51	23877.03	402.35	23410.63	23415.77	651.42	24 840.83	1608.00
G-H-08	27529.83	27704.67	1001.25	27410.59	27691.33	1279.05	27 710.66	2584.00
G-H-09	1309.41	1314.18	388.67	1304.09	1306.57	476.03	1346.03	814.00
G-H-10	1553.28	1573.59	741.45	1540.66	1544.10	975.70	1575.82	1332.00
G-H-11	2207.44	2216.93	979.98	2179.20	2198.61	1194.02	2218.91	2140.00
G-H-12	2455.97	2489.76	970.28	2420.02	2488.38	1682.73	2510.07	2970.00
G-H-13	2216.13	2221.17	368.16	2209.29	2214.35	438.71	2253.45	733.00
G-H-14	2698.15	2722.66	559.88	2602.33	2652.28	732.21	2711.81	1246.00
G-H-15	3144.65	3149.45	950.21	3101.90	3126.67	1132.09	3156.93	1895.00
G-H-16	3629.03	3644.28	735.69	3623.12	3635.92	856.93	3649.09	2785.00
G-H-17	1673.07	1702.71	400.80	1661.96	1668.01	548.05	1705.48	762.00
G-H-18	2732.36	2740.59	593.50	2718.41	2733.38	809.24	2759.99	1299.00
G-H-19	3478.48	3488.30	807.61	3408.60	3480.26	991.69	3517.48	1892.00
G-H-20	4382.12	4391.83	971.68	4322.97	4382.63	1182.87	4413.82	2733.00
%deviation	0.79%	0.99%	326.805	0%	0.24%	418.92	1.20%	791.90

Table 5. Best known solution for the homogeneous limited fleet instances

Instances	Best solution		SRI		RIP		IDEA/2-opt	
	Z		Z	CPU	Z	CPU	Z	CPU
CE-01	1119.47		1199.99	0.00	1119.47	25.00	**1119.47**	7.49
CE-02	1814.52		1890.33	0.00	1814.52	73.00	**1814.52**	3.73
CE-03	1920.36		2050.33	1.00	1937.23	107.00	**1920.36**	22.98
CE-04	2511.63		2694.72	1.00	2528.36	250.00	**2511.63**	37.72
CE-05	3087.95		3228.67	3.00	3107.04	474.00	**3087.95**	172.31
CE-06	1204.56		1282.94	0.00	1207.47	25.00	**1204.56**	131.21
CE-07	2004.02		2092.32	0.00	2006.52	71.00	**2004.02**	159.14
CE-08	2045.63		2163.32	1.00	2052.05	110.00	**2045.63**	48.09
CE-09	2427.99		2526.82	1.00	2436.02	260.00	**2427.99**	53.26
CE-10	3391.23		3511.02	3.00	3407.13	478.00	**3391.23**	112.86
CE-11	2329.01		2375.71	1.00	2332.21	195.00	**2329.01**	26.41
CE-12	1950.64		2037.54	0.00	1953.55	128.00	**1950.64**	18.21
CE-13	2857.03		2916.21	1.00	2858.94	188.00	**2857.03**	31.78
CE-14	2214.14		2220.77	1.00	2216.68	110.00	**2214.14**	19.57
G-01	14 160.77		14675.33	4.00	**14 160.77**	651.00	14206.51	81.12
G-02	19169.84		20108.84	9.00	19 234.03	1178.00	**19169.84**	220.98
G-03	24 646.79		26046.80	16.00	**24 646.79**	2061.00	24763.11	1145.91
G-04	34601.79		36234.51	27.00	34 607.12	3027.00	**34601.79**	1985.27
G-05	14249.82		15751.31	5.00	**14 249.82**	589.00	**14249.82**	654.09
G-06	21550.39		23255.65	8.00	21 703.54	1021.00	**21550.39**	612.99
G-07	23525.15		25298.48	13.00	23 549.53	1628.00	**23525.15**	958.36
G-08	30025.13		30899.74	18.00	30 173.53	2419.00	**30025.13**	1072.75
G-09	1316.53		1378.67	4.00	1336.91	832.00	**1316.53**	103.91
G-10	1583.10		1646.91	8.00	1598.76	1294.00	**1583.10**	204.30
G-11	2123.98		2238.57	14.00	2179.71	2004.00	**2123.98**	1020.36
G-12	2485.00		2597.14	17.00	2503.71	2900.00	**2485.00**	825.66
G-13	2266.21		2339.93	5.00	2268.32	802.00	**2266.21**	78.56
G-14	2688.31		2825.76	8.00	2704.01	1251.00	**2688.31**	698.71
G-15	3104.68		3269.96	12.00	3171.20	1862.00	**3104.68**	1249.13
G-16	3595.22		3784.63	19.00	3654.20	2778.00	**3595.22**	1180.49
G-17	1631.29		1732.70	5.00	1677.22	806.00	**1631.29**	65.92
G-18	2686.54		2821.82	8.00	2742.72	1303.00	**2686.54**	125.97
G-19	3413.56		3614.59	11.00	3528.36	1903.00	**3413.56**	988.21
G-20	4268.12		4439.45	15.00	4352.95	2800.00	**4268.12**	784.12

SRI: Selection – Routing – Improvement algorithm of Bolduc et al. (2007).
RIP: Randomized construction– Improvement–Perturbation algorithm of Bolduc et al. (2008).
SRI, RIP: Xeon 3.6 GHz processor and 1.00 GB of RAM under Windows XP.
IDEA: PC Pentium 4, 3.2 GHz with 512MB of RAM under Windows XP.

Table 6. Best known solution for the heterogeneous limited fleet instances

instances	Best solution Z	SRI Z	SRI CPU	RIP Z	RIP CPU	IDEA/2-opt Z	IDEA/2-opt CPU
Chu-H-01	387.50	**387.50**	0.00	**387.50**	0.35	**387.50**	0.00
Chu-H-02	586.00	**586.00**	0.02	**586.00**	1.90	**586.00**	0.08
Chu-H-03	826.50	**826.50**	0.03	**826.50**	3.50	**826.50**	1.02
Chu-H-04	1389.00	**1389.00**	0.08	**1389.00**	5.85	**1389.00**	3.00
Chu-H-05	1441.50	1444.50	0.09	1444.50	10.40	**1441.50**	5.70
B-H-01	423.50	**423.50**	0.02	**423.50**	1.85	**423.50**	0.30
B-H-02	476.50	**476.50**	0.02	**476.50**	3.50	**476.50**	1.02
B-H-03	777.00	804.00	0.03	778.50	4.75	**777.00**	2.80
B-H-04	1564.50	**1564.50**	0.09	**1564.50**	15.85	**1564.50**	7.23
B-H-05	1609.50	**1609.50**	0.13	**1609.50**	12.90	**1609.50**	8.29
CE-H-01	1183.91	1 220.72	0.00	1191.70	26.00	**1183.91**	3.58
CE-H-02	1781.59	1858.24	0.00	1790.67	72.00	**1781.59**	12.15
CE-H-03	1918.29	1999.91	1.00	1919.05	105.00	**1918.29**	45.98
CE-H-04	2428.22	2615.95	1.00	2475.16	251.00	**2428.22**	24.76
CE-H-05	2059.86	3248.26	3.00	3146.45	490.00	**2059.86**	321.83
CE-H-06	1202.99	1264.72	0.00	1204.48	25.00	**1202.99**	1.51
CE-H-07	2014.65	2093.48	1.00	2025.98	74.00	**2014.65**	14.60
CE-H-08	1981.17	2058.81	0.00	1984.36	112.00	**1981.17**	77.18
CE-H-09	2409.46	2570.57	2.00	2438.73	267.00	**2409.46**	95.73
CE-H-10	3207.20	3391.25	3.00	3267.85	482.00	**3207.20**	191.78
CE-H-11	2246.87	2334.41	1.00	2303.13	188.00	**2246.87**	93.62
CE-H-12	1902.05	1924.92	0.00	1908.74	130.00	**1902.05**	23.74
CE-H-13	2814.10	2925.27	1.00	2842.18	195.00	**2814.10**	90.47
CE-H-14	1903.01	1957.63	1.00	1907.74	114.00	**1903.01**	12.44
G-H-01	14345.09	14599.16	4.00	14 251.75	647.00	**14345.09**	489.39
G-H-02	18506.39	18945.77	13.00	18 560.07	1254.00	**18506.39**	852.33
G-H-03	25309.61	26151.24	13.00	25 356.93	2053.00	**25309.61**	972.66
G-H-04	34473.26	36519.42	22.00	34 589.11	2049.00	**34473.26**	952.57
G-H-05	15609.30	17173.22	3.00	15667.13	512.00	**15609.30**	376.36
G-H-06	19960.21	21083.42	8.00	19975.32	1005.00	**19960.21**	787.10
G-H-07	23410.63	24854.96	14.00	23510.98	1608.00	**23410.63**	651.42
G-H-08	27410.59	28412.97	21.00	27420.68	2584.00	**27410.59**	1279.05
G-H-09	1304.09	1371.98	5.00	1331.83	814.00	**1304.09**	476.03
G-H-10	1540.66	1599.77	8.00	1561.52	1332.00	**1540.66**	975.70
G-H-11	2179.20	2249.11	14.00	2195.31	2140.00	**2179.20**	1194.02
G-H-12	2420.02	2573.81	19.00	2487.38	2970.00	**2420.02**	1682.73
G-H-13	2209.29	2325.09	5.00	2239.18	733.00	**2209.29**	438.71
G-H-14	2602.33	2783.74	10.00	2682.85	1246.00	**2602.33**	732.21

continued on following page

Table 6. Continued

instances		Best solution	SRI		RIP		IDEA/2-opt	
		Z	Z	CPU	Z	CPU	Z	CPU
G-H-15		3101.90	3224.50	13.00	3131.89	1895.00	**3101.90**	1132.09
G-H-16		3623.12	3740.85	22.00	3629.41	2785.00	**3623.12**	856.93
G-H-17		1661.96	1741.66	4.00	1695.75	762.00	**1661.96**	548.05
G-H-18		2718.41	2787.10	7.00	2740.05	1299.00	**2718.41**	809.24
G-H-19		3408.60	3518.50	11.00	3464.70	1892.00	**3408.60**	991.69
G-H-20		4322.97	4362.31	15.00	4352.35	2733.00	**4322.97**	1182.87

SRI: Selection – Routing – Improvement algorithm of Bolduc et al. (2007);
RIP: Randomized construction– Improvement–Perturbation algorithm of Bolduc et al. (2008);
IDEA/2-opt: Iterated Density Estimation Evolutionary Algorithm with **2-opt** local search.

Figure 4. Comparison results for the homogeneous instances

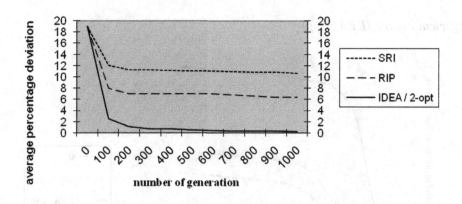

Figure 5. Comparison results for the heterogeneous instances

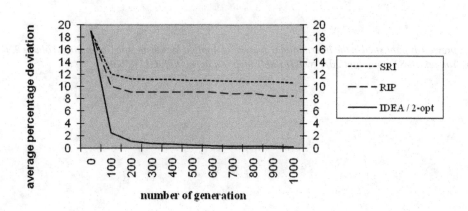

Figure 6. Effect of the number of generation in the fitness value

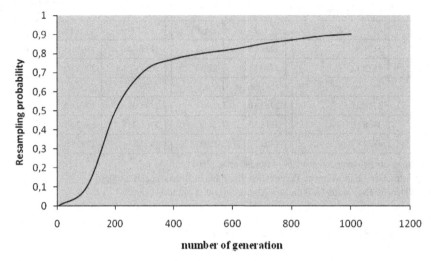

Figure 7. Efficiency of the IDEA/ 2-opt

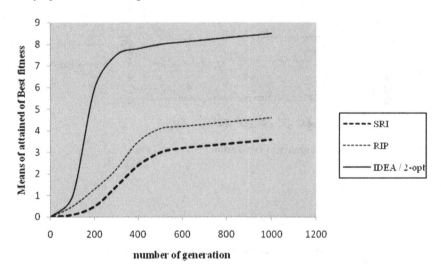

This work was previously published in the International Journal of Applied Metaheuristic Computing, Volume 2, Issue 1, edited by Peng-Yeng Yin, pp. 58-82, copyright 2011 by IGI Publishing (an imprint of IGI Global).

Chapter 10
Scatter Search Applied to the Vehicle Routing Problem with Simultaneous Delivery and Pickup

Gladys Maquera
Universidad Peruana Unión, Peru

Dan Abensur Gandelman
Universidade Federal do Rio de Janeiro, Brasil

Manuel Laguna
University of Colorado, USA

Annibal Parracho Sant'Anna
Universidade Federal Fluminense, Brasil

ABSTRACT

Though its origins can be traced back to 1977, the development and application of the metaheuristic Scatter Search (SS) has stayed dormant for 20 years. However, in the last 10 years, research interest has positioned SS as one of the recognizable methodologies within the umbrella of evolutionary search. This paper presents an application of SS to the problem of routing vehicles that are required both to deliver and pickup goods (VRPSDP). This specialized version of the vehicle routing problem is particularly relevant to organizations that are concerned with sustainable and environmentally-friendly business practices. In this work, the efficiency of SS is evaluated when applied to this problem. Computational results of the application to instances in the literature are presented.

INTRODUCTION

The vigorous industrialization of the modern world and the evolution of consumption habits have resulted in an increase in the demand for efficiently handling the movement of goods as well as managing a growing volume of urban waste. The importance of efficiency of distribution systems becomes evident when their impact on the environment is added to the impact on the operational costs of a firm. Managing urban waste involves a set of regulatory, planning, operational

DOI: 10.4018/978-1-4666-2145-9.ch010

and financial activities that require the use of technology compatible with the local environment. To achieve sustainable material cycles, a philosophy commonly called 3R (reduce, reuse and recycle) is generally adopted. These concerns have introduced a new set of features to existing transportation problems that in turn motivated additional operational research work in the area.

The research in this subject focused initially on developing strategies for collecting returning goods without considering the impact on the delivery routes. Some related work appears in Breedam (1995), Carter *et al.* (1998) and Schultmann *et al.* (2006). The main issue to be addressed is that returned materials may be transported together with the orders to be delivered because combining pickup and delivery results in lower transportation costs than employing separate routes and vehicles to achieve the same tasks. We consider integrated strategies for pickup and delivery within our examination of the Vehicle Routing Problem with Simultaneous Delivery and Pickup (VRPSDP). This is a variant of the classical Vehicle Routing Problem (VRP) that is only concerned with the delivery of orders.

VRPSDP is a combinatorial optimization problem and has been shown to be NP-hard (Nagy & Salhi, 2005). Due to its computational complexity, exact solution methods become impractical for other than small-scale instances, leading to the use of various heuristics and meta-heuristics. We describe the application of scatter search (SS) to VRPSDP. SS is an evolutionary meta-heuristic that has demonstrated merit in the solution of discrete optimization problems. The concepts and basic principles of SS were proposed by Glover (1977), based on strategies to combine decision rules and restrictions. Laguna and Marti (2003) and Marti *et al.* (2006) are two relevant SS references where interested readers will find details on basic and advanced SS designs.

In the process applying SS to the VRPSDP, we developed a Diversification Generation Method

based on the Greedy Randomized Adaptive Search Procedure (GRASP) of Feo and Resende (1995) as well as three different procedures for intensifying the search, as mandated by the Improvement Method of the SS framework. Our implementation is capable of obtaining high quality solutions even though it does not include some of the so-called advanced strategies, such as multi-tier reference sets, reference-set rebuilding, path relinking and the like. The article is organized as follows. In the following two sections, brief descriptions of SS and of VRPSDP are presented. Then, the SS approach to deal with the VRPSDP is described in detail. The experimental section presents computational results for the instances of Nagy and Salhi (2005), Dehtloff (2001), and Montané and Galvão (2006), which are an adaptation of those in Solomon (1987) and Gehring and Homberger (1999). We finish with concluding remarks in the last section.

The Scatter Search Methodology

SS is an evolutionary meta-heuristic that operates on a set of solutions, which the SS literature refers to as the reference set ($Refset$). The evolution of the $Refset$ is achieved by way of combining reference solutions to yield trial solutions with combination of attributes (e.g., particular sequence of customers) not present in the previous solutions. The $Refset$ is a collection of "good" solutions found during the search, where the meaning of "good" is not limited to quality as measured by the objective function value. For instance, a solution may be good because it provides diversity with respect to other solutions in the reference set. In fact, some implementations of SS divide the $Refset$ into two subsets, contributing, respectively, with solution quality and diversity.

Although SS shares some common elements with Genetic Algorithms, the fundamental difference between SS and GAs is philosophical: SS emphasizes the use of systematic instead of random strategies. SS offers a flexible framework from which to develop distinct implementations with different degrees of complexity. It has been applied to a fairly large number of combinatorial optimization problems, including the classical VRP (Sosa, Galvão, & Gandelman, 2007), the VRP with time Windows (Russel & Chiang, 2006) and the VRPSDP (Maquera-Sosa, Gandelman, & Sant'Anna, 2007).

SS was introduced by Glover (1977). A template was made available by Glover (1978), in a version customized for nonlinear optimization problems with continuous variables. Laguna and Martí (2003) published the first book on SS, containing introductory tutorials and advanced techniques such as the use of memory and path relinking.

A basic algorithm of SS consists of the following five methods:

- **Method 1:** A set *of* initial solutions is built by a Diversification Generation Method.
- **Method 2:** An Improvement Method is applied to modify trial solutions.
- **Method 3:** High-quality and highly-diverse solutions are chosen by a Reference Set Update Method.
- **Method 4:** A Subset Generation Method determines which subsets of reference solutions will serve as the basis for creating new solutions.
- **Method 5:** New solutions are generated by a Solution Combination Method.

The SS terminology that will be used in the following sections is the same as in Laguna and Marti (2003). P denotes the set of diverse solutions created by the Diversification Generation Method and $Psize$ denotes the size of the population P (i.e., $|P| = Psize$). $Refset$ denotes the reference set and its size is denoted by b (i.e., $|RefSet| = b$). The reference set contains two subsets, one with b_1 "high quality" solutions

denoted by $Refset_1$ and another with b_2 "diverse solutions" denoted by $Refset_2$, hence $b = b_1 + b_2$. Solutions in the reference set are denoted by x^j, for j varying from 1 to b. Within the reference set, we assume that x^1 is the best solution according to the objective function value and that x^b is the worst. $d(x, y)$ denotes the distance between solution x and solution y. S denotes a subset of solutions chosen from $Refset$ and $Pool$ denotes the set of solutions generated by the combination method.

Scatter search procedures proceed iteratively, as shown in Figure 1. The solutions generated either by the diversification generation method (step 1) or the combination method (step 7) are subjected to the improvement method (steps 2 and 8). The use of *Pool* indicates a static update of the reference set (see the Reference Set Update subsection). That is, the reference set solutions are not changed until all solutions combinations are performed as prescribed by the subset generation method. The method ends when no new solution is admitted in the reference set after the execution of steps 5-9 in the outline of Figure 1.

Departures from the basic framework presented in Figure 1 include a mechanism such as "rebuilding" that adds an outer loop to allow the use of the diversification method to generate new reference solutions. These and other variants of the scatter search methodology are found in Laguna and Marti (2003).

The Vehicle Routing Problem with Simultaneous Delivery and Pickup

VRPSDP is a variant of the classical VRP (with capacity restrictions and maximum route time). It consists of determining minimal cost routes of delivery and pickup for a fleet of vehicles in such a way that all customers are visited. The deliveries originate from a central warehouse that is also the final destination of the items picked up. The items supplied by the warehouse are typically

Figure 1. Basic scatter search framework

---

1. **Diversification Generation** — Build a set P of $Psize$ diverse solutions
2. **Improvement Method** — Apply an improvement heuristic to the solutions in P
3. $Refset$ **Update** — Select the b_1 best solutions in P and b_2 solutions in P for diversity
4. Set $Newsolutions = TRUE$
while ($Newsolutions$)
{
     5. $Newsolutions = FALSE$
     6. **Subset Generation** – Create the subsets of reference solutions to be combined
     7. **Solution Combination** – Apply the combination method to all subsets of solutions generated in the previous step and add the new trial solutions to $Pool$
     8. **Improvement Method** — Apply an improvement heuristic to the solutions in $Pool$
     9. $Refset$ **Updating** - Update $Refset$ with b solutions in $Refset \cup Pool$
     if ($Refset$ changed) set $Newsolutions = TRUE$
}

---

different from those brought from the customers to the warehouse. The fleet is assumed to be homogeneous. Each customer i is identified by delivery and pickup demands and sometimes a service time. It is assumed that the total demand of a single customer must be satisfied by a single vehicle in a single visit. The total load of any vehicle is bounded by the vehicle capacity. The total time to complete any route, including the travel times between locations and the service times at each location, is bounded by a time limit previously set. The objective is to minimize the total time (equivalently, distance traveled) to complete all deliveries and pickups, under the constraint that the vehicles must have enough capacity to carry all the delivery and pickup orders along the route.

Works in the literature on VRPSDP include Min (1989), Galvão and Guimarães (1990), Mosheiov (1998), Salhi and Nagy (1999), Dethloff (2001), Nagy and Salhi (2005), Montané and Galvão (2006), Ganesh and Narendran (2007), Bianchessi and Righini (2007), Wassan *et al.* (2008), Zachariadis *et al.* (2009), and Gajpal and Abad (2009). Min (1989), in the context of a distribution system for public libraries, is the first one to formulate the VRPSDP. The proposed heuristic, based on

clustering and routing, is applied to real problem with 22 libraries and 2 identical vehicles. Galvão and Guimarães (1990) extend the VRPSDP to add a restriction on the number of customers per route as well as a limit on the maximum distance between two adjacent customers. They considered heterogeneous vehicles and applied their procedure to problems of transporting personnel in the oil industry. Mosheiov (1998) investigates the properties of the special case of the VRPSDP with unit demand values for pickup and delivery. A mathematical formulation and two heuristics were developed and analyzed as part of this work. Dethloff (2001) proposes four insertion heuristics based on different insertion criteria. Nagy and Salhi (2005) develop a local search that allows certain degrees of infeasibility, introducing the concept of weak and strong infeasibility. Montané and Galvão (2006) describe the application of tabu search to the VRPSDP. The procedure uses a neighborhood search composed of three different neighborhoods. Long-term memory based on frequency is used to induce a balance between search intensification and diversification. Bianchessi and Righini (2007) also implement tabu search with several neighborhoods of variable complexity. Wassan *et al.* (2008) apply reactive tabu search, while

Zachariadis *et al.* (2009) suggest a hybrid tabu search and guided local search procedure. Gajpal and Abad (2009) develop construction and local search procedures within the framework of ant colony optimization. The procedure is capable of tackling instances of the VRPSDP and the VRP with backhaul and mixed load.

Regarding the application of SS in the context of vehicle routing, the work of Rochat and Taillard (1995), which present a probabilistic technique to diversify and intensify a local search, should be mentioned. Such technique is applied to two different kinds of VRPs: the classical VRP and the VRP with time windows. It is composed of two phases: (1) generation of initial solutions and (2) identification of (good and bad) tours, common to the initial solutions, to generate new solutions. A local search starts from the set of new solutions, followed by a post-optimization.

Rego (2000) applies SS to the classical VRP and combines solutions from the reference set by means of a common arcs mechanism. Corberán *et al.* (2002) developed an efficient SS procedure to solve a multiobjective school bus routing problem in a rural area. Russell and Chiang (2006) applied SS to a VRP with time windows, studying the influence of parameters of the reference set such as total size, as well as the mix of quality and diverse solution. The common arcs mechanism and an adjusted model with a recovery feature are employed to combine solutions. Reactive tabu search is employed as the improvement method. Computational results were obtained showing that the SS implementation is competitive with other solution methods for this problem.

Sosa, Galvão, and Gandelman (2007) developed an application of SS to the classical VRP. Their work tests two diversification generation procedures and three improvement methods. The combination method operates on pairs of solutions, which are combined by a mechanism that identifies customers that are common in routes of the solutions being combined. Computational experiments on four datasets available in the literature were performed and results for several parameter values for the reference set, as well as for different updating criteria, were obtained. One of the main results was the confirmation that SS exhibits a robust behavior when the *RefSet* is relatively small ($b \approx 10$) and it is initially constructed with equal number of high-quality and diverse solutions (i.e., $b_1 = b_2$).

Scatter Search for the VRPSDP

This section describes our application of SS to the VRPSDP. Of the five scatter search methods shown in Figure 1, we discuss four in detail (diversification generation, improvement, reference set update and combination). The subset generation method (step 6 in Figure 1) that we employ is the so-called "type 1", which consists of all pairs of solutions (see page 107 in Laguna & Martí, 2003). The method is such that it generates all pairs of solutions that contain at least one new element. That is, assume that the reference set a given iteration consists of five solutions (two of which were added in the previous iteration). Then, the subset generation method would create a list of 7 pairs of reference solutions. Note that the total number of pairs that can be generated with a set of five reference solutions is 10. However, if the set contains only two new solutions, then there are only 7 pairs with at least one new solution in them. The other three pairs do not contain any new solutions and therefore they were already examined in the previous iteration.

For the diversification generation method, we tested four constructive heuristics before choosing one. We provide a detailed description of the three improvement methods that we implemented and that employ several types of inter-routes and intra-routes exchanges. In addition, we describe and illustrate the distance function that is used to measure solution diversity, which is applied when building the first reference set (step 3 in Figure 1).

Figure 2. Outline of the diversification method

---

1. Set $k = 1$ and $i = 0$
2. If all customers are already assigned then stop, otherwise identify all feasible customers for the current route. If there are no feasible customers then make $k = k + 1$ and $i = 0$ and all currently unassigned customers are feasible.
3. Calculate $c(j)$ for all feasible customers and randomly choose one for which $c(j) \leq c_{min} + \alpha(c_{max} - c_{min})$
4. Update capacity and time limits for the current route. Go to 2.

Diversification Method

The GRASP methodology is employed as the basis for generating diverse solutions. GRASP is a multi-start procedure, with each iteration consisting of a construction and an improvement step. Solutions are constructed following the guidance of a greedy function and controlled randomization. Solutions are subsequently improved by the improvement method. Details on GRASP and numerous applications as well as hybridized approaches can be found in http://www.research. att.com/~mgcr/papers.html.

For the purpose of this study, we created a GRASP based on a simple greedy function that measures the incremental distance of adding a customer to an existing or new route. New routes are created when capacity or time limits are reached. The procedure starts by opening the first route (i.e., setting $k=1$) and then calculating the value of the greedy function for all customers in the problem. Let i be the index of the last customer visited by vehicle k in the current partial solution. Also, let j be a candidate customer to be added to the route, immediately after customer i. Then, the greedy function value associated with customer j is given by:

$$c(j) = d_{ij} + d_{j0} - d_{i0}$$

where d_{ij} is the distance between the ith and the jth customer and 0 denotes the warehouse. For new routes, $i=0$ and the value of the greedy function is simply the round trip distance from the warehouse to the customer that is being considered as the first stop. Once the first customer has been assigned to a route, the procedure considers only those unassigned customers for which its addition to the current route does not violate either the load or the time constraint. From those "feasible" candidate customers, the procedure calculates $c(j)$ as well as c_{min} and c_{max}, the minimum and maximum values of c in the set of feasible customers. Then, a customer is randomly chosen among the feasible candidates for which:

$$c(j) \leq c_{min} + \alpha(c_{max} - c_{min})$$

If no customers can be added without keeping the current route feasible, then a new route is added (by increasing the value of k). An outline of the procedure is shown in Figure 2. This procedure is called $Psize$ times in order to generate the solutions to be included in the initial population P.

Preliminary experiments showed that a good balance between diversity and quality of the solutions generated with this method could be achieved by setting α to 0.5. With the goal of testing the robustness of the GRASP implementation with

Figure 3. Diversity and quality of solutions obtained as a function of α

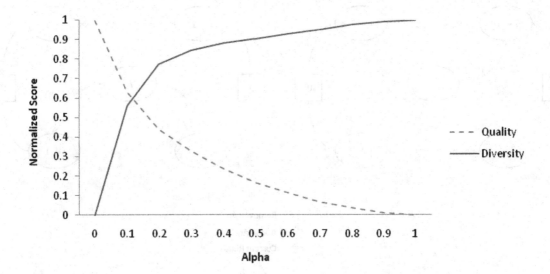

respect to the value of α, we performed more than 28,000 runs. The experiments consisted of trying values of α from 0.0 to 1.0 in increments of 0.1, performing 30 runs on 86 problem instances. We then measured the diversity (with the metric described in the Reference Set Update subsection) and quality (as the objective function value gap against the best known) of the solutions obtained with the different values of this parameter. Gap values here and elsewhere are calculated as the percentage deviation from the best known solution. Figure 3 shows the result of this experiment, in which we have normalized the diversity measures and the gap values to be in the same scale (0 to 1).

As expected, the quality of the solutions found deteriorates as the value of alpha increases. The best quality (with a normalized score of 1) is achieved with $\alpha = 0$ and has an average gap value of 16.2%, while the worst quality (with a normalized score of 0) has an average gap value of 27.0%. On the other hand, diversity increase with the value of α, as additional randomness is allowed in the process. Since diversity is more important than quality at the beginning of the

search, we selected a value of $\alpha = 0.5$ for subsequent experimentation, which is the first point where the normalized diversity score is above 90%.

Improvement Method

This section describes the mechanisms employed to explore the neighborhoods of solutions within the improvement method. Three variants of the improvement method were studied. Each variant includes three inter-routes moves (relocation, interchange and crossover) and three intra-route moves (2-opt, relocation and 1-interchange). The inter-route moves known as relocation and interchange were defined in articles describing "first generation" heuristics for the VRP. See, for instance, Ball, Magnanti, Monma, and Nemhauser (1995). Also, Psaraftis (1983) defined *k*-interchange moves for local search algorithms in VRP with precedence relations.

Relocation, as an inter-route move, consists of moving one customer from one route to another, as shown in Figure 4. The example contains two routes, where the warehouse is represented

Figure 4. Relocation movement

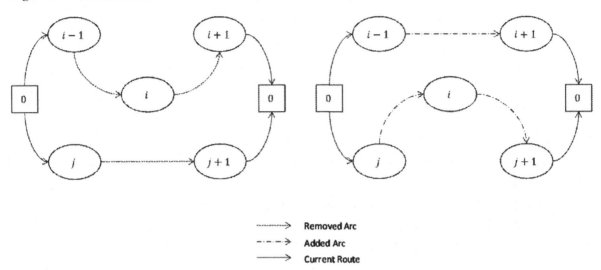

---------->	Removed Arc
--- --->	Added Arc
———>	Current Route

by an 0 inside of a square. In the figure, customer i is changed from one route to another. This move may also be viewed in terms of the changes in the corresponding arcs. Specifically, when customer i is moved to the position immediately after customer j, arcs $(i-1, i)$, $(i, i+1)$ and $(j, j+1)$ are replaced, respectively, with arcs (j, i), $(i, j+1)$ and $(i-1, i+1)$.

1-*Interchange*, as an inter-route move, consists of exchanging two customers belonging to two different routes. The two routes in Figure 5 show how customers i and j are interchanged. This move causes arcs $(i-1, i)$, $(i, i+1)$, $(j-1, j)$ and $(j, j+1)$ to be replaced, respectively, with arcs $(i-1, j)$, $(j, i+1)$, $(j-1, i)$ and $(i, j+1)$.

Crossover is an inter-route move that is applied to a pair of routes. Each route is divided into two

Figure 5. 1-Interchange movement

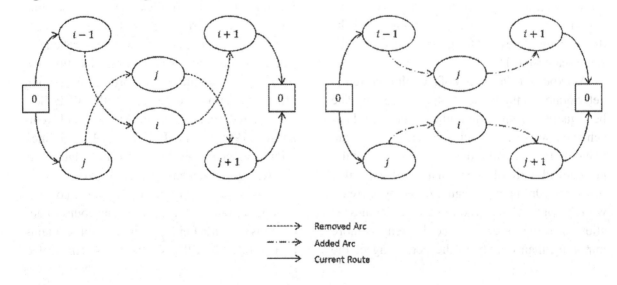

---------->	Removed Arc
--- --->	Added Arc
———>	Current Route

Figure 6. Crossover movement

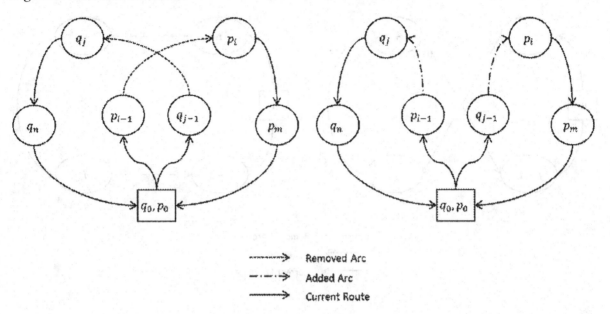

sections by removing an arc in each route and substituting it with a new arc; these two new arcs connect, respectively, the initial section of the first route with the final section of the second and the initial section of the second with the final section of the first. Figure 6 illustrates how the move operates on a pair of routes. The move may also be described in terms of adding and deleting arcs. In particular, route $p_0 \rightarrow p_{i-1} \rightarrow p_i \rightarrow p_m \rightarrow p_0$ is replaced with route $p_0 \rightarrow p_{i-1} \rightarrow q_j \rightarrow q_n \rightarrow p_0$ and route $q_0 \rightarrow q_{j-1} \rightarrow q_j \rightarrow q_n \rightarrow q_0$ is replaced with route $q_0 \rightarrow q_{j-1} \rightarrow p_i \rightarrow p_m \rightarrow q_0$, where $|m|$ and $|n|$ are the lengths of the original routes and m and n are the indices of the last nodes before the depot.

The 2-opt move, proposed by Lin (1965), is an intra-route procedure that consists of substituting two nonadjacent arcs belonging to the route with two other arcs that improve the solution without breaking connectivity. Figure 7 shows how it works, with arcs (i, j) and $(i+1, j+1)$ substituted respectively with arcs $(i, i+1)$ and $(j, j+1)$. Intra-route relocation and 1-interchange

moves are similar to the inter-route counterpart (described above) with the difference that they are applied to a single route.

We use these neighborhoods to implement and test three variants of the improvement method: best overall neighbor, best current neighbor and first improvement. The best overall neighbor variant consists of searching for the best neighbor solution by examining all the neighbors produced by the move mechanisms described above with the exception of 2-opt. The best overall neighbor is the feasible solution with the best objective function value among all the neighbors that can be reached by the five move mechanisms. If no improvement of the objective function is possible by moving to any of the neighbors, then 2-opt is applied until a local optimum is reached.

The best current neighbor variant consists of searching five neighborhoods sequentially in the following order: inter-route relocation, inter-route 1-interchange, crossover, intra-route relocation and intra-route 1-interchange. In other words, the local search starts with the first neighborhood until no improvement is possible. Then it switches to

Figure 7. 2-opt movement

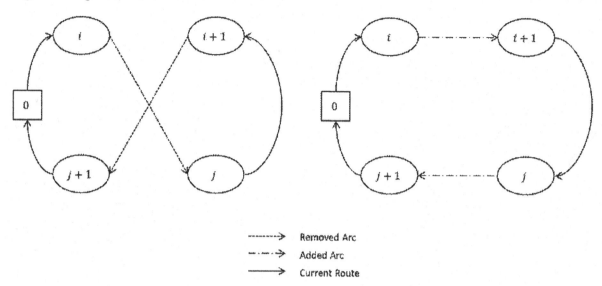

```
--------->   Removed Arc
-·--·-->     Added Arc
--------->   Current Route
```

the second neighborhood, and so on and so forth. When the last neighborhood is not capable of producing an improving move, the 2-opt procedure is applied until a local optimum is reached.

The first improvement variant is similar to best current neighbor except that when examining a neighborhood the procedure executes the first improving move. That is, the procedure does not examine the entire neighborhood to determine the best improving move. Instead, it executes the first improving move and continues the examination of the remaining moves in a way that mimics a circular list. When the entire list of moves is examined and no move improves the objective function, the procedure switches to the next neighborhood. When all five mechanisms have been examined, the 2-opt neighborhood is explored until a local optimum is reached.

We perform an experiment consisting of generating 10 initial solutions for the 28 instances in Salhi and Nagy (1999). We measured the solution quality and the computational effort associated with the three variants of the improvement method. We also tested the variants with and without the use of the crossover neighborhood. The results of this experiment are summarized in Table 1.

Considering both quality and computational effort, we chose as our best option the "best overall neighbor" method without crossover. In addition, this preliminary experimentation allowed us to determine that the crossover method increases the computational time with minimal objective function improvement.

Table 1. Quality and computational effort of several combination methods

Variant	Quality (GAP)	Computational effort (average in seconds)
Best overall neighbor without crossover	0.52%	2.14
Best overall neighbor with crossover	0.62%	1.32
Best current neighbor without crossover	0.69%	1.98
Best current neighbor with crossover	0.72%	1.35
First improvement with crossover	0.54%	2.27
First improvement without crossover	0.75%	1.28

Reference Set Update Method

The SS method of building the *Refset* (line 3 in Figure 1) starts by choosing — among the *Psize* solutions in the set P resulting from the two first steps in Figure 1 — the b_1 solutions with the best objective function values. These solutions are added to *Refset* and eliminated from P. Then, additional b_2 solutions are chosen and added to the *Refset* in such a way as to maximize diversity. To measure diversity, the distance between two solutions, x and y, must be calculated. In order to do this, we create a square matrix *ClientCount* of size $r \times r$, where r is the maximum number of routes in either x or y. We then populate the matrix in such a way that $ClientCount(x_i, y_j)$ contains the number of clients that are common to the ith route in solution x and the jth route in solution y. Once the matrix is populated, a process begins that consists of finding the largest *ClientCount* value, accumulating this value and then deleting the corresponding column and row. The process finishes when all columns and rows have been eliminated. The cumulative count of common clients in routes is then subtracted from the total number of clients and the difference gives a measure of distance between the two routes.

To illustrate this distance calculation, consider the following two solutions:

- x : (3, 7) (4, 5, 8, 6) (10, 2, 9, 1)
- y : (7, 9, 6, 8, 5) (1, 2, 3, 4, 10)

The value of r is 3 and the resulting *ClientCount* matrix has the following form:

	y_1	y_2	y_3
x_1	1	1	0
x_2	3	1	0
x_3	1	3	0

After selecting elements (x_3, y_2) and (x_2, y_1), in any order, and eliminating their corresponding rows and columns, the only element left is (x_1, y_3), which has a count of zero. The total count of common clients in routes is then 6 and the distance $d(x, y)$ between solutions x and y is given by $10 - 6 = 4$.

The distances between all solutions x in P and all solutions y in *Refset* are calculated as described above. Then, the solution x in P that maximizes the minimum distance between itself and all the solutions in *Refset* is added to the *Refset* and eliminated from P. This is repeated until b_2 solutions are selected and the *Refset* consists of a total of b solutions, b_1 of them selected by quality and b_2 selected by diversity.

We use a standard method for the reference set update that occurs in step 9 of Figure 1, the so-called static update (see page 91 in Laguna & Martí, 2003). In this updating, a pool of solutions is constructed by including the current reference solutions plus all the trial solutions generated by the combination method. The new reference set is chosen as the best solutions (according to the objective function value) from the pool of solutions. The method is called "static" because the reference solutions are not replaced until all trial solutions are generated.

Combination Method

The combination method operates in a pair of solutions that is taken from the list of solution pairs generated by the subset generation method. The list of solution pairs is such that at least one of the two solutions being combined must be a new solution (i.e., a solution that has been admitted in the reference set in the last iteration). Initially, all solutions are new and therefore all pairs of solutions are processed by the combination method. The combination method starts by finding a "route match" and then extracts the common clients to

Figure 8. Illustration of the combination method

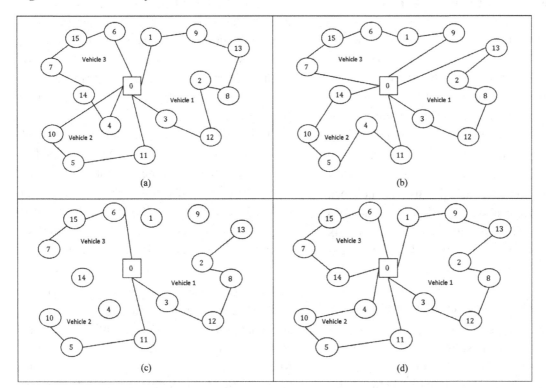

form partial routes or segments. A route in one solution is matched with the one in another solution that has the largest number of common clients, where ties are broken arbitrarily. Consider, for instance, the two reference solutions depicted in Figures 8(a) and 8(b). For ease of illustration, we have that vehicles 1, 2 and 3 in 8(a) match with the corresponding vehicles in 8(b). The common clients for these vehicles are (2, 3, 8, 12, 13), (5, 10, 11) and (6, 7, 15), respectively. The resulting partial solution formed with these common clients is shown in Figure 8(c).

In order to create a complete solution, the unassigned clients — in this case (1, 4, 9, 14) — are inserted in the existing (partial) routes. The insertion is greedy in the sense that we attempt to minimize the incremental distance of adding the unassigned clients to an existing route. In other words, each unassigned client is tentatively inserted in each position in the partial solution. The feasible insertion that results in the least in-

crease of the objective function is chosen. This is done until all unassigned clients have been inserted.

By the nature of the process, it may happen that unassigned clients cannot be inserted in any of the existing routes without violating one or more constraints. This may force the creation of a new route where the unassigned clients will be inserted. In the illustration of Figure 8, clients 1 and 9 are inserted in route 1, client 4 is inserted in route 2 and client 14 is inserted in route 3. The resulting solution is shown in Figure 8(d).

The trial solution that results from the combination of two reference solutions is always feasible. However, by insisting on preserving feasibility, the resulting solution may be inefficient with respect to the use of available vehicles. This is why the scatter search framework includes the application of the improvement method to the trial solutions that result from the execution of the combination method.

Computational Results

To evaluate the merit of our SS implementation we compared our results with the best reported in the literature. Three datasets available in the literature were employed. The first has 50 to 199 customers and consists of 28 instances introduced by Salhi and Nagy (1999). Originally, these are the 14 instances of the classical VRP—introduced by Christofides *et al.* (1979) — that include capacitated vehicles, service times at the customer sites, and a time limit for each route. According to these characteristics, the entire set may be divided into four subsets: 1) time limit and no service time, 2) time limit and service time, 3) no time limit and no service time and iv) no time limit and service time. For the VRPSDP, the instances in Salhi and Nagy (1999) are divided into two subsets (X and Y), where X includes the instances without time limit and Y includes the instances with time limit. See in Nagy and Salhi (2005) for additional details.

The second data set consists of the 40 instances introduced by Dethloff (2001) with 50 customers each. The third data set consists of 18 instances proposed by Montané and Galvão (2006), derived from the Solomon instances and the Extended Solomon instances — see Solomon (1987) and Homberger (1999). The size of these instances ranges between 100 and 400 customers.

The instances of Salhi and Nagy (1999) were employed for fine-tuning purposes to analyze the effect of the different components of our SS implementation. These components include the diversification generation method as well as the improvement methods. The size of the population P and the reference set was also studied. The fine-tuning process resulted in the following settings:

$$Psize = \min(30, n)$$

$$Refset: (b_1 = 5, b_2 = 5) \text{ and } (b_1 = 10, b_2 = 10)$$

$$\alpha = 0.5$$

- **Improvement Method:** Best overall neighbor.

Our results are reported using Euclidean distances without rounding (unless otherwise specified). The codes were implemented in C++ and all experiments were run on a personal computer with a 2 GHz Athlon processor. The distance calculation with or without rounding has a major impact on the performance comparison. Therefore, we limit our comparisons to those results for which the authors explicitly state that their distance calculations are done without rounding.

We start our performance assessment with the results from the Salhi and Nagy instances. For this set, we detected another complicating factor. Namely, some authors have solved these instances without considering the time limit on the routes (Nagy & Salhi, 2005; Dethloff, 2001; Crispim & Brandão, 2005; Chen & Wu, 2006; Montané & Galvão, 2006), while other have solved them with the time limit in place (Nagy & Salhi, 2005; Dethloff, 2001; Montané & Galvão, 2006; Wassan *et al.*, 2008). Furthermore, some authors include the service time (Nagy & Salhi, 2005; Dethloff, 2001) and some do not (Montané & Galvão, 2006; Wassan *et al.*, 2008). To complicate matters even more, Nagy and Salhi (2005) allow for two visits to each customer, one for delivery and one for pickup. Therefore, for this data set, we limit our comparison to the results of Montané and Galvão (2006) and Wassan *et al.* (2008), both of which use a time limit on the routes, no service time and a single visit to each customer (i.e., simultaneous delivery and pickup).

Before making comparisons with existing methods, we ran an experiment with the Salhi and Nagy instances to determine the best values for b_1 and b_2. This experiment resulted in the setting of $b_1 = b_2 = 5$, which when compared to a setting of $b_1 = b_2 = 10$ produced almost identical solution quality but the computational time favored the setting with the smaller *Refset*. Since our scatter search implementation contains stochastic ele-

Table 2. Comparison against published results of the Salhi and Nagy (1999) dataset

Instance	Rounded Data					Unrounded Data				
	Scatter Search			Montané and Galvão	P(match or improve)	Scatter Search			Wassan. et al	P(match or improve)
	Min	Max	Average			Min	Max	Average		
CMT1X	472	483	472.49	472	8.2%	473.13	483.55	475.47	468.30	0.0%
CMT1Y	463	485	473.20	470	54.8%	458.96	498.22	475.14	458.96	3.3%
CMT2X	689	702	691.95	695	38.2%	690.37	702.33	691.34	668.77	0.0%
CMT2Y	668	699	683.92	700	100.0%	663.25	705.09	690.51	663.25	2.8%
CMT3X	728	744	728.37	721	0.0%	729.04	752.88	730.94	729.63	1.6%
CMT3Y	719	755	720.61	719	32.2%	743.71	758.12	745.27	745.46	28.9%
CMT4X	855	880	861.37	880	100.0%	857.39	881.26	861.23	876.5	79.9%
CMT4Y	857	882	866.64	878	92.4%	859.14	894.84	864.19	870.44	55.3%
CMT5X	1040	1068	1041.95	1098	100.0%	1041.71	1075.87	1042.12	1044.51	85.3%
CMT5Y	1050	1085	1068.28	1083	94.1%	1051.21	1100.05	1058.82	1054.46	42.7%
CMT6X	472	498	473.11	476	28.9%	473.77	501.41	475.45	471.89	0.0%
CMT6Y	474	488	474.89	474	18.1%	470.40	490.42	473.62	467.7	0.0%
CMT7X	669	712	688.96	695	68.3%	669.51	727.15	686.32	663.95	0.0%
CMT7Y	667	701	688.25	700	97.4%	668.11	710.43	670.82	662.5	0.0%
CMT8X	729	756	730.84	720	0.0%	730.61	769.02	733.67	726.88	0.0%
CMT8Y	720	754	721.93	721	43.1%	720.72	760.42	741.96	741.96	63.7%
CMT9X	858	899	862.55	885	91.2%	858.16	901.53	880.79	880.61	44.7%
CMT9Y	866	905	885.46	900	95.2%	866.34	915.85	875.93	886.84	71.2%
CMT10X	1061	1165	1083.17	1100	31.2%	1062.25	1174.88	1082.66	1079.99	60.7%
CMT10Y	1059	1086	1070.28	1083	97.3%	1061.26	1093.04	1071.73	1058.09	0.0%
CMT11X	866	899	883.08	900	100.0%	867.99	904.14	880.60	861.97	0.0%
CMT11Y	857	884	877.18	910	100.0%	858.95	897.87	874.34	830.39	0.0%
CMT12X	677	705	687.10	675	0.0%	677.56	707.65	681.78	644.70	0.0%
CMT12Y	675	691	678.81	689	84.1%	677.27	705.90	681.48	659.52	0.0%
CMT13X	883	945	900.51	918	78.3%	883.77	959.96	888.06	858.48	25.6%
CMT13Y	861	889	866.04	910	100.0%	858.99	894.20	860.87	880.56	91.3%
CMT14X	679	702	681.24	675	0.0%	681.18	708.30	681.24	644.70	0.0%
CMT14Y	662	695	677.45	689	87.3%	664.03	703.06	672.86	659.52	0.0%
Average	**760**	**791**	**769.27**	**780**	**67.8%**	**761.38**	**799.20**	**769.62**	**759.30**	**17.8%**

ments, our experimentation was based on executing SS 30 times on each problem instance. Tables 2 through 4 reporting our results include the minimum (Min), the maximum (Max) and the average values of the objective functions associated with the 30 runs. The tables also include the benchmark objective function value that we are using for comparison. Additionally, we calculate the empirical probability that any given run would match or improve upon the benchmark. This is done by simply counting the number of times that SS matches or improves upon the benchmark

Table 3. Results for the Dethloff (2001) dataset

Instance	Scatter Search $b_1 = b_2 = 5$			Montané and Galvão (2006)	P(match or improve)
	Min	Max	Average		
SCA3-0	637.10	640.74	640.30	640.55	98.8%
SCA3-1	697.14	704.63	698.64	697.84	5.3%
SCA3-2	655.15	655.15	655.15	659.34	100.0%
SCA3-3	674.58	680.19	678.98	680.04	96.9%
SCA3-4	690.19	692.07	690.63	690.50	3.7%
SCA3-5	656.74	657.53	656.82	659.90	100.0%
SCA3-6	650.12	653.57	651.24	653.81	100.0%
SCA3-7	659.17	674.77	674.50	659.17	2.8%
SCA3-8	702.20	712.31	708.91	719.47	100.0%
SCA3-9	681.00	683.74	683.57	681.00	7.5%
SCA8-0	980.75	997.62	985.05	981.47	18.5%
SCA8-1	1077.44	1126.37	1117.90	1077.44	3.7%
SCA8-2	1019.60	1054.28	1038.63	1050.98	93.8%
SCA8-3	983.34	991.7	987.23	983.34	7.6%
SCA8-4	1064.56	1069.12	1067.39	1073.46	100.0%
SCA8-5	986.63	999.83	988.26	1047.24	100.0%
SCA8-6	943.93	959.95	948.08	995.59	100.0%
SCA8-7	1068.56	1082.11	1079.40	1058.56	0.0%
SCA8-8	1080.58	1090.76	1089.12	1080.58	5.0%
SCA8-9	1078.82	1090.18	1085.04	1084.80	53.7%
CON3-0	621.82	627.79	624.93	631.39	100.0%
CON3-1	554.47	563.56	562.94	554.47	6.7%
CON3-2	515.73	521.38	516.43	522.86	100.0%
CON3-3	591.19	596.84	595.11	591.19	3.9%
CON3-4	589.73	591.42	590.74	591.12	78.4%
CON3-5	558.49	568.33	566.62	563.70	47.2%
CON3-6	490.91	505.49	493.36	506.19	100.0%
CON3-7	577.68	588.22	587.05	577.68	2.7%
CON3-8	508.71	513.92	510.03	523.05	100.0%
CON3-9	578.37	588.91	583.90	580.05	29.5%
CON8-0	860.48	905.09	894.63	860.48	4.9%
CON8-1	740.85	776.37	759.69	740.85	2.9%
CON8-2	691.42	706.62	695.34	723.32	100.0%
CON8-3	811.23	841.47	838.25	811.23	2.8%
CON8-4	772.25	823.51	818.84	772.25	6.0%
CON8-5	756.91	766.52	764.98	756.91	3.9%

continued on following page

Table 3. Continued

Instance	Scatter Search $b_1 = b_2 = 5$			Montané and Galvão (2006)	P(match or improve)
	Min	Max	Average		
CON8-6	646.12	652.27	646.73	678.92	100.0%
CON8-7	814.50	874.19	870.38	814.50	4.2%
CON8-8	728.90	751.72	737.12	775.59	100.0%
CON8-9	809.00	816.82	815.85	809.00	2.7%
Average	**755.16**	**769.93**	**764.94**	**764.00**	**49.8%**

objective value (i.e., the objective value produced by the method used for comparison) and dividing it by the number of runs (i.e., 30). This empirical probability is labeled as "P(match or improve)" in the tables.

The results in Table 2 show that our procedure performs much better on rounded data than unrounded data. For the rounded data, we calculated a probability of almost 68% of at least matching the benchmark. This includes the estab-

Table 4. Results for the Montané and Galvão (2006) dataset

Instance	Scatter Search $b_1 = b_2 = 5$			Montané and Galvão (2006)	P(match or improve)
	Min	Max	Average		
r101	1032.8	1054.26	1040.55	1042.60	47.9%
r201	673.01	705.41	688.17	671.03	0.0%
c101	1234.3	1267.32	1239.77	1259.79	73.4%
c201	663.09	706.15	673.31	666.01	29.3%
rc101	1066.5	1100.77	1074.50	1094.15	61.7%
rc201	672.92	708.98	677.11	674.46	33.4%
r1_2_1	3391.6	3580.24	3419.78	3447.20	69.1%
r2_2_1	1676.5	1703.67	1686.48	1690.67	66.3%
c1_2_1	3699.5	3757.87	3712.67	3792.62	100.0%
c2_2_1	1741.9	1797.93	1781.29	1767.58	18.7%
rc1_2_1	3446.71	3497.92	3458.87	3427.19	0.0%
rc2_2_1	1570.14	1687.33	1601.32	1645.94	62.4%
r1_4_1	9963.43	10273.50	10024.69	10027.81	61.2%
r2_4_1	3667.1	3794.40	3721.14	3695.26	55.5%
c1_4_1	11745	11983.70	11764.45	11676.27	0.0%
c2_4_1	3661.2	3778.72	3701.91	3732.00	61.4%
rc1_4_1	10005.87	10233.30	10102.75	9883.30	0.0%
rc2_4_1	3539.1	3628.51	3584.60	3603.50	69.1%
Average	**3525.04**	**3625.55**	**3552.97**	**3544.30**	**45.0%**

lishment of 21 new benchmarks in this set of 28 problem instances. The SS also found 11 new best-known solutions to the 28 instances in unrounded data set. We note that the worst solutions obtained in the 30 runs have an average deviation of about 5.25% from the benchmark values for the unrounded data and 1.4% for the rounded data.

For the Dethloff (2001) and Montané and Galvão (2006) datasets, we compare the $(b_1 = 5, b_2 = 5)$ version of our SS procedure against the results of Montané and Galvão (2006) given that no results from Wassan *et al.* (2008) are available for these problem instances. Therefore the best results available for these problem instances are those by Montané and Galvão (2006). Table 3 shows the results associated with the Dethloff problems instances and Table 4 shows the results for the Montané and Galvão instances.

The results in Table 3 show that SS is able to find 24 new best solutions. The probability of at least matching the benchmark solutions is almost 50% and in 14 cases all runs found a better solution than the one previously published. The average deviation of the worst solutions found is only 0.78% from the benchmark values. Since the average time to find the solutions reported in this table is 3.12 CPU seconds, with no time exceeding 6.5 seconds, a practical strategy would be to run the procedure more than once to increase the probability of finding better solutions.

Table 4 shows the results of comparing the best known solutions for the problems introduced by Montané and Galvão (2006) and the outcomes from our SS implementation.

The results show that for any given run the probability of matching the benchmark values is only 45%. However, the table also shows that the experiment yielded 14 new best solutions in this set of 18 problem instances. The average deviation from the benchmark solutions of the worst solutions is only 2.29%. This data set contains 6 instances with 100 customers, 6 instances with 200 customers and 6 instances with 400 customers. The average time to find the SS solutions reported in Table 4 is 48.6 seconds, 1217.7 seconds and 2754.3 seconds, for each of the problem sizes respectively.

It is difficult to compare the computational effort across the procedures for the VRPSDP because the results have been obtained employing equipment of different characteristics and the experiments themselves have not been homogenous (i.e., some considered time restrictions, service time, and fractional or integer values for distance and demand). Galvão (2006) and Gajpal and Abad (2009) consider a distance limit on the routes. Gajpal and Abad (2009) consider service times. Crispim and Brandão (2005), Chen and Wu (2006) and Gajpal and Abad (2009) use fractional distance values and Chen and Wu (2006) and Tang and Galvão (2006) use rounded demand values.

Aside from the difficulties associated with producing an "exact" comparison of the computational times required by each procedure, the relevant question is whether these implementations are able to provide their reported best solutions within the timeframe that the context requires. In other words, if in practice the available time to provide a high quality solution to a particular instance of the VRPSDP is say 5 minutes, then there is no practical difference between a procedure that does it in 1 minute and another one that requires 2 minutes. Time differences (even small ones) are critical in situations where the problem being solved is embedded in a system that requires real-time response and a large number of instances must be solved continuously. This is not the case for the VRPSDP and all the procedures in the literature and our SS are certainly capable of providing their best solutions within the time that practical context requires.

CONCLUSION

We have described the development of a SS procedure to deal with the VRPSDP. The importance of the VRPSDP has been well documented in the areas of Operational Research, Reverse Logistics and Transportation. The difficulty of this version of the general vehicle routing problem is also known. Our development includes new ways of generating diverse solutions as well as a mechanism to measure distance between two solutions. Our method also includes a new way of integrating several mechanisms to produce trial solutions out of the combination of two reference solutions. Computational experiments were performed for three data sets available in the literature and SS was shown to be a robust alternative for solving VRPSDP instances.

Through the application of the SS method that we developed, 70 new best solutions were found out of the 114 instances studied (counting the rounded and unrounded instances in Table 2 as two separate data sets). When SS failed to improve upon published results, the solutions obtained yielded small relative percent deviations.

ACKNOWLEDGMENT

The authors thank CNPq for its financial support.

REFERENCES

Ball, M. O., Magnanti, T. L., Monma, C. L., & Nemhauser, G. L. (1995). *Network routing: Handbooks in operations research and management science* (*Vol. 8*). Amsterdam, The Netherlands: North Holland.

Bianchessi, N., & Righini, G. (2007). Heuristic algorithms for the vehicle routing problem with simultaneous pick-up and delivery. *Computers & Operations Research, 34*, 578–594. doi:10.1016/j.cor.2005.03.014

Breedam, A. V. (1995). Vehicle routing: Bridging the gap between theory and practice. *Belgian Journal of Operations Research . Statistics and Computer Science, 35*, 63–80.

Campos, V., Glover, F., Laguna, M., & Martí, R. (2001). An experimental evaluation of a scatter search for the linear ordering problem. *Journal of Global Optimization, 21*, 397–414. doi:10.1023/A:1012793906010

Carter, C., & Ellram, L. (1998). Reverse logistics: A review of the literature and framework for future investigation. *Journal of Business Logistics, 19*, 85–102.

Chen, J. F., & Wu, T. H. (2006). Vehicle routing problem with simultaneous deliveries and pickups. *The Journal of the Operational Research Society, 57*, 579–587. doi:10.1057/palgrave.jors.2602028

Christofides, N., Mingozzi, A., & Toth, P. (1979). The vehicle routing problem . In Christofides, N., Mingozzi, A., Toth, P., & Sandi, C. (Eds.), *Combinatorial optimization* (pp. 315–338). Chichester, UK: John Wiley & Sons.

Corberán, A., Fernández, E., Laguna, M., & Martí, R. (2002). Heuristic solutions to the problem of routing school buses with multiple objectives. *The Journal of the Operational Research Society, 53*, 427–435. doi:10.1057/palgrave.jors.2601324

Crispim, J., & Brandão, J. (2005). Metaheuristics applied to mixed and simultaneous extensions of vehicle routing problems with backhauls. *The Journal of the Operational Research Society, 56*, 1296–1302. doi:10.1057/palgrave.jors.2601935

Dethloff, J. (2001). Vehicle routing and reverse logistics: The vehicle routing problem with simultaneous delivery and pick-up. *OR-Spektrum, 23*, 79–96. doi:10.1007/PL00013346

Feo, T., & Resende, M. G. C. (1995). Greedy randomized adaptive search procedures. *Journal of Global Optimization, 2*, 1–27.

Gajpal, J., & Abad, P. L. (2009). Multi-ant colony system (MACS) for a vehicle routing problem with backhauls. *European Journal of Operational Research, 196*, 102–117. doi:10.1016/j.ejor.2008.02.025

Galvão, R. D., & Guimarães, J. (1990). The control of helicopter operations in the Brazilian oil industry: Issues in the design and implementation of a computerized system. *European Journal of Operational Research, 49*, 266–270. doi:10.1016/0377-2217(90)90344-B

Ganesh, K., & Narendran, T. (2007). CLOVES: A cluster-and-search heuristic to solve the vehicle routing problem with delivery and pick-up. *European Journal of Operational Research, 178*, 699–717. doi:10.1016/j.ejor.2006.01.037

Gehring, H., & Homberger, J. (1999). A parallel hybrid evolutionary metaheuristic for the vehicle routing problem with time windows. In *Proceedings of the EUROGEN Conference on Evolutionary Algorithms in Engineering and Computer Science* (pp. 57-64).

Glover, F. (1977). Heuristics for integer programming using surrogate constraints. *Decision Sciences, 8*, 156–166. doi:10.1111/j.1540-5915.1977.tb01074.x

Glover, F. (1998). A template for scatter search and path relinking. In J. K. Hao, E. Lutton, E. Ronald, M. Schoenauer, & D. Snyers (Eds.), *Proceedings of the Third European Conference on Artificial Evolution* (LNCS 1363, pp. 3-54).

Laguna, M., & Martí, R. (2003). *Scatter search: Methodology and implementations*. Boston, MA: Kluwer Academic.

Lin, S. (1965). Computer solutions of the traveling salesman problem. *The Bell System Technical Journal, 44*, 2245–2269.

Maquera Sosa, N. G., Gandelman, D., & Sant'Anna, A. (2007). Logística inversa y ruteo de vehículos: Búsqueda dispersa aplicada al problema de ruteo de vehículos con colecta y entrega simultanea. In *Proceedings of the 1er Congreso de Logística y Gestión de la Cadena de Suministro*.

Martí, R., Laguna, M., & Glover, F. (2006). Principles of scatter search. *European Journal of Operational Research, 169*, 359–372. doi:10.1016/j.ejor.2004.08.004

Min, H. (1989). The multiple vehicle routing problem with simultaneous delivery and pickup points. *Transportation Research, 23A*, 377–386.

Montané, F. A., & Galvão, R. D. (2006). A tabu search algorithm for the vehicle routing problem with simultaneous pick-up and delivery service. *Computers & Operations Research, 33*, 595–619. doi:10.1016/j.cor.2004.07.009

Mosheiov, G. (1994). The traveling salesman problem with pick-up and delivery. *European Journal of Operational Research, 79*, 299–310. doi:10.1016/0377-2217(94)90360-3

Nagy, G., & Salhi, S. (2005). Heuristic algorithms for single and multiple depot vehicle routing problems with pickups and deliveries. *European Journal of Operational Research, 162*, 126–141. doi:10.1016/j.ejor.2002.11.003

Psaraftis, H. (1983). K-interchange procedures for local search in a precedence-constrained routing problem. *European Journal of Operational Research, 13*, 391–402. doi:10.1016/0377-2217(83)90099-1

Rego, C. (2000). *Scatter search for vehicle routing problem.* Paper presented at the INFORMS National Meeting, Salt Lake City, UT.

Rochat, Y., & Taillard, E. D. (1995). Probabilistic diversification and intensification in local search for vehicle routing. *Journal of Heuristics, 1,* 147–167. doi:10.1007/BF02430370

Russell, R. A., & Chiang, W. (2006). Scatter search for the vehicle routing problem with time windows. *European Journal of Operational Research, 169,* 606–622. doi:10.1016/j.ejor.2004.08.018

Salhi, S., & Nagy, G. (1999). A cluster insertion heuristic for single and multiple depot vehicle routing problems with backhauling. *The Journal of the Operational Research Society, 50,* 1034–1042.

Scheuerer, S., & Wendolsky, R. (2006). A scatter search heuristic for the capacitated clustering problem. *European Journal of Operational Research, 169,* 533–547. doi:10.1016/j.ejor.2004.08.014

Shultmann, F., Zumkeller, M., & Rentz, O. (2006). Modeling reverse logistic task within closed-loop supply chains: An example from the automotive industry. *European Journal of Operational Research, 171,* 1033–1050. doi:10.1016/j.ejor.2005.01.016

Solomon, M. M. (1987). Algorithms for the vehicle routing and scheduling problems with time window constraints. *Operations Research, 35,* 254–265. doi:10.1287/opre.35.2.254

Sosa, N. G. M., Galvão, R. D., & Gandelman, D. A. (2007). Algoritmo de busca dispersa aplicado ao problema clássico de roteamento de vehiculos. *Pesquisa Operacional, 27,* 293–310. doi:10.1590/S0101-74382007000200006

Toth, P., & Vigo, D. (2002). *The vehicle routing problem: SIAM monographs on discrete, mathematics and applications.* Philadelphia, PA: Society for Industrial & Applied Science.

Wassan, N. A., Wassan, A. H., & Nagy, G. (2008). A reactive tabu search algorithm for the vehicle routing problem with simultaneous pickups and deliveries. *Journal of Combinatorial Optimization, 15*(4), 368–386. doi:10.1007/s10878-007-9090-4

Zachariadis, E., Tarantilis, C., & Kiranoudis, C. (2009). A hybrid metaheuristic algorithm for the vehicle routing problem with simultaneous delivery and pick-up service. *Expert Systems with Applications, 36*(2), 1070–1081. doi:10.1016/j.eswa.2007.11.005

This work was previously published in the International Journal of Applied Metaheuristic Computing, Volume 2, Issue 2, edited by Peng-Yeng Yin, pp. 1-20, copyright 2011 by IGI Publishing (an imprint of IGI Global).

Chapter 11
Parallel Scatter Search Algorithms for Exam Timetabling

Nashat Mansour
Lebanese American University, Lebanon

Ghia Sleiman-Haidar
Lebanese American University, Lebanon

ABSTRACT

University exam timetabling refers to scheduling exams into predefined days, time periods and rooms, given a set of constraints. Exam timetabling is a computationally intractable optimization problem, which requires heuristic techniques for producing adequate solutions within reasonable execution time. For large numbers of exams and students, sequential algorithms are likely to be time consuming. This paper presents parallel scatter search meta-heuristic algorithms for producing good sub-optimal exam timetables in a reasonable time. Scatter search is a population-based approach that generates solutions over a number of iterations and aims to combine diversification and search intensification. The authors propose parallel scatter search algorithms that are based on distributing the population of candidate solutions over a number of processors in a PC cluster environment. The main components of scatter search are computed in parallel and efficient communication techniques are employed. Empirical results show that the proposed parallel scatter search algorithms yield good speed-up. Also, they show that parallel scatter search algorithms improve solution quality because they explore larger parts of the search space within reasonable time, in contrast with the sequential algorithm.

DOI: 10.4018/978-1-4666-2145-9.ch011

INTRODUCTION

Wren (1999) defined timetabling as "the allocation, subject to constraints, of given resources to objects being placed in space time, in such a way as to satisfy as nearly as possible a set of desirable objectives". The university timetabling problem is of two types: exam timetabling and course timetabling. In this work, we are concerned with the exam timetabling problem. Exam timetabling consists of allocating periods and rooms to a set of exams, subject to constraints, which can be hard or soft (Qu et al., 2009). An example of hard constraints is room capacity. Examples of soft constraints are number of students having consecutive exams and number of students with multiple exams per day. In our case, we consider a predefined number of periods and a limited number of rooms as resources for allocation. Despite the similarity between exam and course timetabling, they are different; they have different objectives and constraints. For example, in course timetabling, a hard constraint is not to schedule courses taught by the same instructor at the same time, whereas this might be requested by some instructors for exam timetabling and in the same room whose capacity must allow such assignment. In addition, even in exam timetabling alone, different problem versions have been reported and addressed in a variety of ways. Furthermore, it is not possible to schedule exams at the same time as the corresponding course, since courses normally last one hour whereas exams last 2-3 hours. Also, students might take 4 or 5 courses on the same days whereas this is considered unacceptable for exams.

The university timetabling problem is an NP-hard optimization problem (Schaerf, 1999). A number of heuristic algorithms have been developed for finding sub-optimal exam timetabling solutions. Examples of these algorithms and techniques are: case-based reasoning technique in combination with a simple simulated annealing algorithm (Burke et al., 2003); hybridization within a graph based hyperheuristic framework (Qu & Burke, 2009); genetic and evolutionary algorithms (Cheong et al., 2009; Cote et al., 2005); clustering and clique algorithms (Carter & Johnson, 2001; Lotfi & Cerveny, 1991); tabu search (Kendall & Hussin, 2005); simulated annealing approaches (Burke et al., 2004; Mansour et al., 2003; Thompson & Dowsland, 1998); scatter search algorithm (Mansour et al., 2009); hybrid heuristics (Azimi, 2005; Bilgin et al., 2007; Pillay & Banzhaf, 2009); heuristic ordering based method combined with backtracking technique (Carter & Johnson, 2001); great-deluge hyperheuristic (Ozcan et al., 2010). A recent survey of exam timetabling techniques appears in Qu et al. (2009). Clearly, the exam timetabling literature includes a wide variety of approaches and algorithms and presents a variety of objectives and constraints for the problem (Qu et al., 2009) and, thus, different ways to evaluate solutions. This variety is based on theoretical assumptions as well as diverse requirements found, in reality, among different institutions even within the same country (Burke et al., 1996). Hence, methods that address real-world problems based on real-world data are still needed for bridging the gap between research and practice (McCollum, 2007).

For large numbers of students and exams, sequential algorithms are likely to be slow. In this paper, we present a parallel scatter (PSS) search algorithm for exam timetabling to improve both execution time and solution quality. Not much work has been reported on parallel scatter search algorithms. Adenso-Diaz et al. (2006) have proposed parallelization strategies for scatter search for the 0-1 knapsack problem by dividing the scatter search algorithm into 2 phases. Phase I includes tasks related to initial population generation and reference set creation. Phase II includes all tasks related to Subsets creation and combination, solution improvement reference set update. For each one of these phases, they implemented three different methods of processors communication. The first method is single walk strategy, where every processor works independently from other processors. The second method is multiple walk with independent threads strategy, which is a simple replication on processors. The third method is multiple walk with cooperative threads strategy where processors communicate together and share generated information. For all phases, they used both static and dynamic reference set update methods to update the reference set solutions. Emulation results using threads showed a low parallel efficiency and no clear increase in the solution quality. Garcia-Lopez et al. (2003) have developed a parallel scatter search algorithm to solve the p-median problem. They proposed three different parallel techniques: a low-level synchronous parallel scatter search model by using parallel search instead of local search, a replicated combination scatter search model by distributing multiple subsets on the processors, and a natural replication of parallel scatter search. Bozejko and Wodecki (2008) have presented a parallel algorithm based on scatter search and path re-linking methods to solve a flow shop scheduling problem. In this algorithm, the root processor creates the starting solutions set S of size n, while calculations of path re-linking procedures are executed by all processors on local data. On every processor, n/2 randomly chosen pairs from S undergo path re-linking to produce another set S' of n/2 solutions. The non-root processors send back the S' solutions to the root processor and the root will next create a new set of starting solutions. Another parallel scatter search algorithm has been developed by Garcia-Lopez et al. (2006) to solve the classification subset selection feature. The parallelism is based on replacing the combination method by parallel execution of two greedy methods on every processor. However, none of these strategies has addressed the exam timetabling problem. Our purposed PSS is based on distributing computations over different processors with limited

amount of communication. Most inter-processor communication is implemented using an efficient tree-based scheme. In order to assess the merits of PSS, we have also implemented two other parallel algorithms: (a) Farmer-Worker Parallel Scatter Search (FWPSS), where the root processor is assigned the role of a master processor, and (b) Embarrassingly Parallel Scatter Search (EPSS), where scatter search computations are replicated on different processors that run independently with minimal communication.

This paper is organized as follows. Section 2 describes the exam timetabling problem. Section 3 presents the parallel scatter search algorithms for exam timetabling. In Section 4, we present the empirical results. Conclusions are given in Section 5.

EXAM TIMETABLING PROBLEM DESCRIPTION

Given:

- A list of students STD: (STD_1, STD_2, ..., STD_S), where S denotes the total number of students,
- A list of course exams CRS: (CRS_1, CRS_2, ..., CRS_C), where C denotes the total number of courses,
- A pre-defined number of classrooms R, each with a maximum capacity Ψ_r for $r = 1, 2, ..., R$, and
- A pre-defined number of Exam Periods $\prod$ ($\prod = D * E$, where D is the number of exam days and E is the number of exam period per day), exam timetabling consists of assigning time periods and rooms to the given set of exams. These exams refer to courses in which students are enrolled based on choices allowed by their curricula. We consider exam timetabling with the following constraints:

- Eliminate / minimize the number of students having simultaneous exams (S_{SE}).
- Minimize the number of students having consecutive exams per day (S_{CE}).
- Minimize the number of students having 3 multiple exams per day (S_{3E}).
- Eliminate/minimize the number of students having 4 multiple exams per day (S_{4E}).
- No exam is assigned to a room whose maximum capacity is less than the number of students taking the exam.

We define the following objective function (*OF*), which determines the quality of timetabling solutions:

$$OF = \alpha_* S_{SE} + \varphi_* S_{CE} + \sigma_* S_{3E} + \beta_* S_{4E} + \gamma_* (\Sigma_{1<=x<=\Pi} \Sigma_{1<=y<=R} \rho_{xy})$$

where the last summation term gives the total number of rooms that violate the room capacity constraint. That is, $\rho_{xy}=1$ if room y exceeds its allocated capacity for exam period x. The coefficients of the objective function (*OF*) represent penalties or weights that determine the importance of the respective terms. They also provide the user flexibility in shifting emphasis between constraints.

PARALLEL SCATTER SEARCH ALGORITHMS FOR EXAM TIMETABLING

Overview

Scatter search is a population-based meta-heuristic algorithm that can be applied to hard optimization problems (Glover et al., 2000; Laguna & Marti, 2003). The major steps in sequential scatter search

Figure 1. Basic sequential scatter search design

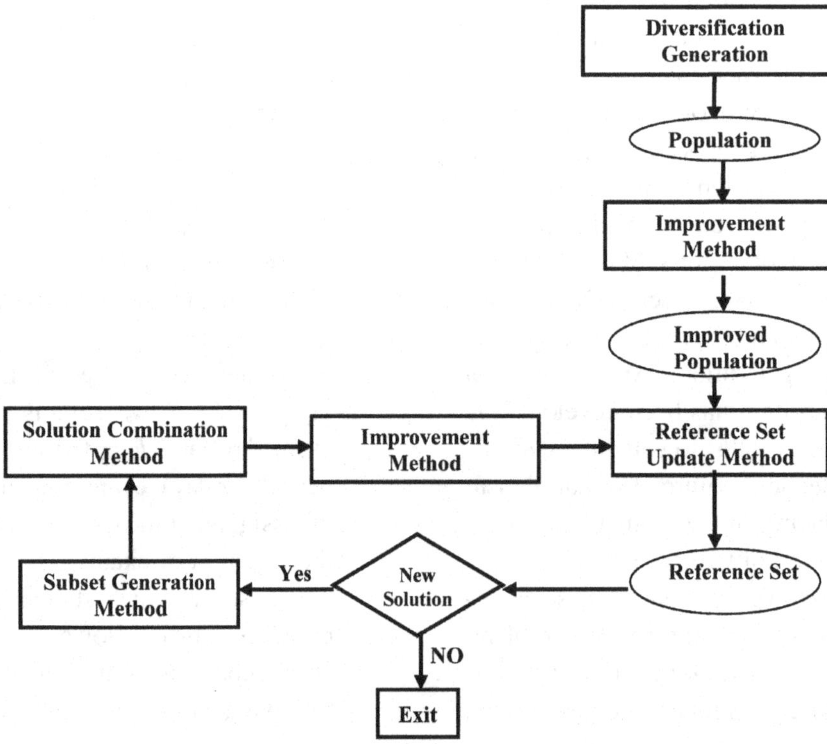

are shown in Figure 1. We have recently developed a scatter search algorithm for exam timetabling (Mansour et al., 2009). Scatter search starts with the generation of an initial population of candidate exam timetables by the diversification generation method that involves diversification and randomization. Each timetable is represented by a list of scheduled course-exams, *CRS*. Each scheduled course-exam, $CRS_i(t, r)$, is associated with an exam period $t \in \{1, 2, \ldots, \Pi\}$ and an assigned room $r \in \{1, 2, \ldots, R\}$, for $i = 1, 2, \ldots, C$. From this initial population set, a smaller subset, called the reference set, is created by selecting the highest quality and diverse solutions by the reference set update method. Then, we select solutions subsets and apply, to every subset, both the combination and improvement methods to generate new solutions. The subset generation method generates subsets from the reference set solutions. The solution combination method combines subset solutions to yield new solutions. The improvement method attempts to improve the solution quality. The reference set update method maintains the reference set with high quality and diverse solutions. These methods are repeated until no new solutions can be added to the reference set.

In this section, we propose a parallel scatter search algorithms (PSS) to solve the exam timetabling problem. We also describe two more parallel algorithms: Farmer-Worker Parallel Scatter Search (FWPSS) and Embarrassingly Parallel Scatter Search (EPSS) for the purpose of comparing the three different paradigms. FWPSS and EPSS are based on classical strategies for inter-process communication. It is well known that the cost of interprocess communication is an overhead, in parallel processing, which slows down execution and leads to a reduction in the required speed. PSS is proposed with a different, tree-based, communication strategy that aims to reduce the communication overhead and, thus, improve speed up. We present the basic design of the three parallel algorithms. Then, we describe the constituent methods of PSS.

PARALLEL SCATTER SEARCH ALGORITHM

The Parallel Scatter Search (PSS) algorithm applies tree-based communication among all processors $Processor_p$, where $0 \leq p \leq (n\_pr\text{-}1)$, when unifying the RefSet in every iteration. Figure 2 shows the PSS algorithm steps. First, $Processor_0$ (or *RootProcessor*) broadcasts the input data files to all other processors; the input data files are courses file, student enrolments file, conflict file and rooms file. Then, each processor determines its own population size, $ProcPopSize_p$, which is defined as a function of the total *PopSize* and the number of processors (n_pr). If *PopSize* is divisble by n_pr, then $ProcPopSize_p$ will be equal for all $p=0, 2, \ldots, n\_pr\text{-}1$; otherwise, the reminder of the division is distributed evenly among the lower rank processors.

In Step 4 of Figure 2, each processor, $Processor_p$, runs the diversification generation method to create its local sub-population. This is followed by the improvement method locally (Step 5).

After the generation and improvement of $ProcPopSize_p$, every $Processor_p$ generates its complete local reference set $ProcRefSet_p$. Once all reference sets are defined, PSS merges all $ProcRefSet_p$ in a bottom-up tree-based communication way starting from leaf processors and moving up toward *RootProcessor*. Each $Processor_p$ ($0 \leq p \leq n\_pr\text{-}1$) creates a local $ProcRefSet_p$ of size equal to *GlobalRefSet*. Then, every two processors unify their corresponding *GlobalRefSet* in a bottom-up tree-based communication toward the *RootProcessor*. Unification means that the two local Reference Sets are compared and the solutions of higher quality and higher diversity are selected from both. The tree-based communication system in PSS is implemented to reduce the communication time. Real reduction in communication cost is obtained when the inter-processor connection network is compatible with the hierarchical tree-based communication scheme. In the best case, the communication cost

Figure 2. Parallel scatter search algorithm

Algorithm PSS
1. *Broadcast InputData (*from *RootProcessor* to *n_pr processors);*
2. *Determine ProcPopSize$_p$(n_pr processors);*
3. for each *processor p = 0, ..., n_pr – 1,* do in parallel
4.   *DiversificationGeneration(ProcPopSize$_p$);*
5.   *Improvement(ProcPopSize$_p$);*
6.   *RefSetUpdate(ProcRefSet$_p$);*
    end for;
7. *Gather GlobalRefSet(ProcRefSet$_p$, p = 0, ..., n_pr – 1) in Buttom-up Tree-based Communication;*
8. if *(RootProcessor)* then
9.   *Broadcast GlobalRefSet;*
    end if;
10. repeat
11.   for each processor *p = 0, ..., n_pr – 1,* do in parallel
12.   *SubSetGeneration(SubsetSize, Subsets);*
13.   *Select ProcSubsets(SubsetSize$_p$, ProcSubset$_p$);*
14.   *while (ProcSubset$_p$ ≠ Ø) do*
15.   *SelectSubset s from ProcSubset$_p$;*
16.   *SolutionCombination(s) to produce solution x;*
17.   *Improvement (x) to produce x';*
18.   *RefSetUpdate(ProcRefSet$_p$) with x';*
            end while;
        end for;
19.   if *(at least 1 ProcRefSet$_p$ is improved)* then
20.   *Gather GlobalRefSet(ProcRefSet$_p$, p= 0 ...,n_pr – 1) in Bottom-up Tree-based Comm.;*
        end if;
21.   if *(RootProcessor)* then
22.   *Broadcast GlobalRefSet;*
        end if;
23. until *(no new solution in GlobalRefSet);*

will be of the order of *log(n_pr)*, which is the height of the binary tree, in comparison with *n_pr* for the other two communication schemes. After having a complete unified *GlobalRefSet* from all *ProcRefSet$_p$*, the *RootProcessor* broadcasts it to all other processors.

In Step 12, the subset generation method creates all possible 2-element subsets. In Step 13, each *Processor$_p$ (0 ≤ p ≤ n_pr-1)* selects it own local *ProcSubset$_p$* of size *SubsetSize$_p$*. We note that these subsets are evenly or near-evenly distributed among processors and that the selected local *ProcSubsets* are, thus, different for different processors, *p*. After *SubsetSize$_p$ ProcSubsets* are determined, combination and improvement

methods are applied to the subsets in order to yield improved solutions. The reference set update method is then applied on every generated solution to check whether this new candidate solution can be added to the processor reference set *ProcRefSet$_p$*.

This procedure is applied on every processor. If *ProcRefSet$_p$* is changed at any processor, all processors re-unify all *ProcRefSet$_p$* in bottom-up tree-based communication towards the *RootProcessor*, where the *RootProcessor* re-generates a *GlobalRefSet* and broadcasts it to all processors. Then, another iteration is initiated. PSS ends when no new solution is added to *GlobalRefSet*.

The communication performance of tree-based PSS could be more promising for appropriate interconnection networks that support tree communication than the cluster results.

Farmer-Worker Parallel Scatter Search Algorithm

The Farmer-Worker Parallel Scatter Search (FW-PSS) is described in Figure 3. FWPSS differs from PSS in that the master processor, *RootProcessor,* is responsible for dividing the work and delegating tasks to the remaining processors. FWPSS applies the same diversification generation method, subset generation method, and combination and improvement methods used in PSS. For the reference set generation method, FWPSS generates a partial reference set at each *Processor$_p$*, denoted as *ProcLocalRefSet$_p$* such that the grouping of all *ProcLocalRefSet$_p$* makes up the complete

Figure 3. Farmer-worker parallel scatter search algorithm

Algorithm FWPSS
1. *Broadcast InputData (*from *RootProcessor* to *n_pr processors);*
2. *Determine ProcPopSize$_p$(n_pr processors);*
3. for each *processor p* = 0, ..., *n_pr* – 1, do in parallel
4.   *DiversificationGeneration(ProcPopSize$_p$);*
5.   *Improvement(ProcPopSize$_p$);*
6.   *RefSetUpdate(ProcLocalRefSet$_p$);*
    end for;
7. if *(RootProcessor)* then
8.   *Receive ProcLocalRefSet(ProcLocalRefSet$_p$, p = 0, ..., n_pr – 1);*
9.   *Create GlobalRefSet;*
10.   *Broadcast GlobalRefSet;*
11. else
12.   *Send ProcLocalRefSet$_p$ to RootProcessor;*
13.   *ReceiveGlobalRefSet* from *RootProcessor;*
    end if;
14. repeat
15.   for each processor *p* = 0, ..., *n_pr* – 1, do in parallel
16.   *SubSetGeneration(SubsetSize, Subsets);*
17.   *Select ProcSubsets(SubsetSize$_p$, ProcSubset$_p$);*
18.   while (*ProcSubset$_p$* ≠ Ø) do
19.   *SelectSubset s from ProcSubset$_p$;*
20.   *SolutionCombination(s) to produce solution x;*
21.   *Improvement (x) to produce x';*
22.   *RefSetUpdate(ProcLocalRefSet$_p$) with x';*
        end while;
    end for;
23.   if (*at least 1 ProcLocalRefSet$_p$ is improved*) then
24.   if *(RootProcessor)* then
25.   *Receive ProcLocalRefSet(ProcLocalRefSet$_p$, p = 0, ..., n_pr – 1);*
26.   *Create GlobalRefSet;*
27.   *Broadcast GlobalRefSet;*
28.   else
29.   *Send ProcLocalRefSet$_p$ to RootProcessor;*
30.   *ReceiveGlobalRefSet* from *RootProcessor;*
        end if;
31. until (*no new solution in GlobalRefSet*);

requested *GlobalRefSet*. Once all partial local reference sets are defined, the non-root processors send *ProcLocalRefSet$_p$* to the *RootProcessor*, which then creates *GlobalRefSet* and broadcasts it back to all other processors.

Embarrassingly Parallel Scatter Search Algorithm

Embarrassingly Parallel Scatter Search (EPSS), described in Figure 4, implements a replicated SS algorithm on all processors where every processor creates its own initial population, generates its related *ProcLocalRefSet$_p$*, and applies combination and improvement methods to all subsets generated from corresponding *ProcLocalRefSet$_p$*. This algorithm is also referred to as replicated parallel SS in Alba (2005). In the EPSS algorithm, no communication takes place except at the beginning where *RootProcessor* broadcasts

the input data files to all other processors and at the end where it gathers all *ProcLocalRefSet$_p$* from n_pr-1 processors in *RootProcessor* to create *GlobalRefSet* and select the best solution out of it. The communication cost is reduced in the EPSS model.

Diversification Generation Method

The diversification generation method in PSS is implemented in full parallelism over all processors. That is, the initial population size is divided equally or near-equally among the n_pr processors.

The diversification generation method combines diversification with randomization. For this purpose, we use controlled randomization and a frequency-based memory technique to generate an initial set of diverse solutions. In our implementation, we divide the range $\prod$ into 4 sub-ranges. To assign a period to a given exam, we first select the

Figure 4. Embarrassingly parallel scatter search algorithm

Algorithm EPSS
1. *Broadcast InputData (*from *RootProcessor* to n_pr *processors);*
2. *Determine ProcPopSize$_p$(n_pr processors);*
3. for each *processor p = 0, ..., n_pr – 1, do in parallel*
4.   *DiversificationGeneration(ProcPopSize$_p$);*
5.   *Improvement(ProcPopSize$_p$);*
6.   *RefSetUpdate(ProcLocalRefSet$_p$);*
   end for;
7. repeat
8.   for each *processor p = 0, ..., n_pr – 1, do in parallel*
9.   *SubSetGeneration(SubsetSize, Subsets);*
10.   *Select ProcSubsets(SubsetSize$_p$, ProcSubset$_p$);*
11.   *while (ProcSubset$_p$ ≠ Ø) do*
12.   *SelectSubset s from ProcSubset$_p$;*
13.   *SolutionCombination(s) to produce solution x;*
14.   *Improvement (x) to produce x';*
15.   *RefSetUpdate(ProcLocalRefSet$_p$) with x';*
          end while;
       end for;
16. until (*MaxIterationNumber*);
17. if *(RootProcessor)* then
18.   *Receive ProcLocalRefSet(ProcLocalRefSet$_p$, p = 0, ..., n_pr – 1);*
19.   *Create GlobalRefSet;*
20. else
21.   *Send ProcLocalRefSet$_p$ to RootProcessor;*
    end if;

sub-range that is the least-frequently selected one so far, and then randomly select a period value within this sub-range. We also assign a randomly selected room. This method is replicated on all the n_pr processors, so that each processor is responsible for creating its $ProcPopSize_p$ initial solutions.

Improvement Method

The improvement method in PSS is also implemented in full parallelism for improving the quality of the local solutions generated on the individual processors.

Our improvement method sorts the exams in terms of decreasing number of enrolled students. Then, it considers every exam starting with the one that has the largest number of enrolments down to the exam with the lowest enrolments. For each exam, we apply a better-move heuristic procedure. In this procedure, we reassign an exam to other randomly selected period and randomly selected room. This reassignment is repeated a limited number of times, say 5, and the assignment that yields the greatest decrease in the value of *OF* is accepted. If none of these 5 moves leads to a decrease in *OF*, the original period and room assignments are kept.

Reference Set Update Method

The reference set update method is invoked in one of two ways. One way corresponds to generating the initial reference set and the second corresponds to updating a previous reference set. After updating the local reference set, inter-processor communication takes place in order to produce a global reference set for the next iteration.

The reference set is composed of two equal-size subsets: the subset of the highest quality solutions, *HQRefSet*, and the subset of diverse solutions, *DivRefSet*. The initial reference set on a processor p is denoted as $ProcLocalRefSet_p$. It is selected from the improved initial population

of solutions generated on this processor (of size $ProcPopSize$). The local reference set, $ProcLocalRefSet_p$, has a size, $RefSetSize_p$, which is typically less than 20% of $ProcPopSize$. $ProcLocalRefSet_p$, is the union of $HQRefSet_p$ and $DivRefSet_p$. $HQRefSet_p$ consists of the best solutions, selected from the local population, as determined by the objective function values. $DivRefSet_p$ items are, then, selected from the initial population based on diversity. To measure the diversity of a solution, we use a function $\partial(x', x'')$ that gives the number of different exam assignments (to periods and rooms) that appear in solutions x' and x''. To select the $DivRefSet_p$ solutions, we define $\partial_{min}(x)$ as the minimum distance between a solution x and all the solutions x' in the $HQRefSet_p$. Then, $DivRefSet_p$ will consist of the solutions that have the greatest ∂_{min} values.

The second type of the reference update method occurs at the end of every scatter search iteration, where the preceding local reference set is updated with newly generated solutions, produced by the combination and improvement methods. The criterion for accepting a newly generated solution, x, is based on the following. The first case is: if the *OF* of x is less than that of x_{worse}, the worst solution in the $HQRefSet_p$, then x replaces x_{worse}. The second case is: if the *OF* of x is greater than that of x_{worse}, we check if it can be added to the $DivRefSet_p$. If the distance $\partial_{min}(x)$ to all the solutions in the $HQRefSet_p$ is greater than that of the least diverse solution in the $DivRefSet_p$, we replace the current worst solution with x.

When all n_pr $ProcLocalRefSet$s are created, the global reference set update method gathers all n_pr $ProcLocalRefSet$s on the root processor to generate the global reference set, $GlobalRefSet$, and broadcast it to all processors as a pre-requisite stage required by the generation method. In PSS, the gather step is implemented by a tree-based interprocessor communication where the leaves of the binary tree correspond to the local processor reference sets. Adjacent pairs of processors compare their $ProcLocalRefSet$s and unify the

Table 1. Subject Problems

Subject Problem	Π	# Rooms R	# Exams E	# Students	# Enrolments
SP1	32	38	473	3652	13662
SP2	32	38	472	3704	13455
SP3	32	38	537	3911	15392
SP4	36	38	634	3794	15858

HQRefSet and *DivRefSet* parts in a similar way to what is described above except that the comparison is made on the union of the 2 local reference sets of the pair of processors. The resulting reference set is forwarded to one higher level in the binary tree so that the two adjacent unified reference sets will also be compared and unified. This tree traversal continues up the tree until the

GlobalRefSet is produced at the root processor. Then, *GlobalRefSet* is broadcast to all processors.

Subset Generation Method

The Subset Generation method generates subsets of the reference set solutions on *n_pr* processor. We choose to generate subsets of size 2 (subset type 1) and subsets of size 3 (subset type 2) by

Table 2. Results for PSS, PopSize=1024, RefSize=32

Subject Problem	No. of Procs	S_{SE}	S_{CE}	S_{3E}	S_{4E}	R_V	OF	No. of Iterations	Execution Time (Mins)	Time/ Iteration	PEff
SP1	1	0	357	16	0	0	517	11	919	83.5	-
	2	1	405	11	0	0	715	17	704	41.4	65
	4	1	398	22	0	0	818	15	314	20.9	73
	8	0	352	10	0	0	452	13	139	10.6	83
	16	0	403	8	0	0	483	6	37	6.1	155
SP2	1	0	329	12	0	0	449	6	545	90.8	-
	2	0	314	6	0	0	374	13	562	43.2	49
	4	0	213	0	0	0	213	4	143	35.7	95
	8	0	203	1	0	0	213	3	56	18.6	122
	16	0	398	14	0	0	538	21	117	5.5	29
SP3	1	1	328	7	0	0	598	6	812	135.3	-
	2	1	224	6	0	0	484	23	1527	66.3	27
	4	1	268	3	0	0	498	15	476	31.7	43
	8	1	266	7	0	0	536	9	157	17.4	65
	16	1	219	6	0	0	479	16	145	9.0	35
SP4	1	0	276	2	0	0	296	10	1929	192.9	-
	2	0	272	1	0	0	282	10	990	99	98
	4	0	260	1	0	0	270	17	836	49.1	58
	8	0	260	2	0	0	280	31	754	24.3	32
	16	0	261	1	0	0	271	10	134	13.4	90

Table 3. Results for FWPSS, PopSize=1024, RefSize=32

	No. of Procs	S_{SE}	S_{CE}	S_{3E}	S_{4E}	R_V	OF	No. of Iterations	Execution Time (Mins)	Time/ Iteration	PEff
SP1	1	0	357	16	0	0	517	11	919	83.5	-
	2	0	363	9	0	0	453	4	192	48	239
	4	0	385	15	0	0	535	11	234	21.2	98
	8	0	368	12	0	0	488	20	209	10.4	55
	16	0	399	13	0	0	529	16	90	5.6	64
SP2	1	0	329	12	0	0	449	6	545	90.8	-
	2	0	311	7	0	0	381	9	398	44.2	69
	4	0	311	7	0	0	381	4	98	24.5	139
	8	0	313	9	0	0	403	10	112	11.2	61
	16	0	314	8	0	0	394	10	60	6	114
SP3	1	1	328	7	0	0	598	6	812	135.3	-
	2	1	242	4	0	0	482	10	690	69	59
	4	1	235	4	0	0	475	18	609	33.8	33
	8	1	212	5	0	0	462	13	222	17.0	46
	16	1	176	0	0	0	376	18	162	9	31
SP4	1	0	276	2	0	0	296	10	1929	192.9	-
	2	0	271	1	0	0	281	9	915	101.6	105
	4	0	244	1	0	0	254	11	557	50.6	87
	8	0	269	1	0	0	279	11	279	25.3	87
	16	0	261	1	0	0	271	18	235	13.0	51

augmenting 2-elements subset with the best solution outside this subset. Thus, for subset type 1, we pair all possible solutions (*x, x'*) from the *GlobalRefSet* generated by the parallel reference set update method.

The subset generation method is a fast operation; we replicate this method on all *n_pr* processors.

Solution Combination Method

The Solution Combination Method combines the solutions grouped in subsets of size 2 in the preceding step. Our combination method approach employs a voting technique depending on the *OF* values and applies a random selection of Room and Exam Period in case we have the same scores.

First, we initiate an empty new solution array *s* with all courses ordered in decreasing order according to the number of student enrolments; the course corresponds to the array index. Then, for each course *c* in the ordered solution, we consider the partial *OF* values that result from the period-room assignments found in each of the solutions to be combined. Then, we select the room and the exam period that correspond to the lower *OF* value. For example, if solutions 1 and 2 consist of [a,b,c] and [x,y,z], respectively. First, we include either a or x in the new combined solution depending on which one yields the smaller *OF* value. Next, we select between b and y to include in the new solution depending on which gives the lower *OF* value, up to this point (i.e., using the

Table 4. Results for EPSS, PopSize=1024, RefSize=32

Subject Problem	No. of Procs	S_{SE}	S_{CE}	S_{3E}	S_{4E}	R_V	OF	No. of Iterations	Execution Time (Mins)	Time/ Iteration	PEff
SP1	1	0	357	16	0	0	517	11	919	83.5	-
	2	0	163	0	0	0	163	17	770	45.2	60
	4	0	129	1	0	0	139	15	341	22.7	67
	8	0	163	2	0	0	183	13	149	11.4	77
	16	0	161	2	0	0	181	6	37	6.1	155
SP2	1	0	329	12	0	0	449	6	545	90.8	-
	2	0	314	7	0	0	384	13	622	47.8	44
	4	0	128	0	0	0	128	4	105	26.2	130
	8	3	432	15	0	0	1182	3	41	13.6	166
	16	4	418	17	0	0	1388	21	128	6.0	27
SP3	1	1	328	7	0	0	598	6	812	135.3	-
	2	1	146	4	0	0	386	23	1650	71.7	25
	4	1	96	3	0	0	326	15	542	36.1	37
	8	1	123	0	0	0	323	9	167	18.5	61
	16	1	139	2	0	0	359	16	151	9.4	34
SP4	1	0	276	2	0	0	296	10	1929	192.9	-
	2	0	121	2	0	0	141	10	1084	108.4	89
	4	0	93	0	0	0	93	17	902	53.0	134
	8	0	81	0	0	0	81	31	814	26.2	30
	16	0	93	0	0	0	93	10	142	14.2	85

earlier selections too). Then, the same is repeated to select either c or z.

The Solution Combination Method is applied in parallel by distributing the subsets generated (in the Subset Generation Method) over the n_pr processors. That is, each processor combines a different part of the generated subsets.

Table 5. Results for PSS on SP1 and SP4, PopSize=200, RefSize =20

Program	SP	No. of Procs	RefSize	S_{SE}	S_{CE}	S_{3E}	S_{4E}	R_V	OF	No. of Iterations	Execution Time (Mins)	Time/ Iteration	PEff
PSS	SP1	1	20	0	496	16	0	0	656	11	321	29.1	-
		2	20	1	543	18	0	0	923	10	149	14.9	108
		4	20	0	444	20	0	0	644	18	134	7.4	60
		8	20	0	498	13	0	0	628	16	61	3.8	66
	SP4	1	20	0	295	3	0	0	325	14	900	64.2	-
		2	20	0	320	1	0	0	330	13	426	32.7	106
		4	20	0	324	3	0	0	354	12	199	16.5	113
		8	20	0	349	5	0	0	399	19	156	8.2	72

Table 6. Results for PSS on SP1and SP4, PopSize =100, RefSize =10

Program	SP	No. of Procs	RefSize	S_SE	S_CE	S_3E	S_4E	R_V	OF	No. of Iterations	Execution Time (Mins)	Time/ Iteration	PEff
PSS	SP1	1	10	2	489	15	0	0	1039	6	45	7.5	-
		2	10	1	484	16	0	0	844	3	13	4.3	173
		4	10	2	416	15	0	0	966	8	16	2	70
	SP4	1	10	0	350	3	0	0	380	5	89	17.8	-
		2	10	0	300	2	0	0	320	24	191	7.9	23
		4	10	0	319	1	0	0	329	13	56	4.3	40

EMPIRICAL RESULTS

Procedure

We apply our parallel scatter search algorithms to four real-world-university subject problem in-stances described in Table 1. The solution quality is evaluated by the objective function (*OF*) and its 5 components. Based on our previous experiences and the specific university requirements, the objective function user-defined weights are used as follows: $\alpha = 200$, $\varphi = 1$, $\sigma = 10$, $\beta = 100$, and γ

Table 7. Results for PSS, PopSize=1024, RefSize=32, 8 iterations

Subject Problem	No. of Procs	S_SE	S_CE	S_3E	S_4E	R_V	OF	No. of Iterations	Execution Time (Mins)	Time/ Iteration	PEff
SP1	1	0	357	16	0	0	517	8	670	84	-
	2	1	398	7	1	0	768	8	394	49.2	85
	4	2	384	9	0	0	874	8	180	22.5	93
	8	1	353	10	0	0	653	8	94	11.7	89
	16	0	403	8	0	0	483	8	40	5	105
SP2	1	0	329	12	0	0	449	8	735	92	-
	2	0	315	7	0	0	385	8	384	48	96
	4	0	213	0	0	0	213	8	289	36.1	64
	8	0	203	1	0	0	213	8	150	18.7	61
	16	0	364	10	1	0	564	8	46	5.7	100
SP3	1	1	328	7	0	0	598	8	1095	137	-
	2	1	301	8	0	0	581	8	532	66.5	102
	4	1	270	7	0	0	540	8	254	31.7	108
	8	1	268	7	0	0	538	8	140	17.5	98
	16	1	220	2	1	0	540	8	73	9.1	94
SP4	1	0	276	2	0	0	296	8	1564	196	-
	2	0	272	1	0	0	282	8	801	100.1	98
	4	0	259	2	0	0	279	8	409	51.1	96
	8	0	254	1	1	0	364	8	208	26.2	94
	16	0	261	1	0	0	271	8	120	15	81

Table 8. Results for FWPSS, PopSize=1024, RefSize=32, 8 iterations

Subject Problem	No. of Procs	S_{SE}	S_{CE}	S_{3E}	S_{4E}	R_v	OF	No. of Iterations	Execution Time (Mins)	Time/ Iteration	PEff
SP1	1	0	357	16	0	0	517	8	670	84	-
	2	0	323	8	0	0	403	8	416	52	81
	4	0	385	15	0	0	535	8	236	29.5	71
	8	0	405	14	0	0	545	8	105	12.8	80
	16	0	436	14	0	0	576	8	52	6.5	81
SP2	1	0	329	12	0	0	449	8	735	92	-
	2	0	312	8	0	0	392	8	372	46.9	99
	4	0	302	5	0	0	352	8	201	25	91
	8	0	313	10	0	0	413	8	105	13.1	87
	16	0	314	8	0	0	394	8	56	7	82
SP3	1	1	328	7	0	0	598	8	1095	137	-
	2	1	249	9	0	0	539	8	576	72	95
	4	1	264	8	0	0	544	8	297	37.1	92
	8	1	231	6	1	0	591	8	168	21.3	81
	16	1	281	5	0	0	531	8	87	10.9	79
SP4	1	0	276	2	0	0	296	8	1564	196	-
	2	0	286	1	0	0	296	8	832	104	94
	4	0	246	3	0	0	276	8	429	53.6	91
	8	0	250	1	1	0	360	8	227	28.4	86
	16	0	285	2	0	0	305	8	114	14.2	86

= 100 to indicate the relative importance of each *OF* term as required by the pertinent university. The performance of the parallel algorithms is measured by their parallel efficiency (*PEff*) over different numbers of processors (*n_pr*), where *PEff* is given by: [sequential time / (parallel time * *n_pr*)]. The behavior of the parallel algorithms is further explored by varying the *PopSize* and the *RefSize* values. *PopSize* values used are: 100, 200 and 1024. *RefSize* values used are: 10, 20 and 32. We run the parallel algorithms for as many iterations as they take to reach the situation where the Reference Set Update method does not produce a new solution that can be added to the reference set. But, in order to verify our observations and comparative analysis, we also run the parallel

algorithms for a fixed number of 8 iterations and record the *PEff* and OF results.

We execute our parallel programs on a cluster of PCs running Linux operating system and connected by an Ethernet network. Each PC has 2.33 GHz CPU and 2 GByte RAM memory. The programs are implemented in C++ and MPI-2 software.

RESULTS AND DISCUSSIONS

Tables 2 through 4 show the results of executing PSS, FWPSS and EPSS, respectively, for as many iterations as needed for convergence using *PopSize* = 1024 and *RefSize* = 32 for SP1-SP4

Table 9. Results for EPSS, PopSize=1024, RefSize=32, 8 iterations

Subject Problem	No. of Procs	S_{SE}	S_{CE}	S_{3E}	S_{4E}	R_V	OF	No. of Iterations	Execution Time (Mins)	Time/ Iteration	PEff
SP1	1	0	357	16	0	0	517	8	670	84	-
	2	0	172	4	0	0	212	8	372	46.5	90
	4	0	154	0	0	0	154	8	180	22.5	93
	8	1	265	2	0	0	485	8	102	12.7	82
	16	0	159	2	0	0	179	8	52	6.5	81
SP2	1	0	329	12	0	0	449	8	735	92	-
	2	0	340	9	0	0	430	8	398	49.7	92
	4	0	118	0	0	0	118	8	232	29	79
	8	1	417	11	0	0	727	8	112	14	82
	16	5	452	19	0	0	1642	8	59	7.3	78
SP3	1	1	328	7	0	0	598	8	1095	137	-
	2	1	216	8	1	0	596	8	587	73.3	93
	4	1	101	5	1	0	451	8	309	38.6	89
	8	1	123	0	0	0	323	8	150	18.7	91
	16	1	159	5	0	0	409	8	79	9.8	87
SP4	1	0	276	2	0	0	296	8	1564	196	-
	2	0	121	2	0	0	141	8	898	112.2	87
	4	0	115	0	1	0	215	8	457	57.1	86
	8	0	105	6	1	0	265	8	233	29.1	84
	16	0	93	1	0	0	103	8	115	14.3	85

on 1-16 *Procs*. Table 5 gives the results of PSS and FWPSS for *PopSize* = 200. Table 6 gives the results of PSS on 1-4 processors for *PopSize* = 100. Tables 7 through 9 are similar to Tables 2 through 4 but for a fixed number of 8 iterations of the scatter search algorithm.

From all these results, we infer the following:

1. Parallel scatter search algorithms PSS, FWPSS, EPSS do reduce the execution time with respect to the sequential SS algorithm. For example, Table 2 shows that sequential SS takes 1,929 min. for SP4 whereas PSS takes 134 min. on 16 processors for the same number of iterations, 10. This reduction in time allows including time demanding scat-

ter search features that are likely to improve the solution quality such as combining subsets with more than 2 solutions, which explores more areas of the search space. However, this significant execution time reduction fails in 4 exceptional cases out of 48 (i.e., in 8% of the time) where parallel algorithms spend several more iterations (13 or 23) than the sequential algorithm (6 iterations).

2. Parallel scatter search algorithms are necessary for handling/solving large timetabling problems in reasonable times by running them on larger number of processors.

3. There is not a clear pattern for PSS, FWPSS, and EPSS as far as *PEff* and *OF* values are

concerned. This demonstrated in both cases of predefined and undefined number of scatter search iterations. However, despite the choppy behavior, the three parallel algorithms are based on adequate parallelization strategies since they yield high efficiency (*PEff*) values in the great majority of the results. This result is more evident in the 8-iteration runs (Tables 7 through 9), where *PEff* values range from 61% on 8 processors to 108% on 4 processors. This is also demonstrated in the decrease in the time per iteration as the number of processors increase. In particular, PSS is expected to yield better efficiency on other parallel architectures that support tree-based interprocessor communication.

4.   In 85% of the cases, the values of *OF* decrease as a result of parallel execution. Moreover, the values of the terms that make up the *OF* expression show that the exam timetables produced are quite adequate. For example, for problem SP3, PSS yields an exam timetable with *OF*=536 (vs. 598 for the sequential algorithm) on 8 processors; this corresponds to 1 simultaneous exam, 266 consecutive exams, and 7 students with 3 exams per day. We note that the very small number of simultaneous exams, obtained in this case, is usually handled in the respective university by giving individual make-up exam for this very small number of concerned students.

5.   Referring to Tables 5 and 6 and inspecting the behavior of PSS' *OF* for different population sizes and reference set sizes, we find that PSS yields better exam timetables for larger population size and reference set size than on smaller population and reference set sizes. Due to parallelization and increase in population size, we have the opportunity to explore larger areas of search space and thus, to find better quality timetabling solutions.

Limitations

Our results indicate the following limitations:

1.   PSS, FWPSS and EPSS do not fully respect the reference set generation and update methods of the sequential scatter search. This is due to the parallelization approaches of these algorithms which result in generating different reference sets on different processors. These reference sets are then aggregated in different ways, to create one overall reference set, unlike what takes place in sequential scatter search.

2.   Due to random selection in a few scatter search operations, PSS runs for different number of iterations than sequential scatter search and in an unpredictable manner. When this happens, it could lead to reduction in *PEff* and/or the solution quality or to unexpected peculiar results that fall outside the pattern implied by the other encountered results.

3.   PSS, FWPSS and EPSS have limited scalability. Over larger number of processors, such as 32 or 64 processors, we need to increase both population size and reference set size. This increase is not consistent with the sequential scatter search and it takes longer time for running large problems.

4.   A threat to the validity of our results is due to the limited number and sizes of the available real-world subject problems.

CONCLUSION

We have proposed three different parallel scatter search algorithms for the exam timetabling problem. These algorithms are: a parallel algorithm (PSS) that involves a bottom-up tree-based communication scheme, an algorithm (FWPSS) based on the Farmer-Worker model, and a replicated

Embarrassingly Parallel Scatter Search algorithm (EPSS). The three algorithms have been executed on a cluster of networked PCs. Parallelism allows us to handle larger exam timetabling problems within reasonable time. Also, parallelism allows us to increase the population size and the reference set size which lead to improving the quality of the exam timetables in comparison with sequential scatter search. The experimental results showed that the proposed parallel scatter search algorithms produce good exam timetables with good parallel efficiency. However, PSS (and EPSS) is more scalable than FWPSS.

Further work should explore the performance of the parallel scatter search algorithms on other parallel architectures and on larger subject problems.

ACKNOWLEDGMENT

This work was partially supported by the Lebanese American University and the Lebanese National Council for Scientific Research. We thank the anonymous reviewers whose questions and suggestions led to improving the paper.

REFERENCES

Adenso-Diaz, B., Garcia-Carbajal, S., & Lozano, S. (2006). An empirical investigation on parallelization strategies for scatter search. *European Journal of Operational Research, 169*(2), 490–507. doi:10.1016/j.ejor.2004.08.011

Alba, E. (2005). *Parallel metaheuristics*. Hoboken, NJ: John Wiley & Sons. doi:10.1002/0471739383

Azimi, Z. N. (2005). Hybrid heuristics for examination timetabling problem. *Applied Mathematics and Computation, 163*(2), 705–733. doi:10.1016/j.amc.2003.10.061

Bilgin, B., Ozcan, E., & Korkmaz, E. E. (2007). An experimental study on hyperheuristics and exam timetabling. In E. K. Burke & H. Rudova (Eds.), *Proceedings of the 6th International Conference on Practice and Theory of Automated Timetabling* (LNCS 3867, pp. 394-412).

Bozejko, W., & Wodecki, M. (2008). Parallel scatter search algorithm for the flow shop sequencing problem. In R. Wyrzykowski, J. Dongarra, K. Karczewski, & J. Wasniewski (Eds.), *Proceedings of the 7th International Conference on Parallel Processing and Applied Mathematics* (LNCS 4967, pp. 180-188).

Burke, E. K., Bykov, Y., Newall, J., & Petrovic, S. (2004). A time-predefined local search approach to exam timetabling problems. *IIE Transactions on Operations Engineering, 36*(6), 509–528.

Burke, E. K., Eckersley, A. J., McCollum, B., Petrovic, S., & Qu, R. (Eds.). (2003, August 13-16). Similarity measures for exam timetabling problems. In *Proceedings of the 1st Multidisciplinary Conference on Scheduling: Theory and Applications*, Nottingham, UK (pp. 120-136).

Burke, E. K., Elliman, D. G., Ford, P. H., & Weare, R. F. (1996). Examination timetabling in British universities: A survey. In E. K. Burke & P. Ross (Eds.), *Proceedings of the 1st International Conference on Practice and Theory of Automated Timetabling* (LNCS 1153, pp. 76-90).

Carter, M., & Johnson, D. G. (2001). Extended clique initialization in examination timetabling. *The Journal of the Operational Research Society, 52*(5), 538–544. doi:10.1057/palgrave. jors.2601115

Cheong, C. Y., Tan, K. C., & Veeravalli, B. (2009). A multi-objective evolutionary algorithm for examination timetabling. *Journal of Scheduling, 12*, 121–146. doi:10.1007/s10951-008-0085-5

Côté, P., Wong, T., & Sabourin, R. (2005). Application of a hybrid multi-objective evolutionary algorithm to the uncapacitated exam proximity problem. In E. Burke & M. Trick (Eds.), *Proceedings of the 5th International Conference on the Practice and Theory of Automated Timetabling* (LNCS 3616, pp. 294-312).

Garcia-Lopez, F., Batista, B. M., & Moreno-Perez, J. A. (2003). Parallelization of the scatter search for the p-median problem. *Parallel Computing, 29,* 575–589. doi:10.1016/S0167-8191(03)00043-7

Garcia-Lopez, F., Torres, M. G., Batista, B. M., Moreno-Perez, J. A., & Moreno-Vega, J. M. (2006). Solving features subset selection problem by a parallel scatter search. *European Journal of Operational Research, 196,* 477–489. doi:10.1016/j.ejor.2004.08.010

Glover, F., Laguna, M., & Marti, R. (2000). Fundamentals of scatter search and path relinking. *Control and Cybernetics, 39,* 575–589.

Kendall, G., & Mohd Hussin, N. (2005). A tabu search hyper-heuristic approach to the examination timetabling problem at the MARA University of Technology. In E. Burke & M. Trick (Eds.), *Proceedings of the 5th International Conference on the Practice and Theory of Automated Timetabling* (LNCS 3616, pp. 270-293).

Laguna, M., & Marti, R. (2003). *Scatter search: Methodology and implementations in C.* London, UK: Springer.

Lofti, V., & Cerveny, R. (1991). A final-exam-scheduling package. *The Journal of the Operational Research Society, 42,* 205–216.

Mansour, N., Isahakian, V., & Galayini, I. (2009). Scatter search technique for exam scheduling. *Applied Intelligence, 34*(2).

Mansour, N., Tarhini, A., & Isahakian, V. (2003, July 14-18). Three-phase simulated annealing algorithms for exam scheduling. In *Proceedings of the ACS/IEEE International Conference on Computer Systems and Applications,* Tunis, Tunisia (p. 90).

McCollum, B. (2007) A perspective on bridging the gap between research and practice in university timetabling. In E. K. Burke & H. Rudova (Eds.), *Proceedings of the 6th International Conference on the Practice and Theory of Automated Timetabling* (LNCS 3867, pp. 3-23).

Ozcan, E., Mısır, M., Ochoa, G., & Burke, E. K. (2010). A reinforcement learning – great-deluge hyper-heuristic for examination timetabling. *International Journal of Applied Metaheuristic Computing, 1*(1), 39–59. doi:10.4018/jamc.2010102603

Pillay, N., & Banzhaf, W. (2009). A study of heuristic combinations for hyper-heuristic systems for the uncapacitated examination timetabling problem. *European Journal of Operational Research, 197*(2), 482–491. doi:10.1016/j.ejor.2008.07.023

Qu, R., & Burke, E. K. (2009). Hybridisation within a graph based hyperheuristic framework for university timetabling problems. *The Journal of the Operational Research Society, 60,* 1273–1285. doi:10.1057/jors.2008.102

Qu, R., Burke, E. K., McCollum, B., Merlot, L. T. G., & Lee, S. Y. (2009). A survey of search methodologies and automated system development for examination timetabling. *Journal of Scheduling, 12,* 55–89. doi:10.1007/s10951-008-0077-5

Schaerf, A. (1999). A survey of automated timetabling. *Artificial Intelligence Review, 13,* 87–127. doi:10.1023/A:1006576209967

Thompson, J., & Dowsland, K. (1998). A robust simulated annealing based examination timetabling system. *Computers & Operations Research*, *25*, 637–648. doi:10.1016/S0305-0548(97)00101-9

Wren, A. (1999). Scheduling, timetabling and rostering - a special relationship? In E. K. Burke & P. M. Ross (Eds.), *Proceedings of the International Conference on Practice and Theory of Automated Timetabling* (LNCS 1153, pp. 46-75).

Chapter 12
Pseudo–Cut Strategies for Global Optimization

Fred Glover
OptTek Systems, Inc., USA

Abraham Duarte
Universidad Rey Juan Carlos, Spain

Leon Lasdon
The University of Texas at Austin, USA

Rafael Marti
Universidad de Valencia, Spain

John Plummer
Texas State University, USA

Manuel Laguna
University of Colorado, USA

Cesar Rego
University of Mississippi, USA

ABSTRACT

Motivated by the successful use of a pseudo-cut strategy within the setting of constrained nonlinear and nonconvex optimization in Lasdon et al. (2010), we propose a framework for general pseudo-cut strategies in global optimization that provides a broader and more comprehensive range of methods. The fundamental idea is to introduce linear cutting planes that provide temporary, possibly invalid, restrictions on the space of feasible solutions, as proposed in the setting of the tabu search metaheuristic in Glover (1989), in order to guide a solution process toward a global optimum, where the cutting planes can be discarded and replaced by others as the process continues. These strategies can be used separately or in combination, and can also be used to supplement other approaches to nonlinear global optimization. Our strategies also provide mechanisms for generating trial solutions that can be used with or without the temporary enforcement of the pseudo-cuts.

DOI: 10.4018/978-1-4666-2145-9.ch012

INTRODUCTION

We consider the constrained global optimization problem (*P*) expressed in the following general form:

(*P*) minimize *f(x)*

subject to:

$G(x) \le b$

$x \in S \subset Rn$

where *x* is an *n*-dimensional vector of decision variables, *G* is an *m*-dimensional vector of constraint functions, and without losing generality the vector *b* contains upper bounds for these functions. The set *S* is defined by simple bounds on *x*, and we assume that it is closed and bounded, i.e., that each component of *x* has a finite upper and lower bound.

We introduce strategies for solving (P) which are based on pseudo-cuts, consisting of linear inequalities that are generated for the purpose of strategically excluding certain points from being admissible as solutions to an optimization problem. The *pseudo* prefix refers to the fact that these inequalities may not be valid in the sense of guaranteeing that at least one globally optimal solution will be retained in the admissible set. Nevertheless, a metaheuristic procedure that incorporates occasional invalid inequalities with a provision for replacing them can yield an aggressive solution approach that can prove valuable in certain settings. The use of pseudo-cuts to create temporary restrictions in a search process was suggested in Glover (1989) in the context of a tabu search procedure. In this approach the cuts are treated in the same way as other restrictions imposed by tabu search, by drawing on a memory-based strategy to cull out certain cuts previously introduced and drop them from the pool of active restrictions. The present approach is particularly motivated by the work of Lasdon et al. (2010), where a simplified instance of such strategies was found to be effective for improving the solution of certain constrained non-convex nonlinear continuous problems.

In the present paper we likewise assume the objective function of (P) is non-convex (hence a local optimum may not be a global optimum), and allow for non-convexity in the constraints. We also allow for the presence of integer restrictions on some of the problem variables under the provision that such variables are treated by means of constraints or objective function terms that permit them to be treated as if continuous within the nonlinear setting. In the case of zero-one variables, for example, a concave function such as $x_j(1 - x_j)$ may be used that is 0 when $x_j = 0$ or 1, and is positive otherwise. See Bowman and Glover (1972) for additional examples.

We make recourse to an independent algorithm to generate trial solutions to be evaluated as candidates for a global optimum, where as customary the best feasible candidate is retained as the overall "winner". The independent algorithm can consist of a directional search (based on gradients or related evaluations) as in Lasdon et al. (2010), or may be a "black box" algorithm as used in simulation optimization as in April et al. (2006) and Better et al. (2007).

PSEUDO-CUT FORM AND REPRESENTATION

Our pseudo-cut strategy is based on generating hyperplanes that are orthogonal to selected rays (half-lines) originating at a point x′ and passing through a second point x″, so that the hyperplane intersects the ray at a point x° determined by requiring that it lies on the ray at a selected distance d from x′. The half-space that forms the pseudo-cut is then produced by the associated inequality that excludes x′ from the admissible half-space. We define the distance d by reference to the Euclidean (L2) norm, but other norms can also be used.

To identify the pseudo-cut as a function of x′, x″ and d, we represent the ray that originates at x′ and passes through x″ by:

$$x = x' + \lambda(x'' - x'), \lambda \geq 0. \quad (1)$$
(Hence x′ and x″ lie on the ray at the points determined by $\lambda = 0$ and 1, respectively.)

A hyperplane orthogonal to this line may then be expressed as.

$$ax = b \quad (2.1)$$

where

$$a = (x'' - x') \quad (2.2)$$

$$b = \text{an arbitrary constant} \quad (2.3)$$

The specific hyperplane that contains a given point x° on the ray (1) results by choosing

$$b = ax°. \quad (2.4)$$

To identify the point x° that lies on the ray (1) at a distance d from x′, we seek a value $\lambda = \lambda°$ that solves the equation

$$d(x',x°) \equiv \|x' - x°\| = d \quad (3.1)$$

where

$$x° = x' + \lambda°(x'' - x'). \quad (3.2)$$

Consequently, by the use of (3.2) the desired value of $\lambda°$ is obtained by solving the equation

$$(\sum(x_j' - x_j°)^2)^{.5} = d \quad (3.3).$$

For the value of $\lambda°$ and the hyperplane thus determined, the associated half-space that excludes x″ (and x′) is then given by

$$ax \geq ax°. \quad (4)$$

PSEUDO-CUT STRATEGY

We make use of the pseudo-cut (4) within a 2-stage process. In the first stage x′ represents a point that is used to initiate a current search by the independent algorithm, and x″ is the point obtained at the conclusion of this search phase (e.g., x″ may be a local optimum). The distance d is then selected so that x° lies a specified distance beyond x″.

In the second stage we take x′ to be the point x″ identified in the first stage, and determine x″ by applying the independent algorithm to the problem that results after adding the pseudo-cut generated in the first stage. In this case d is chosen so that x° lies between x′ and x″ at a selected distance from x′.

The value of d in both of these cases may be expressed as a multiple m of the distance between the points currently denoted as x′ and x″, i.e.

$$d = m\|x'' - x'\| \quad (5)$$

The multiple m is selected to be greater than 1 in the first stage and less than 1 in the second. Because the points x′ and x″ change their identities in the two stages, it is convenient to refer to the points generated in these stages by designating them as P0, P1, Q1, etc., as a basis for the following description (We later identify additional variations based on choosing d, x′ and x″ in different ways). The *pseudo-cut pool* (or simply *cut pool*) refers to all pseudo-cuts previously added that have not yet been discarded. The pool begins empty.

Together with the statement of the Pseudo-Cut Generation Procedure, we include parenthetical remarks, underlined and in italics, that identify specific accompanying diagrams to illustrate some of the key steps of the procedure.

Pseudo-Cut Generation Procedure (A Complete Pseudo-Code for this Procedure Appears in the Appendix)

- **Stage 1:**
  - **1.1:** Let $x' = P0$ denote a starting point for the independent algorithm, let $x'' = P1$ denote the best point obtained during the current execution of the algorithm, and let $x^\circ = Q1$ be the point determined by (3) upon selecting a value $m > 1$ in (5) (see Note 1). If x° violates any pseudo-cut contained in the cut pool, remove this cut from the pool.
  - **1.2:** Add the pseudo-cut (4) to the cut pool and apply the independent algorithm starting from the point Q1. Let Q2 denote the best point of the current execution. If Q2 = Q1, then increase the value of m to determine a new Q1 by (3) that replaces the previous cut that was generated for a smaller m value, and then repeat step (1.2) (without increasing an iteration counter). Otherwise, if Q2 differs from Q1, proceed to step (1.3) (see Note 2).
  - **1.3:** If Q2 does not lie on the hyperplane $ax = ax^\circ$ associated with the current pseudo-cut (4) then redefine P0 = Q1, P1 = Q2, and return to step (1.1). (Figure 1 shows this case and Figure 2 shows this case after returning to step (1.1).) Otherwise, if Q2 lies on $ax = ax^\circ$, then proceed to Stage 2 (see Note 3). (Figure 3 shows this case.)
- **Stage 2:**
  - **2.1:** Remove the pseudo-cut (4) just added in step (1.2) and replace it with a new one determined as follows. Let $x' = P1$ and $x'' = Q2$, and determine a point $x^\circ = R1$ by (3) and (5), where m is chosen to satisfy $1 > m > 0$. (See Note 4 for choosing m large enough but less than 1.) If x° violates any pseudo-cut contained in the cut pool, remove this cut from the pool.
  - **2.2:** Add the new pseudo-cut (4) to the cut pool and apply the independent algorithm starting from the point R1. Let R2 denote the best point of the current execution. (a) If R2 = R1, then redefine P0 = Q1, P1 = Q2. Otherwise, (b) if R2 ≠ R1 *(Diagram 2.1 shows this case)*, then whether or not R2 lies on the cut hyperplane, redefine P0 = P1 and P1 = R2. In either case (a) or (b), return to step (1.1) of Stage 1 (see Note 5). *(Diagram 2.1.1 shows this case, inherited from (b), while Figure 6 shows the case inherited from (a). Both of these two diagrams also show the new P0, P1 and Q1, and the new pseudo-cut produced at step (1.1).)*

We observe that each time the method returns to step (1.1) in the Pseudo-Cut Generation Procedure, whether from step (1.3) or step (2.2), the current designation of P0 and P1 is compatible with the original designation, i.e., P0 always represents a point that has been used to start the independent algorithm and P1 represents the resulting best solution found on the current (most recent) execution of the algorithm.

We also remark that when the method specifies that the independent algorithm should start from Q1 in step (1.2) or from R1 in step (2.2), it may be preferable to start the method from a point slightly beyond this intersection with the current pseudo-cut hyperplane, to avoid numerical difficulties that sometimes arise in certain nonlinear methods if starting solutions are selected too close to the boundaries of the feasible region.

Figure 1. Stage 1: Q_2 not on hyperplane

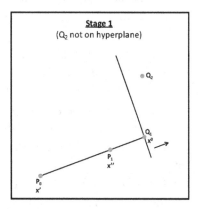

Figure 2. New state 1: start over

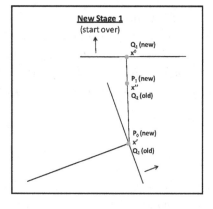

Figure 3. Stage 1: Q_2 on hyperplane

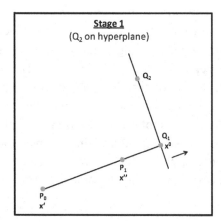

Illustrative Diagrams

The diagrams in Figures 1 through 6 illustrate several main components of the procedure.

A Rule for Dropping Pseudo-Cuts: We allow for pseudo-cuts to be dropped (removed from the cut pool) by a rule that goes beyond the simple provision for dropping cuts already specified in the algorithm. We consider the pseudo-cuts to have the same character as tabu restrictions that are monitored and updated in the short term memory of tabu search. We propose the use of two tabu tenures t_1 and t_2 for using such memory, where t_1 is relatively small (e.g., $1 \leq t_1 \leq 5$) and t_2 is selected to be larger (e.g., $7 \leq t_2 \leq 20$). (The indicated ranges are for illustrative purposes only.) Each pseudo-cut not dropped by the instructions stipulated in the algorithm will be retained for t_1 iterations (executions of step (1.1)) after the cut is created, and then dropped after this number of iterations whenever the cut becomes non-binding (the current solution x" produced by the independent algorithm does not lie on the cut hyperplane). However, on any iteration when no cut is dropped (either directly by the algorithm or by this rule), a second rule is applied by considering the set of all cuts that have been retained for at least t_2 iterations. If this set is non-empty, we drop oldest cut from it (the one that has been retained for the greatest number of iterations).

The following additional observations are relevant.

- **Note 1:** The values chosen for m are a key element of the cut generation strategy in its present variation, and will depend on such things as the sizes of basins of attraction in the class of problem considered. Within step (1.1), m may be chosen to be a selected default fraction greater than 1, but bounded from below by a value that assures x^o will lie a certain minimum distance beyond x".

- **Note 2:** To avoid numerical problems, it is appropriate to require that Q2 differ from

Figure 4. Stage 2: Q_2 on stage 1 hyperplane, R2 $\neq$R1

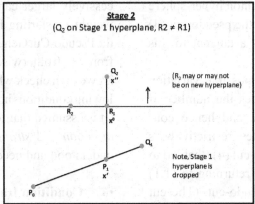

Q1 by a specified amount in step (1.2) in order to be considered "not equal" to Q1. Also, the increase in the value of m in step (1.2) can be chosen either as a default percentage increase or as an amount sufficient to assure that d grows by a specified value independent of this percentage. This value of m drops back to its original value whenever the method re-visits step (1.1), but if a succession of increases in step (1.2) causes the distance separating Q1 from P1 to exceed a specified threshold (anticipated to render all feasible solutions for the original problem inadmissible relative to the pseudo-cut (4) at step (1.2)), then the

procedure may be terminated or re-started from scratch from a new initial starting solution x' = P0 produced by a multi-start procedure, e.g., as described in Ugray et al. (2009).

- **Note 3:** In step (3.3) we require the point Q2 to lie a certain minimum distance from the hyperplane ax = ax° in order to be considered as not lying on the hyperplane.

- **Note 4:** The value of m in step (2.1) is assumed to be chosen to prevent the point Q1 from satisfying the pseudo-cut (4) produced in step (2.2). It suffices to choose m so that the distance of R1 from P1 is as least as great as the distance of Q1 from

Figure 5. New stage 1: start over Figure 6. Stage 2: followed by new stage 1, $R_2 = R_1$

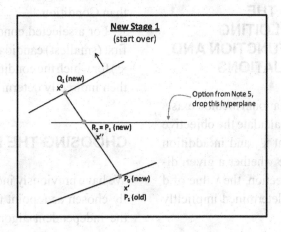

P1. (If this distance is the same, then R1 and Q1 will lie on a common hyper-sphere whose center is P1, and the pseudo-cut (4) of (2.2) is produced by a tangent to this hyper-sphere.)

- **Note 5:** An interesting possible variation in Step (2.2) that reduces the number of pseudo-cuts maintained, and hence constrains the search space less restrictively, is to drop the latest pseudo-cut (4) (that led to determining R2) before returning to (1.1) to generate the new pseudo-cut. (The cut thus dropped is not immediately relevant to the next step of the search in any event.) Another variation is to make sure that d is large enough to render the most recent Q2 infeasible relative to the pseudo-cut. This variation will avoid cases where sometimes Q2 may be revisited as a local optimum (The procedure may be monitored to see if multiple visits to the same Q2 point occur, as a basis for deciding if the indicated variation is relevant).

Finally, we observe that a simplified version of the Pseudo-Cut Generation Procedure can be applied that consists solely of Stage 1, with the stipulation in step (1.3) that the pseudo-cut (4) is generated and the method returns to (1.1) in all cases.

DETERMINATION OF THE DISTANCE D BY EXPLOITING QUICK OBJECTIVE FUNCTION AND DIRECTIONAL EVALUATIONS

In a context where a computational method exists that can relatively quickly calculate the objective function value for the point x^o, and in addition can fairly quickly calculate whether a given direction is an improving direction, the value of d that determines x^o can be determined implicitly rather than explicitly.

This is done by generating a number of successively larger candidate values for the scalar weight λ^o, starting from $\lambda^o > 1$ for Step (1.1) of the Pseudo-Cut Generation Procedure, and starting from $\lambda^o > 0$ otherwise. For each candidate value of λ^o, we then check whether one or more of the following conditions hold for the associated x^o vector (It is assumed that terms like *feasible improving direction* and *stronger improving direction* are understood and need not be defined).

- **Condition 1(a):** There exists a feasible improving direction from x^o that lies in the region satisfying the pseudo-cut (4).
- **Condition 1(b):** The direction from x^o on the ray for $\lambda > \lambda^o$ is a feasible improving direction.
- **Condition 2(a):** The improving direction from Condition 1 (for a given choice of 1(a) or 1(b)) is stronger than any feasible improving direction that does not lie in the region satisfying the pseudo-cut (4).
- **Condition 2(b):** The improving direction from Condition 1 (for a given choice of 1(a) or 1(b)) is stronger than the direction from x^o on the ray for $\lambda < \lambda^o$ (automatically satisfied the latter is not a feasible improving direction).

The conditions 1(b) and 2(b) are more restrictive than 1(a) and 2(a), respectively, but are easier to check. Condition 2 is evidently more restrictive than Condition 1.

For a selected condition, we then choose the first (smallest) candidate λ^o value (and associated x^o) for which the condition is satisfied. This choice then indirectly determines the distance d.

CHOOSING THE POINTS X' AND X"

We have previously indicated that x' is customarily chosen as a point that initiates the search of the independent algorithm, and x" denotes the best point determined on the current pass of the algorithm, as where x″ may denote a local optimum. We now consider other choices that can be preferable under various circumstances.

It is possible, for example, that an effort to determine a point x° according to Condition 1 or 2 of the preceding section will not be able to identify a feasible point that qualifies. In this case, it may be preferable to reverse the roles of x′ and x″ to seek a qualifying x° on the ray leading in the opposite direction. Moreover, it may be worthwhile to examine the option of reversing the roles of x′ and x″ in any event, where the ultimate choice of which point qualifies as x′ will depend on the evaluation of the point x° that is generated for each case.

Still more generally, the collection of candidate points from whose members a particular pair of points x′ and x″ will be chosen can be generated by a variety of considerations, including those used in composing a Reference Set in Scatter Search (see, for example, Glover, Laguna, & Marti, 2000; Marti, Glover, & Laguna, 2006). Likewise the criteria for selecting x′ and x″ from such a collection can also incorporate criteria from Scatter Search. Here, however, we suggest three alternative criteria.

- **Criterion 1:** Let x′(i) and x″(i), i = 1,...,i*, identify the points used to determine previous pseudo-cuts (i.e., those successfully generated and introduced at some point during the search). Let x*(i) identify the point on the ray from x′(i) through x″(i) that lies a unit distance from x′(i). Finally for a candidate pair of points x′ and x″, let x* denote the point on the ray from x′ through x″ that lies a unit distance from x′. From among the current pairs x′ and x″, we select the one such that x* maximizes the minimum distance from the points x*(i), i = 1,...,i*.

- **Criterion 2:** Choose the candidate pair x′ and x″ by the same rule used in Criterion

1, except that x*(i) is replaced by the point x°(i) (the "x° point" previously determined from x′(i) and x″(i)), and x* is likewise replaced by the point x° determined from the currently considered x′ and x″.

- Criterion 2 allows for the possibility that x′ and x″ may lie on the same ray as generated by some pair x′(i) and x″(i), provided the point x° lies sufficiently distant from the point x°(i). This suggests the following additional criterion.

- **Criterion 3:** Employ Criterion 1 unless the minimum distance of the selected point x* from the points x*(i), i = 1,...,i* falls below a specified threshold, in which case employ Criterion 2.

A variant on Criterion 3 is to employ Criterion 1 except where the minimum distance determined from Criterion 2 exceeds a certain lower bound, where this latter may be expressed in terms of the minimum distance obtained for Criterion 1.

ADDITIONAL CONSIDERATIONS FOR CHOOSING X°

To this point we have assumed that x° will lie beyond x″ on the ray leading from x′ through x″, on each execution of Step (1.1) of the Pseudo-Cut Generation Procedure. However, in some case, as in the customary application of Scatter Search, it may be preferable to select a point x° that lies between x′ and x″. We add this possibility as follows.

First, we stipulate that the candidate values for $\lambda°$ lie in the interval $0 < \lambda° < 1$. Second, we apply Condition 1 or Condition 2 (in either the (a) or (b) form)) to determine a value $\lambda°_{min}$ which is the least $\lambda°$ value that satisfies the condition (assuming such a value exists in the interval in the interval $0 < \lambda° < 1$). Next, we examine the candidate $\lambda°$ values in the reverse direction (from larger to smaller) in the interval $\lambda°_{min} < \lambda° < 1$, and choose one of the following Reverse Conditions as a basis for choosing a particular candidate value.

- **Reverse Condition 1(a):** There exists a feasible improving direction from x^o that lies in the region not satisfying the pseudo-cut (4).
- **Reverse Condition 1(b):** The direction from x^o on the ray for $\lambda < \lambda^o$ is a feasible improving direction.
- **Reverse Condition 2(a):** The improving direction from Reverse Condition 1 (for a given choice of 1(a) or 1(b)) is stronger than any feasible improving direction that lies in the region satisfying the pseudo-cut (4).
- **Reverse Condition 2(b):** The improving direction from Reverse Condition 1 (for a given choice of 1(a) or 1(b)) is stronger than the direction from x^o on the ray for $\lambda > \lambda^o$ (automatically satisfied if the latter is not a feasible improving direction).

Finally, we identify the first (largest) λ^o candidate value satisfying the selected Reverse Condition, denoted by λ^o_{max} (provided such a value exists in the indicated interval), and choose $\lambda^o = (\lambda^o_{min} + \lambda^o_{max})/2$. This final λ^o value is the one used to find a point strictly between between x' and x'' from which to launch a new search. This search can optionally be constrained by adding a pseudo-cut (4) for x^o determined from $\lambda^o = \lambda^o_{min}$ (or from a "reverse" pseudo-cut determined from $\lambda^o = \lambda^o_{max}$).

From among the various candidate values x^o identified for launching a new search as above, and also from among those that may be identified from applying Condition 1 or 2 for $\lambda^o > 1$ (allowing x' and x'' to be interchanged), one may ultimately choose the option such that x^o receives a highest evaluation. This evaluation can be in terms of objective function value (possibly considering directional improvement), or in terms of maximizing the minimum distance of x^o from points

in a Reference Set. By such a use of a Reference Set, the approach can foster diversity in conjunction with the search for improvement. In fact, the indicated strategies can be used to create rules for a version of Scatter Search that differs from more customary forms of the method.

It should be noted that these strategies for choosing x^o vectors can be used without bothering to introduce pseudo-cuts. For example, such a strategy can be employed for some initial duration of search to produce x^o trial solutions, and then the pseudo-cuts can subsequently be invoked to impose greater restrictiveness on the search process.

CONCLUSION

The proposed collection of pseudo-cut strategies for global optimization expands the options previously available for guiding solution processes for non-convex nonlinear optimization algorithms. These strategies can be used to supplement other approaches for solving such problems, or can be used by themselves. The mechanisms proposed for generating trial solutions can similarly be used in a variety of ways, and may even be used independently of the pseudo-cuts themselves. The demonstration that an exceedingly simplified instance of a pseudo-cut strategy succeeded in enhancing a non-convex optimization method in Lasdon et al. (2010) suggests the potential value of more advanced pseudo-cut strategies as described here, and of empirical studies for determining which combinations of these strategies will prove most effective in practice. The use of pseudo-cuts reinforces the theme of joining mathematically based exact methods for convex problems with special strategies capable of modifying these methods to enable them to solve non-convex problems. In this guise, the proposals of this paper offer a chance to create a wide range of new hybrid algorithms that marry exact and metaheuristic procedures.

REFERENCES

April, J., Better, M., Glover, F., Kelly, J., & Laguna, M. (2006). Enhancing business process management with simulation-optimization. In *Proceedings of the Winter Simulation Conference* (pp. 642-649).

Better, M., Glover, F., & Laguna, M. (2007). Advances in Analytics: Integrating dynamic data mining with simulation optimization. *IBM Journal of Research and Development*, *51*(3-4), 477–487. doi:10.1147/rd.513.0477

Bowman, V. J., & Glover, F. (1972). A note on zero-one integer and concave programming. *Operations Research*, *20*(1), 182–183. doi:10.1287/opre.20.1.182

Glover, F. (1989). *Tabu Search - Part I. ORSA Journal on Computing*, *1*(3), 190–206.

Glover, F., Laguna, M., & Marti, R. (2000). Fundamentals of scatter search and path relinking. *Control and Cybernetics*, *29*(3), 653–684.

Lasdon, L., Duarte, A., Glover, F., Laguna, M., & Marti, R. (2010). Adaptive memory programming for constrained global optimization. *Computers & Operations Research*, *37*, 1500–1509. doi:10.1016/j.cor.2009.11.006

Martí, R., Glover, F., & Laguna, M. (2006). Principles of scatter search. *European Journal of Operational Research*, *169*, 359–372. doi:10.1016/j.ejor.2004.08.004

Ugray, Z., Lasdon, L., Plummer, J., & Bussieck, M. (2009). Dynamic filters and randomized drivers for the multi-start global optimization algorithm MSNLP. *Optimization Methods and Software*, *24*, 635–656. doi:10.1080/10556780902912389

APPENDIX

Figure 7. Pseudo-code for the Pseudo-Cut Method (Initial Simplified Version)

Initialization

1. Let *TabuCutList* be the memory list of pseudo-cuts with *TabuTenure* size
2. Let $m = 0.1$, *MaxIter* = 5 and *GlobalIter*= 20
   Maxpert = 5 and *Iter1* = 0. % Maxpert is a limit on *pert*
   % The preceding values in 2. are suggestive only
3. Let *P0* be a *random point and Best_f = f(P0)*

While (*Iter1* < *GlobalIter*)

4. *TabuCutList* = $\varnothing$
5. *improved* = FALSE
6. *pert* = *m* and *Iter2* = 0
7. *P1* = LS(*P0, f(x),G,S*)
If (*f(P1)* < *Best_f*)
   8. *Best_f = f(P1)*

While (*Iter2* < *MaxIter* **and** $\|P1 - P0\| > \underline{minDist}$)

   //**Begin Stage 1**
   9. *Q1 = P0 + (1 + pert)*(P1 − P0)*
   10. Remove from *TabuCutList* the cuts violated at *Q1*. If no cuts are removed and if
       there are any cuts retained more than t_2 iterations, drop the oldest. (Disregard this
       last instruction if 9 and 10 are reached from 16, below.)
   11. Add to *TabuCutList* the cut *pcut(Q1)*: $(P1 − P0) x \geq (P1 − P0) Q1$
   12. *Q2 = LS(Q1, f(x),G$\cup$TabuCutList, S*)
   If (*f(Q2)* < *Best_f*)
       13. *improved* = TRUE
       14. *Best_f = f(Q2)*; *P0 = Q1*; *P1 = Q2*; *pert = m*
       15. go to 22
   If ($\|Q1 − Q2\| < minDist$)
       16. Drop the cut just added in 11, set *pert = pert + m*. If *pert* > *Maxpert*, go to 23.
           Otherwise, return to 9.
   Else
       17. *pert = m*
   If (*Q2* does not satisfy *pcut(Q1)* with equality (**and** $\|Q2 − Q1\| > minDist$))
       18. *P0 = Q1*; *P1 = Q2*; Proceed to 22 (increase Iter2 and repeat Stage 1)
   //**Begin Stage 2**
   (Here *Q2* satisfies *pcut(Q1)* with equality and $\|Q2 − Q1\| > minDist$)
   19. Drop the cut just added in 11 and record *P01 = P1* and *Q01 = Q1*.
   20. *P0 = P1*; *P1 = Q2*; *Q1 = P0 + (1 - pert)*(P1 - P0)*
   21. Execute instructions 10 − 15, but without dropping any cuts in 10 other than
       those violated at Q1.
       If ($\|Q1 − Q2\| < minDist$)
           21.1. *P0 = Q01* (P1 is unchanged);
       Else ($\|Q1 − Q2\| \geq minDist$)
           21.2. *P0 = P01*; *P1 = Q2*
   22. *Iter2 = Iter2 + 1* (Return to 9 if *Iter2* < *MaxIter*.)

23. Generate a new point *P0* by a diversification step, *Iter1 =Iter1+1* (and return to 4 if
    Iter1 < *GlobalIter*)

This work was previously published in the International Journal of Applied Metaheuristic Computing, Volume 2, Issue 4, edited by Peng-Yeng Yin, pp. 1-12, copyright 2011 by IGI Publishing (an imprint of IGI Global).

Chapter 13
Management of Bus Driver Duties Using Data Mining

Walid Moudani
Lebanese University, Lebanon

Félix Mora-Camino
Ecole Nationale de l'Aviation Civile, France

ABSTRACT

This paper presents Bus Driver Allocation Problem (BDAP) which deals with the assignment of the drivers to the scheduled duties so that operations costs are minimized while its solution meets hard constraints resulting from the safety regulations of public transportation authorities as well as from the internal company's agreements. Another concern in this study is the overall satisfaction of the drivers since it has important consequences on the quality and economic return of the company operations. This study proposes a non-dominated bi-criteria allocation approach of bus drivers to trips while minimizing the operations cost and maximizing the level of satisfaction of drivers. This paper proposes a new mathematical formulation to model the allocation problem. Its complexity has lead to solving the BDAP by using techniques such as: Genetic algorithms in order to assign the drivers with minimal operations cost and Fuzzy Logic and Rough Set techniques in order to fuzzyfied some parameters required for the satisfaction module followed by applying the Rough Set technique to evaluate the degree of satisfaction for the drivers. The application of the proposed approach to a medium size Transport Company Bus Driver Allocation Problem is evaluated.

INTRODUCTION

The management of planning problems confronting logistics service providers frequently involves complex decisions. For more than four decades now the Bus Driver Allocation Problem (BDAP)

in the Road Passenger Transport Company (RPTC) has retained the attention of the Management and Operations Research community since bus drivers management is extremely complex and may generate many problems that hinder the smooth operation, influencing the total transport

DOI: 10.4018/978-1-4666-2145-9.ch013

company's operations profit. Therefore, RPTC considers that the efficient management of their drivers is a question of the highest economic relevance. Many studies have tried to solve the BDAP problem either using optimization methods or complex type of heuristic methods. Unfortunately, the exact numerical solution of the associated large scale combinatorial optimization problem is very difficult to obtain. Early rules of thumb have been quickly overrun by the size of the practical problems encountered (hundreds or thousands of drivers to be assigned to at least as many duties) and by the complexity of the set of constraints to be satisfied, leading very often to poor performance solutions. More recently, with the enhancement of computer performances, optimization approaches have been proposed to solve this problem: mathematical programming methods (large scale linear programming and integer programming techniques) (Desrochers, Gilbert, Sauve, & Soumis, 1992; Curtis, Smith, & Wren, 1999; Smith, Layfield, & Wren, 2000), artificial intelligence methods (logical programming, simulated annealing, neural networks, fuzzy logic and genetic algorithms) as well as heuristic approaches and their respective combinations (Brusco, 1993; Clement & Wren, 1995; Beasley, 1996; Sengoku & Yoshihara, 1998; Yoshihara & Sengoku, 2000; Aickelin & Dowsland, 2003). Many studies refer to the BDAP which is a static decision problem, based on a monthly table of trips, and devoted exclusively to the minimization of RPTC operations costs. This problem is in general split in two sub problems: the first problem consists to generate the set of trips (or rotations) which cover optimally all the scheduled duties. A trip is composed of a set of duties that start and take end in the same base. The second problem is called BDAP where the retained trips are assigned to the drivers.

In this paper, we introduce the BDAP as a bi-criterion decision problem where the main decision criterion is the drivers operations cost of the RPTC and the secondary decision criterion is

relative to the drivers overall degree of satisfaction. A set of hard and soft constraints are considered while solving the BDAP. The solution produced some knowledge which is used with the previous satisfaction level reached in the latest scheduling in order to get the current individual satisfaction level for the next planning by applying a classification process. The presence of multiple objectives in a problem, gives rise to a set of Pareto-optimal solutions. Several approach proposed in literature aims to detect the Pareto optimal solutions and capture the shape of the Pareto front. Over the past decade, a number of multi-objective evolutionary algorithms have been suggested (Horn, Nafploitis, & Goldberg, 1994; Srinivas & Deb, 1995; Deb, 2001). The primary reason for this is their ability to find multiple Pareto-optimal solutions in one single simulation run. Evolutionary Algorithms (EA) are good candidate to multiple objective optimization problems due to their abilities to search simultaneously for multiple Pareto-optimal solutions and, perform better global search of the search space. In recent years, many evolutionary techniques for multi-objective optimization have been proposed (Deb, 2001). Several multi-objective evolutionary algorithms are presented in literature such as: Non-dominated Sorting Genetic Algorithm (Zitzler, Deb & Thiele, 2000; Deb, Pratap, Agarwal, & Meyarivan, 2002), Multi-Objective Particle Swarm Optimization, and Multi-Objective Bacteria Foraging Optimization algorithm.

This paper is organized as follows: an overview of research found in literature about this problem is presented; followed, by a brief description of an efficient classification technique such as the Rough Sets technique. Our proposed solution approach deals with the BDAP based firstly, on cost minimization using Genetic Algorithms technique and secondly, on maximization of the bus drivers satisfaction in RPTC using the fuzzy logic and rough set techniques in order to enhance the quality of the "*line of work*". We describe our solution approach through a numerical example applied to a medium size BDAP followed by discussion and analysis of the results obtained. Finally, we ended by a conclusion concerning this new approach and the related new ideas to be tackled in the future.

THE BUS DRIVERS ALLOCATION PROBLEM IN RPTC

The Bus Driver Allocation Problem in RPTC is treated in general once the schedule of the trips has been established for the next month and once the available buses have been assigned to the scheduled trips. Two classes of constraints are considered in order to produce the "*line of work*" for the drivers over the planning period: hard constraints whose violation impair the security of the trip (driver qualifications, national regulations concerning duration of work and rest times, training requirements) and soft constraints (internal company rules, agreements with unions regarding the driver's working and remuneration conditions, holidays and declared claims by the drivers) which are relevant to build the drivers schedule but whose relaxation may lead to lower cost solutions. While some of these soft constraints are common to most RPTC, others are only relevant for some classes of RPTC and some few are specific to a given RPTC. The primary objective sought by RPTC at this level of decision making is to minimize the drivers related operations costs, so in most research studies, the BDAP has been formulated as a mono-criterion minimization problem. It is known that mathematical programming formulations of this problem are generally too complex to be exactly solved for real-world applications. A sub optimal but widely accepted approach to tackle more efficiently the BDAP, which is of the NP-hard computational complexity class (Fischetti, Martello, & Toth, 1989; Desrochers, Gilbert, Sauve, & Soumis, 1992), consists in decomposing it in two sub-problems of lower difficulty. In the first sub-problem, the RPTC involves the construction of an efficient set of trips (a trip is a sequence of duties which starts and ends at the same base of RPTC base while meeting all relevant legal regulations) which covers the whole programmed duties. In the second sub-problem, the RPTC considers the assignment of the drivers to the generated set of trips so that an effective "*line of work*" is obtained for each driver (Curtis, Smith, & Wren, 1999).

In literature, a set of various approaches have been proposed to tackle with these two sub-problems related to RPTC. The several approaches proposed in the literature which are detailed and discussed here, can be classified as:

- Approaches for solving set covering and/or set partitioning problems. These approaches are based on classical techniques of optimization in graphs;
- The column generation method which consists in generating a subset of trips (or rotations) since the number of possible trips is extremely high and cannot be treated globally; and subsequently, generating eligible workloads for each crew member;
- Evolutionary Algorithms and Heuristic methods that generate solutions directly hoped to be effective while overcoming the computer constraints (CPU time, memory space).

Regarding the research found in literature, the BDAP problem is very rich and can be divided into two main classes, namely theoretical and application-based papers. The first class of theoretical papers deals with formulations and algorithms for simplified crew scheduling problems with only a few constraints. In the second class those algorithms are applied to solve problems which arise from practical applications. The set of papers gathered from literature dealt not only with the bus transport field, but also with the field of railways (Caprara, Fischetti, Guida, Toth, & Vigo, 1999b; Freling, Lentink, & Odijk, 2001;

Kroon & Fischetti, 2001) and airlines (Desaulniers, Desrosiers, Dumas, Marc, Rioux, Solomon, & Soumis, 1997; Desaulniers, Desrosiers, Lasry, & Solomon, 1999). Thus, our presentation for the related works in literature will cover all these fields while illustrating the main related research activities.

Solving the Problem of Construction of the Trips (Rotations)

In this section, we present the problem of BDAP which aims to define the set of trips that cover all scheduled duties and assign the crew to the generated trips (rotations). The results issued from literature show that this problem is formulated and solved based on: set covering/partitioning problem, Column generation approach, and the heuristics and its combinations. We present in details the research found in literature dealing with this problem:

Set Covering and Partitioning Problems

Rubin, (1973), Ball and Roberts, (1985), Balas and Ho, (1980), and Crainic and Rousseau (1987) are among the first having proposed to deal with the problem of building the crew rotations in airlines, as set covering/partitioning problem. The choice of a formalism corresponding to this approach will lead to the definition of two sub-problems: the generation of the constraint matrix where each column corresponds to an acceptable rotation; and, the optimization of this matrix by using different appropriate techniques. Numerous studies have subsequently been developed based on this resolution principle. We note the following work:

- Marsten (1979) presented a heuristic solution to combine the simple flights in order to form flight services. This leads to reduce the size of the matrix constraints. Then, based on this set of flight services found, a model of set partitioning problem

is adopted. A branch and bound method is also used to generate solutions with integer values where the values of bounds are obtained by linear relaxation. Shepardson (1981) proposed to obtain the values of bounds by using the method of Lagrangian relaxation and subgradient method. The numerical tests performed had seen that the computation time taken by this approach is much lower.

- Ball (1985) proposed a solution approach leading to generate a subset of feasible rotations by finding disjoint paths in a directed graph where each path corresponds to a feasible rotation. Then, a coupling method is used to improve the quality of initial rotations. This process is continued until no improved solutions can be found.

- Anbil (1991) proposed a solution aiming to consider the problem in a global approach rather than local. A heuristic is used to generate a large number of rotations instead of just covering some scheduled flights. The primal simplex method is used to solve the set partitioning problem using the linear relaxation. Dual variables obtained to assess the trips with a negative marginal cost thereby reducing the overall cost associated of the problem. This process is repeated until no new rotations can integrate the constraint matrix. The heuristic of Ryan and Foster (1981) is used to find the integer solution since the method of branch and bound presents enormous difficulties.

- Hoffman (1993) presented a solution approach using the method of branch and cut to solve the problems of set partitioning of large dimensions. The basic idea is to analyze the impact of the size of the considered problem on the quality of the solution. An application of this method has provided a gain of 0.5% in terms of operating cost associated with the construction of trips with respect to the method of Rubin (1973).

This technique uses a heuristic method to find a solution with integer values. Cuts (or constraints) are added to reduce the search space in the polyhedron of solutions.

- Smith and Wren (1988) formulated the driver scheduling problem as an integer linear programming (ILP) model. An ILP solution approach becomes complex since the problem size increased. Therefore, Smith and Wren (1988) developed a set of heuristics which reduced the problem size by first eliminating relief opportunities which were unlikely to be useful, then restricting the potential shifts generated to those which satisfied not only the legal requirements but also some parameters relating to the precise work content, and finally filtering out some of the generated shifts which compared unfavorably with others. The ILP model was solved by first relaxing the 0-1 conditions, and then using a specially developed branch and bound process which took advantage of the authors' knowledge of driver scheduling.

Column Generation Approach

Optimizing the construction of trips with a formulation of set covering/partitioning problems requests to enumerate all eligible trips and then solve the problem where each column corresponds to a decision variable. The resolution of this problem becomes impassable when it reaches a large size. Hence, the importance of the method of column generation. The new variables (columns) are generated by solving a sub-problem. So, unlike the traditional approach of set covering/partitioning, the column generation approach solves the problem by repeatedly generating the matrix of constraints. It has the advantage of offering the user a choice between obtaining an optimal solution or a feasible solution at an acceptable cost. On the other hand, a prohibitively time becomes necessary when the problem size increases. Col-

umn generation is a technique used for problems with a huge number of variables. The general idea, introduced by Dantzig and Wolfe (1960), is to solve a sequence of reduced problems, where each reduced problem contains only a small portion of the set of variables (columns). After a reduced problem is solved, a new set of columns is obtained by using dual information of the solution. The column generation algorithm converges once it has established that the optimal solution based on the current set of columns cannot be improved upon by adding more columns. The set of columns can be enumerated or generated heuristically. Then the optimal solution of the reduced problem is the optimal solution of the overall problem. Many authors such as Crainic and Rousseau (1987) and Desrosiers (1997) have used the technique of column generation to optimize the construction of crew rotations in airlines. Using this technique leads to a decomposition of the main problem into: firstly, a master problem for solving a problem of set covering or set partitioning, and secondly, a sub-problem for constructing the columns associated with rotations. The master problem involves solving a relaxed linear program to find a solution associated with branch and bound method to find a solution with integer values. The sub-problem in making columns allows the construction of columns associated with rotations at minimum marginal cost. It may be in the form of a research problem of constrained shortest path in a graph. If the marginal cost of a column (a path found eligible in the graph) is negative, the column is added to the master problem. If we cannot generate negative marginal cost columns, the resulting solution is optimal.

If (some of) the decisions variables values need to be integer, different algorithms, exact or heuristically, can be used. Branch-and-bound algorithms are most often used to solve the resulting (mixed) integer programming problem (Barnhart, Johnson, Nemhauser, Savelsbergh, & Vance, 1998). Hoffman and Padberg (1993) describe a branch-and-cut algorithm where cutting planes are generated based on the underlying structure of the polytope. In Caprara, Fischetti, and Toth (1999a), Lagrangian heuristics and variable fixing techniques are used to find integer solutions. Mingozzi, Boschetti, Ricciardelli, and Bianco (1999) suggest an approach to compute a dual solution of the LP-relaxation combined with Lagrangian relaxation (Desrochers & Soumis, 1989; Desrochers, Gilbert, Sauve, & Soumis, 1992; Carraresi, Girardi, & Nonato, 1995). Afterwards, the dual solution is used to reduce the number of primal variables such that the resulting problem can be solved by a branch and bound algorithm. Furthermore, the different procedures use lower bounds of previous procedures and improve them. In some of these procedures Lagrangian relaxation and column generation are used. A completely different formulation is used by Fischetti, Lodi, Martello, and Toth (2001), namely a single commodity flow formulation. The disadvantage of such a formulation is the poor LP relaxation, however, they solve this weakness by adding valid inequalities to strengthen the formulation. Experiments were carried out by (Lavoie, 1988) showing that the best results, in computing time, are obtained by integrating multiple columns in the matrix of constraints. This reduces the number of calls to the column generation process that generates a waste of time much larger than a few pivoting operations in the simplex method. Steinzen, Gintner, Suhl, and Kliewer (2010) discuss the integrated vehicle-and crew-scheduling problem in public transit with multiple depots. It is well known that the integration of both planning steps discloses additional flexibility that can lead to gains in efficiency, compared to sequential planning. He proposed a new modeling approach based on a time-space network representation of the underlying vehicle-scheduling problem. The integrated problem is solved with column generation in combination with Lagrangian relaxation. The column generation sub-problem is modeled as a resource-constrained shortest-path problem based on a novel time-space network formulation.

Feasible solutions are generated by a heuristic branch-and-price method that involves fixing service trips to depots.

TRACS II, was developed specifically to satisfy the needs of rail operation, while still being capable of scheduling bus drivers. It uses a set covering model, but both the heuristics which generate potential shifts, and the ILP process, are completely new. The former generate far more potential shifts than previously, while the latter incorporates a new column generation approach as well as better optimization procedures (Fores, Proll, & Wren, 1999; Kwan, Parker, & Wren, 1999). Layfield, Smith, and Wren (1999) have used constraint programming to build several alternative sets of shifts covering critical periods of the day. Relief opportunities which do not appear in any of the generated sets are then removed, and the resultant smaller problem is speedily and efficiently solved by TRACS II. To date tests have only been carried out on bus scheduling problems from a single depot, and it is not yet known whether this is a suitable approach for more complex problems.

Heuristics Methods

Some of the earliest work on bus driver scheduling was undertaken in the early 1960s and it was based on developing an exhaustive tree search approach, which failed to produce good solutions even on small problems. Elias (1966) and Ward and Deibel (1972) developed a series of heuristics which produced a large number of different schedules quite quickly, then chose the best. Reasonable results were obtained in some practical, lightly constrained, situations. Thus, heuristic approaches have provided an alternative to overcome the constraints associated with the drivers scheduling. These methods are often inspired by the knowledge gained by practitioners who have developed empirically simplified methods of solving the problem. Among these methods, is retained mainly:

- **The Method "day by day":** This method is proposed, for the case of airlines, in which the assignment of the crew is performed, chronologically, based on the day unit time (Marchettini, 1980; Sarra, 1988). The set of assignments defined for the day "*j*" and which are assigned to crew based on their availability and respecting the various constraints, become constraints in the processing of the assignment process for the next day "*j +1*". This process continues until all the trips, arranged chronologically, are covered by the required crew. Moreover, it is considered in their approach to solving the pre-allocation of activities such as training, medical and flight crew desires. These activities are pre-assigned to different crew and become hard constraints during the process of crew scheduling. This method leads to daily assignments more and more degraded and can generate a lot of uncovered rotations. Using this method makes it difficult to predict the consequences of previous decisions on the difficult problems to solve in the coming days. This can be seen especially in peak periods (summer and holidays).

- **The Method "pilot by pilot ":** This method is proposed to perform the assignment by considering the crew members one after the other (Byrne, 1988). In this approach, we select an arbitrary crew "*i*" and assigning to him a monthly workload. This process continues until all crew is assigned a monthly workload. The disadvantage of this method is relative to the unequal treatment which is subject to different crew, as the first crew selected will be favored over the latter. Here too, at the end of the assignment process, in order to cover the residual rotations, it will be necessary to assign overtime, and/or to over-size the crew. Moreover, both methods offer a limited calculation in difficulty and in volume.

On the other hand, misuse of the crew may leave uncovered rotations.

- **The Combination of the Two Methods "day by day" and "pilot by pilot":** Another solution approach aims to combine the two previous methods was proposed by Antosik (1978). This approach of resolution can be summarized by the following algorithmic scheme:

Order rotations in chronological order;

Select all rotations from the first day of the period. Let "*j*" the first rotation selected; we consider all available crew members to carry on this rotation "*j*". For every available crew "*i*", we determine an equivalent cost "c_{ij}" associated with the assignment of the rotation "*j*" to crew "*i*". If the pair (i, j) is the lower cost, the rotation "*j*" will be assigned to the crew "*i*".

Go to the next day and return to step 2.

Antosik (1978) has developed a method combining the methods "pilot by pilot" and "day by day". It aims to assign all rotations scheduled to qualified pilots while minimizing the disproportionate workload (flight time, number of working days, number of rotations desired by pilots and assigned to them, number of rotations unwanted by pilots and which do not affect them) between different crew. First, the method "pilot by pilot" is applied where each crew is assigned as many rotations as possible. Its first objective is to cover all rotations scheduled by pilots. Once this step is completed, the method "day by day" is used to reduce the disproportionate workloads, the working hours, the number of working days and the degree of satisfaction related to the crew members.

- **A Method Based on the Attractiveness of Rotation and Staff Seniority:** Glanert (1984) and Byrne (1988) have developed a heuristic approach based on the seniority of the crew. This is to construct, in a sequential approach, the monthly workload for each crew member, ranked in descending order of seniority. Each crew is assigned satisfactorily workload by assigning priority to activities (a particular rotation, a rest period, an attractive destination, etc.) that he seeks. This workload will be complemented by other activities in order to assign a sufficient workload. Then, a validation of the workload set is performed to check if the residual rotations can be covered. This process continues until all the rotations are covered. An improvement of this approach was proposed in which not only the seniority of the crew is considered but also the attractiveness of rotations. This approach has drawbacks. For example, by treating the most attractive activities one after the other, we don't guarantee that the workload is satisfactory (inequality in the distribution of the workload). Thus, we must go back and ensure the admissibility of this workload. On the other hand, the optimization process for the entire crew members may also have to go back because the phase of validation of the workload is incomplete at a current time.

Other approach deals with this problem by finding a set of bus driver duties which covers a given set of bus schedule so as to minimize the number of drivers and satisfy a set of constraints given by the union contract and company regulations (Martello & Toth, 1986). This approach identifies the basic components of the problem and solves each of them either exactly or heuristically. The method is particularly suitable for cases where heavy constraints are present. The package based on this algorithm is currently used by a public transit company. Another approach based on heuristic is developed to analyze bus driver scheduling problems and produce an estimate of the number of drivers required for a bus schedule (Liping, 2006).

Evolutionary Methods

A pure evolutionary algorithm has been proposed in the literature to tackle with this problem while overcoming the complexity structure and calculation. Wren and Wren (1995) used genetic algorithms. They represent a complete schedule as a binary chromosome with one gene position for each shift in the schedule, the value in a cell being the index of a shift in the large set. Thus different schedules have different chromosome lengths, depending on the number of shifts in the schedule. The initial population is generated by a constructive heuristic used to obtain the starting solution to the set covering process. Parents are chosen based on their fitness (number of shifts and cost of schedule), and offspring are obtained by first forming the union of the parents and then eliminating any shifts whose entire work is covered by other shifts. Experiments indicated that three parents are better than two, while using four or five parents is counter-productive. Moreover, results were obtained for small problems, up to 16 shifts. Subsequent research was directed at extending the scope, but did not prove productive in the time available.

Curtis, Smith, and Wren (1999) have presented a genetic algorithms approach which is successfully used to overcome the limitations of the established integer linear programming system. The proposed approach is a hybrid in which all probable potential shifts are generated according to well developed heuristics. This system was designed for urban bus operations, although it was also successfully applied to rural and short interurban services. Selection of such shifts to form a schedule is modeled as a set covering problem, and the relaxation of this problem ignoring integer conditions is solved to optimality. A GA then develops a solution schedule based on some of the characteristics of the relaxed solution. Curtis, Smith, and Wren (2000) have investigated another approach using constraint programming, and subsequently produced promising results on relatively small problems with a mixture of a neural network method and repair heuristics. Forsyth and Wren (1997) reports on the early stages of the development of an ant colony system for driver scheduling. Lucic (1999) proposed a solution approach for crew scheduling in airlines by using the Simulated Annealing method. Firstly, he applied heuristics methods (Byrne, 1988) in order to generate an acceptable initial solution; therefore, he used the Simulated Annealing method in order to improve the solution initially found.

DATA MINING AND THE ROUGH SET THEORY

Introduction

The 1990s has brought a growing data glut problem to many fields such as science, business and government. Our capabilities for collecting and storing data of all kinds have far outpaced our abilities to analyze, summarize, and extract "knowledge" from this data (Hang & Kopriva, 2006; Awad & Latifur, 2009; Vercellis, 2009). Traditional data analysis methods are no longer efficient to handle voluminous data sets. Thus, the way to extract the knowledge in a comprehensible form for the huge amount of data is the primary concern. Data mining refers to extracting or "mining" knowledge from databases that can contain large amount of data describing decisions, performance and operations. However, analyzing the database of historical data containing critical information concerning past business performance, helps to identify relationships which have a bearing on a specific issue and then extrapolate from these relationships to predict future performance or behavior and discover hidden data patterns. Often the sheer volume of data can make the extraction of this business information impossible by manual methods. Data mining is a set of techniques which allows extracting useful business knowledge, based on

a set of some commonly used techniques such as: Statistical Methods, Neural Networks, Decision Trees, Bayesian Belief, Genetic Algorithms, Rough Sets, and Linear Regression, etc. (Clement & Wren, 1995; Hang & Kopriva, 2006).

Presentation of Rough Sets Theory

Pawlak (1982, 1991) introduced the theory of Rough Sets (RS). This theory was initially developed for a finite universe of discourse in which the knowledge base is a partition, which is obtained by any equivalence relation defined on the universe of discourse. Rough Set theory is an efficient technique for knowledge discovery in databases. It is a relatively new rigorous mathematical technique to describe quantitatively uncertainty, imprecision and vagueness. It leads to create approximate descriptions of objects for data analysis, optimization and recognition.

Hu, Lin, and Jianchao (2004) presented the formal definitions of RS theory and described its basic concepts. In RS theory, the data is organized in a table called decision table. Rows of the decision table correspond to objects, and columns correspond to attributes. In the data set, a class label to indicate the class to which each row belongs. The class label is called as decision attribute, the rest of the attributes are the condition attributes. Therefore, the partitions/classes obtained from condition attributes are called elementary sets, and those from the decision attribute(s) are called concepts. Let's consider C for the condition attributes, D for the decision attributes, where $C \cap D = \varphi$, and t_j denotes the jth tuple of the data table. The goal of RS is to understand (construct rules for) the concepts in terms of elementary sets, i.e., mapping partitions of condition attributes to partitions of decision attribute (Vercellis, 2009). However, a rough set is a formal approximation of a crisp set in terms of a pair of sets which give the lower and the upper approximation of the original set. Once the lower and upper approximation is calculated, positive, negative, and boundary regions can be derived from the approximation (Zhong & Skowron, 2001).

A MATHEMATICAL FORMULATION OF THE BDAP

In literature, the optimization approach for solving the driver assignments problem adopted considers unanimously a cost criterion. But the social history of this sector in recent decades has shown that everything was not just a question of economic costs but on the contrary, some elements which are difficult to be quantified could easily disrupt the solutions and lead to spectacular poor economic performance. The BDAP has been formulated as a zero-one integer mathematical programming problem where the driver operations cost is the criterion to be minimized under a finite set of hard constraints (Ryan & Foster, 1981). The classic formulation of the BDAP corresponds to a partition problem manipulating sets of binary decisions variables. The mathematical formulation is given as follows:

$$Minimize \quad f_1 = \sum_{j \in J^t} \sum_{i \in A_j} dh_{ij} x_{ij} + \sum_{i \in I^t} hs_i \tag{1}$$

subject to

$$\sum_{i \in A_j} x_{ij} = 1, \ \forall j \in J^t \tag{2}$$

$$\left(x_{ij_1} + x_{ij_2} \right) \leq 1, \ \forall j_1 \in J^t, j_2 \in O_{j_1}, i \in I^t \tag{3}$$

$$\sum_{j=1}^{m} (dv_j + dh_{ij}) x_{ij} + hs_i \leq LH, \ \forall i \in I^t \tag{4}$$

$$\sum_{j=1}^{m} x_{ij} \leq R_{\max}^i, \ \forall i \in I^t \tag{5}$$

$$x_{ij} \in \{0,1\}, \ \forall i \in I^t, \forall j \in J^t \tag{6}$$

Notations: dh_{ij} represents the number of hours corresponding to the operations of setting up the driver « i » with respect to trip « j », hs_i represents the supplementary driving hours carried on by driver « i » during the assignment period « t », I^t is the set of « n » available drivers during the assignment period « t », J^t is the set of « m » scheduled trips which cover the set of scheduled trips for the assignment period « t », A_j is the set of drivers able to carry on trip « j », O_{j_1} is the set of trips overlapping with trip « j_1 », dv_j represents the number of driving working hours associated to trip « j », x_{ij} are binary decision variables such that it takes « 1 » if trip « j » is assigned to driver « i »; « 0 » otherwise.

Interpretation of Constraints: The estimation by the RPTC of cost associated to the trip « j » and the driver « i » is very approximate, since the cost of operation is evaluated with respect to a set of complex rules concerning conditions of remuneration, overtime, the standby activities (SBY) and other parameters. Constraint (2) ensures that to each trip is assigned a unique driver; the inequality sign allowing driver deadheading (a transfer of driver out of duty to another base in order to carry out a planned trip). Constraint (3) ensures that a given driver cannot be assigned simultaneously to two overlapping trips. Constraint (4) ensures that the workload assigned to each driver complies with the regulations (LH) during the assignment period (a month in general). This limit represents the number of hours corresponding to the minimum guaranteed wage and the permitted number of additional driving hours. Constraint (5), $R_{\max}^i$ represents the maximum number of trips that can be assigned to driver « *i* » over the assignment period « *t* ».

Other constraints considered as not essential may be taken into account in this mathematical formulation. These constraints represent the claims of drivers that the company is willing to consider, as well as other internal rules from the agreements made between the RPTC and union groups. The repeated non-satisfaction of these constraints leads to decrease the degree of satisfaction of each of the driver. The following paragraph applies to define the concept.

Some Important Comments Can be Done:

1. Exact solutions for the resulting large scale combinatorial optimization problems are not available in an acceptable computing time and in general, RPTC have produced their bus driver allocation schedules using simplistic heuristics. However during the last decade, a number of major RPTC have developed optimization-based techniques to solve the BDAP by developing improved heuristics based on column generation techniques and exact solution approaches based on constraints logic programming.
2. The cost resulting from the assignment of a trip to a given driver cannot be determined with precision before the whole workload of this driver has been defined. In many RPTC, driver payment is computed in a complex way from many parameters (total amount of driving hours, standby duties, etc.). So, it appears that the above formulation can result in a poor approximation of the real assignment problem.
3. Other relevant concerns can be added as soft constraints to the mathematical formulation of the BDAP. However, constraints relative to the driver satisfaction levels are difficult to add to the standard mathematical formulation of the BDAP since they depend in a complex way of the considered decision

variables. Another difficulty is relative to the choice of individual versus collective driver satisfaction constraints. So it appears more appropriate to introduce this concern as such criterion to the overall assignment problem.

A BI-CRITERION NON-DOMINATED SOLUTION APPROACH FOR SOLVING THE BDAP

We adopt for solving the BDAP an approach organized in 2 stages. Firstly, we search a set of solutions for the BDAP while maximizing the satisfaction degree of the drivers. We evaluate qualitatively the degree of satisfaction of the drivers while taking into consideration some parameters defined by experts such as: the previous satisfaction of drivers, the claims, the standby activity, the workload, etc. This is done by using a heuristic method developed later. Secondly, we look to find a set of non-dominated sorting assignment solutions while minimizing the associated cost.

A Heuristic to Build Max Satisfaction Assignment Solutions (HAS)

Here, we describe a Heuristic Assignment Solution (*HAS*) of type "Greedy Randomized Adaptive Search Procedure", developed to generate an initial population (size N_{pop}) of assignments while ensuring coverage of all trips and maximizing the satisfaction of the drivers. This approach of resolution is to affect drivers in order by increasing their level of satisfaction. This leads to define priorities among the various drivers in order to equitably increase their level of satisfaction. The proposed greedy heuristic (HAS) is designed in order to generate the initial "max satisfaction" assignment solutions set. It is supposed here that at the beginning of a new planning period the satisfaction level of each driver is available based

Table 1. A set of possible candidates of drivers based on class of satisfaction

	Class C1	Class C2	Class C3	.	.	Class C6
Distribution of drivers by class of satisfaction	$p_1, p_2, \cdots, p_{n_1}$	$p_{n_1+1}, \cdots, p_{n_1+n_2}$	.	.	.	$p_{n_1+n_2+n_3+n_4+n_5+1} \cdots, p_{n_1+n_2+n_3+n_4+n_5+n_6}.$
Different candidates	$p_1, p_2 \cdots\cdots$ $p_2, p_1 \cdots\cdots$.	. . .	. . .	. . .	. .	
# of possible candidates	$h(n_1)$	$h(n_2)$	$\cdots$	$\cdots$	$\cdots$	$h(n_6)$

on historical data. The proposed greedy heuristic is designed so that the claims of drivers of lowest satisfaction levels are considered first. It starts by the elaboration of a set of full arrangements among all the available drivers following an increasing order of satisfaction level. The solution process consists on generating for each satisfaction level a set of sub-arrangements among the corresponding drivers (Table 1), thereafter; these different sub-arrangements are combined together, following in decreasing order, the different satisfaction levels. Then a set of full arrangements is obtained. This method is decomposed of the following steps:

Classification of Drivers based on their Satisfaction Level

The classification approach proposed here considers the different categories of satisfaction qualitatively characterized in $S = \{Very\ weak, Weak, Moderate, Acceptable, Good, Very\ Good\}$. Each driver and each period may be associated with a class of satisfaction. The definition of class of satisfaction can analyze the evolution of the satisfaction of different drivers. Let "n_i" the total number of drivers distributed in class "i" of satisfaction. We have $n = n_1 + n_2 + n_3 + n_4 + n_5 + n_6$ with $n_1 = |Very\ Poor|, \cdots, n_6 = |Very\ Good|$. The classification method used here is decomposed

of 2 stages. Firstly, we consider, separately, a number of possible candidates between the different drivers belonging to different classes of satisfaction (Table 1). This number is defined by the function $h(n_i) = \min\{n_i!, n_i^2\}$ where h is an increasing function, n_i is the number of drivers belonging to the class of satisfaction "i". Secondly, we combine the different candidates found in the previous stage in order of increasing satisfaction. All the candidates found corresponding to different satisfaction class are merged in ascending order. The number of possible candidates depends on the population's size. The number of different assignment solutions is equal to the size of this population by retaining different candidates.

With the use of the concept of satisfaction classes, the total number of different possible classifications becomes equal to $\prod_{i=1}^{6} N_i!$ with $\prod_{i=1}^{6} N_i! \leq n!$. So, reducing the number of candidates is very important.

The application of this method requires the verification of certain assumptions. Firstly, the length of each candidate must be equal to the total number of drivers. Secondly, the number of candidates to generate must be greater than or equal to the population's size (Npop) handled by GAs.

Assignment Algorithm

We choose a candidate A_ℓ and we consider the drivers based on their order in A_ℓ. A driver is assigned to the uncovered activities which fit with his desiderata. This assignment must take into account the availability of the driver regarding the scheduled activities such as: the training, medical visit, etc. After assigning the driver to his favorite activities, we begin to complete his workload by assigning uncovered trips. We begin, first, by the available trips which are not approached by other drivers. If the workload becomes acceptable, i.e. the difference becomes low between the actual driving hours to be performed and the threshold corresponding to paid hours, and then we fix that workload and start with the next driver. If the workload is still not acceptable then we first assign trips requested by drivers classified in descending order of satisfaction. We begin by drivers belonging to the better class of satisfaction, and so on. Once the workload of the driver is built, the allocation process continues with the following driver, and so on until all the trips are covered. However, it is possible that the allocation process generates two similar solutions of assignments. In this case, we retain one and an additional candidate, $A_{\ell+1}$, will be processed and added to the initial set of classifications to ensure a number of assignments equal to the size of the initial population.

Minimizing the Operations Cost by Using Genetic Algorithms

As one of our aim is to reduce the operations cost, so the target considered here is the minimization of overtime working hours performed by the drivers during the operation planning. In an assignment solution, the structure of cost operations is not separable with respect to trips. Indeed, an estimate of the cost associated with a driver is computable only after having developed his *workload* during the entire assignment period. Thus, the technique of GA that handles global solutions in a stochastic manner seems interesting in this context. GA process leads to produce populations of feasible solutions where each solution is similar to a chromosome composed of genes, each gene corresponding to a particular parameter of the problem. The chromosomes can be represented using either a binary or a non-binary codification. We describe our new method, using GA technique (Holland, 1992; Peters & Skowron, 2004), adopted for building a population of assignment solutions with lowest cost of operation. We define a fitness function and the basic operators related to GA such as: coding, selection and reproduction.

In our case, a non-binary codification has been adopted for the representation of the solutions. The ith component of a chromosome indicates which driver is assigned to the ith trip. This codification has been chosen in order to minimize the memory requirements to codify a whole population. In the encoding phase, the length of a chromosome is equal to the number of planned trips and each gene contains the identity of the driver assigned to the corresponding trip. The initial population is generated using the heuristic method (HAS). We assign a fitness score to each assignment solution based on all drivers. Let f_C the fitness function (Equation 7). This leads to define a set of decision vectors and updates the solution set of non-dominated assignments. Classical genetics operators have been adapted to the context of the present assignment problem to produce in a progressive way improved new generations. The genetic operators such as mutation, crossover and inversion are applied using adjusted probabilistic ratio: $P_c = 0.9, P_m = 0.05$ and $P_i = 0.05$. The algorithm is stopped if, after having generated several new populations, no new lower-cost solution is found. An operator selects the chromosomes which become parents according to their evaluation values. A roulette wheel method picks two chromosomes to which classical GA operators like crossover, inversion and mutation are applied according to a chosen probability.

$$f_C = \sum_{i \in I} \left| WL_i - \overline{WL} \right| \tag{7}$$

Where I is the set of index associated to the available drivers during the assignment period under consideration, WL_i is the workload assigned to driver « i », and $\overline{WL}$ is the minimum workload corresponding to the minimum guaranteed wage for drivers.

Non-Dominated Bi-Criteria Solutions Approach

Definition of Non-Dominated Solutions

The multi-objective optimization algorithms are maintained by using a fitness assignment scheme which prefers non-dominated solutions. A non-dominated solution is characterized by an assessment made of two parameters corresponding respectively to the cost associated with the solution and the degree of dissatisfaction generated by this solution. We define the set ($DSND$) which is composed of non-dominated assignment solutions. A solution of ($DSND$) is characterized by an assessment made of two parameters corresponding respectively to the cost associated with the solution and the degree of dissatisfaction generated by this solution:

$$DSND = \left\{ (C_\ell, NS_\ell) \, / \, \ell = 1 \cdots N_{pop} \right\} \tag{8}$$

where "C_ℓ" is the value of the cost criterion, "NS_ℓ" is the overall degree of dissatisfaction of drivers in the solution « ℓ », and N_{pop} is the number of elements belonging to the set ($DSND$) that should be equal to the size of a population of solutions. The iterative application of genetic operators on the various intermediate populations may gradually change the composition of the set ($DSND$) when new non-dominated solutions are generated. Let $X^1 = (C_1, NS_1)$ and $X^2 = (C_2, NS_2)$ two evaluations that are associated with two different assignment solutions where C_ℓ (ℓ=1, 2) represents the cost associated with the solution « ℓ » and NS_ℓ (ℓ=1, 2) represents the overall degree of dissatisfaction reached in the solution « ℓ ». In this case, X^1 is said to dominate X^2 (denoted by $X^1 \, dom \, X^2$), if and only if:

$$(C_1 \prec C_2) \text{ and } (NS_1 \prec NS_2)$$

and

$$(C_1, NS_1) \neq (C_2, NS_2) \tag{9}$$

Application of Non-Dominated Sorting Genetic Algorithm (NSGA-II)

In order to manage the elitism of non-dominated solutions, we have applied in our study one of the most competitive second-generation algorithms, called the Non-dominated Sorting Genetic Algorithm (NSGA-II) approach (Deb, Pratap, Agarwal, & Meyarivan, 2002). This algorithm is stated that is a highly efficient with such a good performance. The characteristic feature of NSGA-II is the implementation of elitism in order to preserve the best individuals for subsequent generations. In the fast non-domination approach, the population is sorted based on the non-domination. Each solution is assigned a fitness (or rank) equal to its non-domination level (1 is the best level, 2 is the next-best level, and so on). The NSGA-II has the following main features: *Ranking based on non-domination sorting, Crowding distance metric as explicit diversity-preserving mechanism, and Diversity preservation and elitism reinforced by a Crowded-Comparison operator*. These features are presented as follows:

Figure 1. Non-dominated sorting ranking in NSGA-II

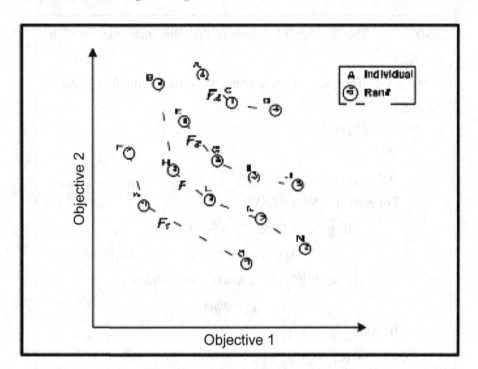

Non-Domination Sorting For Ranking

The method determines for each member of the population its dominance depth FR_i. It then assigns the individual's rank according to the non-domination level. The non-dominated set of the population is identified and made front FR_1 (*rank*=1). It is then disregarded from the population. A new non-dominated set is identified and assigned front FR_2 (*rank*=2). Then this set is also disregarded. The operation continues iteratively until all individuals are assigned to a front FR_i and ranked (Deb, 2001). This ranking method is illustrated in Figure 1.

In order to start the NSGA-II, we start with the population obtained after applying HAS approach followed by identifying the set of solutions of the first non-dominated front in a population of size N_{pop}. Each solution can be compared with every other solution in the population to find if it is dominated. The ranking algorithm is presented in Figure 2. All the solutions in the first non-dominated front will have their domination count as zero. Now, for each solution (C_i, NS_i) with $NSD_{(C_i, NS_i)} = 0$, we visit each member (C_j, NS_j) of its $SSD_{(C_i, NS_i)}$ set reduce its domination count by one. In doing so, if for any member (C_j, NS_j) the domination count becomes zero, we put it in a separate list Q. These members belong to the second non-dominated front. Now, the above procedure is continued with each member of Q and the third front is identified. This process continues until all front are identified. This requires $O(MN)$ comparisons for each solution, where M is the number of objectives (*M=2*). When this process is continued to find all members of the first non-dominated level in the population, the total complexity is *O(MN²)*.

Crowding Distance Density Estimation

The crowding distance is a measure of how close an individual is to its neighbors. Large average

Figure 2. Ranking algorithm in NSGA-II

$NSD_{(C_i, NS_i)}$: The number of solutions that dominate the solution (C_i, NS_i).

$SSD_{(C_i, NS_i)}$: The set of solutions that the solution (C_i, NS_i) dominates.

For each $(C_i, NS_i) \in DSND$

    $NSD_{(C_i, NS_i)} = 0$

    $SSD_{(C_i, NS_i)} = \phi$

    For each $(C_j, NS_j) \in DSND$

        IF $\left((C_i, NS_i) \, dom \, (C_j, NS_j)\right)$ then

            $SSD_{(C_i, NS_i)} = SSD_{(C_i, NS_i)} \cup \{(C_i, NS_i\}$

        Else IF $\left((C_j, NS_j) \, dom \, (C_i, NS_i)\right)$ then

            $NSD_{(C_i, NS_i)} = NSD_{(C_i, NS_i)} + 1$

    IF $NSD_{(C_i, NS_i)} = 0$ {

        $(C_i, NS_i)_{rank} = 1$

        $FR_1 = FR_1 \cup \{(C_i, NS_i)\}$

        }

i=1

while $(FR_i \neq \phi)$ {

    $Q = \phi$

    For each $(C_p, NS_p) \in FR_i$

        For each $(C_r, NS_r) \in SSD_{(C_p, NS_p)}$

            $NSD_{(C_r, NS_r)} = NSD_{(C_r, NS_r)} - 1$

        IF $NSD_{(C_r, NS_r)} = 0$ then {

            $(C_r, NS_r)_{rank} = i + 1$

            $Q = Q \cup \{(C_r, NS_r)\}$

            }

    }

i=i+1

$FR_i = Q$

crowding distance will result in better diversity in the population. Once the non-dominated sort is complete, the crowding distance is assigned. Since the individuals are selected based on rank and crowding distance all the individuals in the population are assigned a crowding distance value. Crowding distance is assigned front wise and comparing the crowding distance between

Figure 3. Crowding distance calculation

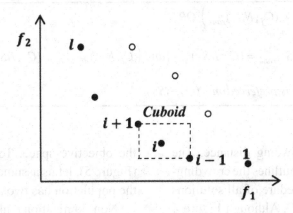

two individuals in different front is meaningless. In Figure 3, the crowding distance $(C_i, NS_i)_{distance}$ of the ith solution in its front is the average side length of the cuboid. The measure $(C_i, NS_i)_{distance}$ represents a measure of population density around i (Deb, 2001). Therefore, the smaller $(C_i, NS_i)_{distance}$ is the more crowded its surrounding environment. To get an estimate of the density of solutions surrounding a particular solution in the population, we calculate the average distance of two points on either side of this point along each of the objectives k. The algorithm as shown below outlines the crowding-distance computation procedure of all solutions in a non-dominated set

FR. The crowding-distance computation requires sorting the population according to each objective function value in ascending order of magnitude. Thereafter, for each objective function, the boundary solutions (solutions with smallest and largest function values) are assigned an infinite distance value. All other intermediate solutions are assigned a distance value equal to the absolute normalized difference in the function values of two adjacent solutions. This calculation is continued with other objective functions. The overall crowding-distance value is calculated as the sum of individual distance values corresponding to each objective. Each objective function is normalized

Figure 4. Crowding distance procedure

$$L = |ND|$$

For each $(C_i, NS_i) \in ND$

$$(C_i, NS_i)_{distance} = 0$$

For each $k \in \{C_i, NS_i\}$

$$ND_k = sort(ND, k)$$

$$(C_1, NS_1)_{distance} = (C_{|L|}, NS_{|L|})_{distance} = \infty$$

For $i = 2$ to $|L| - 1$

$$(C_i, NS_i)_{distance} = (C_i, NS_i)_{distance} + \left(\left((C_i, NS_i)^k - (C_{i-1}, NS_{i-1})^k \right) \Big/ (f_k^{max} - f_k^{min}) \right)$$

Figure 5. Crowding-comparison operator procedure

$$\textbf{If } \left((C_i, NS_i)_{rank} < (C_j, NS_j)_{rank} \right) \textbf{ OR}$$

$$\left(\left((C_i, NS_i)_{rank} = (C_j, NS_j)_{rank} \right) and \left((C_i, NS_i)_{distance} > (C_j, NS_j)_{distance} \right) \right)$$

$$\textbf{then } (C_i, NS_i) \quad is\ preferred\ on \quad (C_j, NS_j)$$

before calculating the crowding distance. The algorithm as shown below outlines the crowding-distance computation procedure of all solutions in a non-dominated set *ND*. Although Figure 4 illustrates the crowding-distance computation for two objectives, the procedure can be applied to more than two objectives as well.

The index ND_k denotes the solution index based on the kth objective member in the sorted list. Here, $(C_i, NS_i)^k$ refers to the kth objective function value of the ith individual in the set *ND* and the parameters f_k^{max} and f_k^{min} are the maximum and minimum values of the kth objective function.

Crowded-Comparison Operator

After all population members in the set *ND* are assigned a distance metric, we can compare two solutions for their extent of proximity with other solutions. A solution with a smaller value of this distance measure is, in some sense, more crowded by other solutions. The crowded-comparison operator guides the selection process at the various stages of the algorithm toward a uniformly spread-out Pareto-optimal front. However, besides comparing couples of individuals by ranking, it compares their crowded distance for tie-breaking. The algorithm is called Crowded Tournament Selection. It works as follows: When comparing two solutions *i* and *j*, solution *i* is preferred over *j* if it has a better (lower) rank. If both *i* and *j* have the same rank, *i* is preferred over *j* if it has a higher crowding distance $((C_i, NS_i)_{distance} > (C_j NS_j)_{distance})$. The crowding distance is normally calculated in

the objective space. To explain this algorithm (Figure 5), let us assume that every individual in the population has two attributes:

Non-domination rank $(C_i, NS_i)_{rank}$ and Crowding distance $(C_i, NS_i)_{distance}$

The complexity of this procedure is governed by the sorting algorithm. Since *M* independent sorting of at most *N* solutions are involved, the above algorithm has $O(MN \log N)$ computational complexity.

The Complete Algorithm

We present the complete NSGA-II algorithm as proposed in Deb, Pratap, Agarwal, and Meyarivan (2002). The algorithm works with a population of fixed size N_{pop}. Initially, a random parent population P_0 is created. The population is sorted into different non-domination levels. Each solution is assigned a fitness (or rank) equal to its non-domination level. For the population size of *N* and for *M* objective functions, we describe a fast non-dominated sorting approach. Thus, minimization of fitness is assumed. Binary tournament selection (with a crowded tournament operator), recombination, and mutation operators are used to create an offspring population Q_0 of size *N*. The NSGA-II procedure is outlined, in Figure 6. The offspring population is combined with the current generation population and selection is performed to set the individuals of the next generation. Since all the previous and current best individuals are added in the population, elitism is ensured. Population is now sorted based on

Figure 6. Crowding-comparison operator procedure

Step 1:	* Combined population and offspring and create $R_t = P_t \cup Q_t$. * Perform a non-dominated sorting to R_t and identify different non-dominated fronts FR_i, $i = 1, \ldots ,$etc.
Step 2:	* Set new population $P_{t+1} = \Phi$ * Set a counter $i = 1$. * Until $|P_{t+1}| + |FR_i| < N_{pop}$, perform $P_{t+1} = P_{t+1} + FR_i$ and $i = i + 1$.
Step 3:	* Perform the Crowding-sort (FR_i, $\leq_n$) procedure and include the most widely spread ($N_{pop} - |P_{t+1}|$) solutions by using the crowding distance values in the sorted FR_i to P_{t+1}.
Step 4:	* Create offspring population Q_{t+1} from P_{t+1} by using the crowded tournament selection, crossover and mutation operators. * Goes back to step 1.

non-domination. The new generation is filled by each front subsequently until the population size exceeds the current population size. If by adding all the individuals in front FR_j the population exceeds N_{pop} then individuals in front FR_j are selected based on their crowding distance in the descending order until the population size is N_{pop}.

AN APPROACH FOR THE EVALUATION OF THE DRIVER SATISFACTION LEVEL

Evaluation of Satisfaction Level for Drivers

In this section, we propose an approach for assessing individual satisfaction of the drivers using RS technique. Thus, a new mathematical formulation for this problem is proposed by considering a set of related constraints. However, the level of satisfaction at any given moment of any of the driver depends on a complex of several parameters, that have an impact on the satisfaction level and that may be incorporated in the resolution of the problem, such as the number of driving hours, the number of occurrences of standby activities

(SBY) and the percentage of performed claims reported by the driver, etc. It recalls here that SBY activity is when the driver is put on hold to make a possible replacement of another driver would be unable to achieve or finish his trip for an unforeseen reason such as: disease, the failure of a bus, regulatory issues, etc. However, the occurrence of SBY activities provided for each driver is generally limited (regarding the collective agreements) to some units, over a given assignment period. Indeed, this activity is not attractive to drivers because the pay rate is significantly lower than that of trips and it is an activity often considered tedious. So, assessing the level of satisfaction for each driver, such as assessing the overall average satisfaction of the set of drivers, appears as a matter of great difficulty.

Mathematical Formulation Integrating the Satisfaction in the Optimization Problem

Thus, in this section we try to define and quantify the concepts of individually and collectively satisfaction, so as to be considered in the BDAP solutions. We can write symbolically:

$$S_i^t = f(S_i^{t-1}, V_t^i) \forall \ i \ \in \ I^t, \quad S_i^t \ \in \ S$$

(10)where $V_t^i = (x_{i1}^t, ..., x_{iJ}^y)'$ with $x_{ij}^t = 1$ if the duty « *j* » is assigned to driver « *i* » during the assignment period « *t* »; otherwise, $x_{ij}^t = 0$. Therefore, « *f* » represents the evaluation process using RS technique described later. It is then possible to formulate the constraints relative to the degree of satisfaction of the driver. These can be:

From individual character:

$$S_i^t : d(S_i^t) > S_i, \forall i \in I^t \tag{11}$$

with δ_i is an integer comprises between 1 and $|S| - 1$

From collective character:

$$\min_{i \in I^t} \left\{ d(S_i^t) \right\} > \delta \tag{12}$$

with δ is an integer comprises between 1 and $|S| - 1$

Le also the subsets $\wp_k$ consisting of drivers whose satisfaction degree is resulting '*k*', we can also consider collective constraints such as:

$$\rho_{\min}^n \cdot |I| \leq |\wp_n| \leq \rho_{\max}^n \cdot |I| \tag{13}$$

where $\rho_{\min}^n et \ \rho_{\max}^n$ are parameters comprises between 0 and 1.

$$\sum_{\ell=1}^k |\wp_\ell| \leq \rho_k \cdot |I| \tag{14}$$

or

$$\sum_{\ell=h}^{|S|} |\wp_\ell| \geq \rho_h |I| \ h \in \{1, ..., |S|\}, \ k \neq h \tag{15}$$

where ρ_k and ρ_h are parameters comprises between 0 and 1.

It is up to the manager to choose the nature and number of qualitative constraints to include in the assignment problem of the drivers.

Given the multiplicity of possible constraints regarding the distribution of levels of satisfaction between the drivers, it is useful to introduce an overall assessment of the degree of satisfaction of drivers. One of the simplest examples is the sum of weights as follows:

$$DS^t = \sum_{k=1}^{|S|} \alpha_k |\wp_k^t| / |I| \tag{16}$$

where $\alpha_k, k = 1$ to $|S|$ are weights such as:

$$\alpha_k \in [0,1] \tag{17}$$

$$\alpha_{k+1} \geq \alpha_k \tag{18}$$

This overall satisfaction index will be such as:

$$0 \leq DS^t \leq 1 \tag{19}$$

It will be the more satisfying if it is close to unity, and a global constraint imposed by the manager may be such that:

$$DS^t \geq DS_{\min} \text{ with } 0 \leq DS_{\min} \leq 1 \tag{20}$$

It will be up to the manager to choose the weights α_k so that the overall evaluation outcomes of the relationship (16) are coherent with its own assessments. This can be achieved using an auxiliary linear quadratic optimization problem. For example, if the manager is given a set of situations (i, j) as it considers that the distribution $(\wp_k)^i$ is less satisfactory than the distribution $(\wp_k)^j$ which will be noted $(i < j)$, we may introduce inequality constraints, for all pairs (i, j) of given distribution with (i < j), such that:

$$\sum_{k=1}^{|S|} \alpha_k \cdot \left| \wp_{k^i} \right| \leq \sum_{k=1}^{|S|} \alpha_k \left| \wp_{k^j} \right| \qquad (21)$$

and an overall criterion to maximize regarding the following parameters α_k :

$$\sum_{(i<j)} \left(\sum_{k=1}^{|S|} \alpha_k \cdot \left| \wp_{k^j} \right| - \sum_{k=1}^{|S|} \alpha_k \cdot \left| \wp_{k^i} \right| \right)^2 \qquad (22)$$

associated to constraint (17).

It is clear that the integration of such constraints to the problem leads to a hybrid formulation for which no known technique of Mathematical Programming does get the optimal solution.

Methodology Approach

We attempt to assess qualitatively the evolution of the level of satisfaction of the bus drivers regarding the duties (trips, standby, etc.) assigned to them by the RPTC during the period of assignment. We consider a Rough Set predictive model that can be initially established based on expert knowledge in Human Resources Management for the RPTC to qualify the level of satisfaction (Orlowska, 1998). It is first necessary to identify the main parameters influencing the establishment of driver satisfaction. The different levels associated with these parameters are modeled using fuzzy subsets technique (Zimmermann, 1996; EL Moudani & Mora-Camino, 2000). The RS technique is used to perform this quality evaluation (Jensen & Shen, 2004). However, analyzing the database of historical data containing critical information, related to the satisfaction level, helps to identify relationships which have a bearing on a specific issue and then extrapolate from these relationships to predict future satisfaction level or behavior and discover hidden data patterns. The level of satisfaction of a driver « i » is a qualitative scale values S_i in the universe of discourse $S = \{\sigma_1, ..., \sigma_{|S|}\}$. This universe of discourse is equipped with a linear ordering based on the level of satisfaction. We therefore associate to each element « k » of the finite discrete set an integer d_k that is its degree, with:

$$d : S \rightarrow \{1, 2, ..., |S|\} \text{ with } d_k = k, k = 1\alpha |S|' \qquad (23)$$

The level of satisfaction associated to any driver depends directly on some operational parameters such as the acceptance rate of its claims, his number of standby activities provided, and its workload performed during the assignment period. It is also estimated from the past satisfaction. Suppose that depends essentially on the following parameters:

- The previous degree of satisfaction for driver.
- The number of claims declared by driver « i » over the assignment period « t » (SE_i^t).
- The number of satisfied claims for driver « i » over the assignment period « t » (SR_i^t).
- The number of standby activities assigned to driver « i » over the assignment period « t » (R_i^t).
- The workload assigned to driver « i » over the assignment period « t » (WL_i^t).

The process for evaluating the satisfaction level of a driver is performed in two stages. Firstly, we fuzzify the operations parameters: claims satisfied, standby activities, and workload. Secondly, we integrate the results issued with the previous satisfaction in order to get the new satisfaction degree.

Fuzzification of the Operations Parameters

The first two parameters, (SE_i^t) and (SR_i^t), contribute to calculate the percentage of acceptance of claims for each driver during the assign-

Figure 7. Fuzzy membership function of $\tilde{X}_i^t$

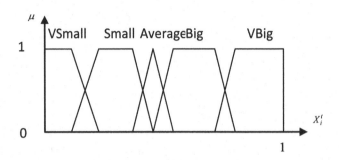

ment period "t". Let (X_i^t) be the variable representing the ratio of satisfied claims $SR_i^t / (SE_i^t)$ for driver "i" over the assignment period "t". The third parameter (R_i^t) helps to determine the degree of SBY activity for a driver during an assignment period "t". Let Y_i^t be the variable representing the normalized SBY activity degree R_i^t / R_{max} performed by driver "i" over the assignment period "t" where R_{max} is the maximum level of SBY activity achievable by a driver over one assignment period. On the other hand, the fourth parameter, (WL_i^t) helps to determine the degree of workload for a driver during an assignment period "t". Let Z_i^t be the normalized variable representing the workload activity degree WL_i^t / WL_{max} performed by driver "i" over the assignment period "t" where WL_{max} is the maximum level of workload activity achievable by a driver over one assignment period.

We aim to get a qualitative assessment of the degree of satisfaction of a driver in a manner to be submitted to the appreciation of managers of driver scheduling. The achievement degree of claims is characterized by a qualitative variable "X_i^t" whose value is based on the definition of fuzzy subsets according to the knowledge gathered from experts. Let $\xi_P = \{VSmall, Small, Average, Big, VBig\}$ the universe of discourse consists of 5 sub-fuzzy sets

to which linguistic variables are associated (Figure 7).

For each value of « X_i^t », a quadruplet $\Psi_i = (V_1, \mu_1, V_2, \mu_2)$ is associated such as:

$$V_i \in \xi_P, \quad \forall i \in \{1, 2, ..., 5\}$$

and

$$0 \le \mu_i \le 1, \forall i \in \{1, 2, ..., 5\} \qquad (24)$$

where a membership function, μ_i, is associated to each fuzzy subset V_i of ξ_P.

In assessing the level of satisfaction, several cases are considered:

First Case:

In this first case, we consider that the qualitative level of claims achieved for each driver is the maximum value given by the membership functions associated with different fuzzy subsets (Figure 8).

IF $\mu_i \succ \mu_j$ **Then***the degree of claims achieved for a driver is qualified by V_i*

IF $\mu_i \le \mu_j$ **Then***the degree of claims achieved for a driver is qualified by V_j*

Second Case:

Figure 8. Defining the degrees of membership in case driver has put claims

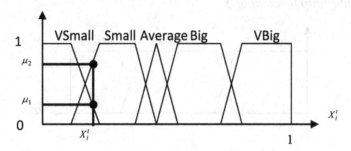

If a driver did not ask for a preference ($SE_i^t = 0$), in this case the quadruplet "ψ_i" associated to this driver is written as follows (Figure 9):

$$\psi_i = (Average, 1, Big, 0) \text{ or}$$

$$\psi_i = (Small, 0, Average, 1). \tag{25}$$

This avoids assigning special priority to the drivers who expressed less preference than others during the previous period of assignment. In this case, the qualitative assessment of the claims achieved is as follows:

IF ($\mu_{Average} \succ \mu_{Small} = 0$) and

($\mu_{Average} \succ \mu_{Big} = 0$)

Then the degree of claims achieved is qualified as Average. (26)

Regarding the third parameter, (R_i^t), a universe of discourse ξ_R is defined by

$\xi_R = \{Many, Average, Few\}$. The achievement degree of SBY activity is characterized by a qualitative variable "Y_i^t" whose value is based on the definition of fuzzy subsets according to the knowledge gathered from experts. It is composed of 3 fuzzy subsets to which linguistic variables are associated (Figure 10). Regarding the fourth parameter, (WL_i^t) a universe of discourse

ξ_{WL}

is defined by

$\xi_{WL} = \{Overloaded, Lightly\ loaded, Moderate, Little\ light, Too\ light\}$

The achievement degree of workload activity is characterized by a qualitative variable "Z_i^t" whose value is based on the definition of fuzzy subsets according to the knowledge gathered from experts. It is composed of 5 fuzzy subsets to which linguistic variables are associated (Figure 11).

Figure 9. Defining the degrees of membership in case driver has no claims

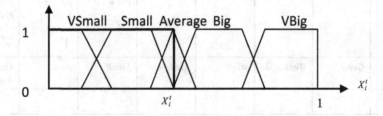

Figure 10. Fuzzy membership function of $\tilde{Y}_i^t$

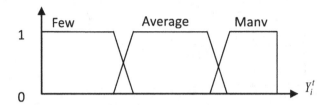

Figure 11. Defining the degrees of membership associated to a ratio of assigned workload

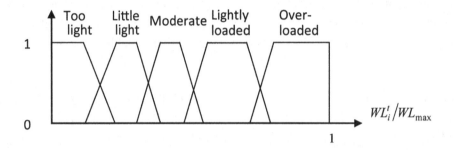

Figure 12. Evaluation of the driver satisfaction using Rough Set technique

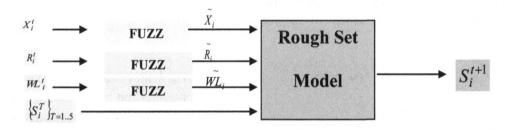

Table 2. Example of evaluation for the level of satisfaction

Satisf. achieved at period « t-5 » S_i^{t-5}	Satisf. achieved at period « t-4 » S_i^{t-4}	Satisf. achieved at period « t-3 » S_i^{t-3}	Satisf. achieved at period « t-2» S_i^{t-2}	Satisf. achieved at period « t-1 » S_i^{t-1}	Satisf. achieved at period « t » S_i^t	Degree of achieved claims at period « t+1 » $\tilde{X}_i^{t+1}$	Degree of standby activities during the period « t+1» $\tilde{Y}_i^{t+1}$	Degree of workload activity during the period « t+1» $\tilde{Z}_i^{t+1}$	Decision S_i^{t+1}
Moderate	Average	Good	Weak	Weak	Moderate	Small	Few	Over-loaded	Weak
…	…	…	…	…	…	…	…	…	…

Evaluation of the Current Satisfaction Using Rough Set Technique

A process of fuzzification is applied to the three critical parameters X_i^t, R_i^t, and WL_i^t in a manner to get fuzzy parameters $\tilde{X}_i^t$, $\tilde{R}_i^t$, and $\tilde{WL}_i^t$. These fuzzyfied parameters are joined with the previous satisfaction degree in order to perform the qualitative assessment of the current satisfaction of each driver during the considered operation period (Figure 12). This uses the universe of discourse "S" already defined. We adopt here that the degree of satisfaction of drivers is developed continually. We may retain such an active scale of 6 assignment periods (the last 5 and the next one). Obtaining an acceptable level of satisfaction for all the drivers may not be feasible on a single period of assignment period because there may be exist some cases where the demands of drivers are competing.

For this reason, an approach of resolution based on rules taking into account the temporal distribution of tasks can be adopted. This is to consider not only the level of satisfaction of drivers for the current assignment period but also the level of satisfaction achieved during the previous assignment periods. For example, if we consider the period assignment "*t+1*", the predictive classification model will incorporate the qualitative satisfaction achieved until the period assignment "*t*" with the critical parameters in order to produce qualitative outputs belonging to set "*S*". This approach is defined in a manner to represent the dynamic process of building the driver satisfaction (Table 2). We have applied our methodology over a rich data gathered and collected from a data warehouse specialized in our case study. The data are grouped and analyzed by month.

Figure 13. Computation time based on GA and Heuristic

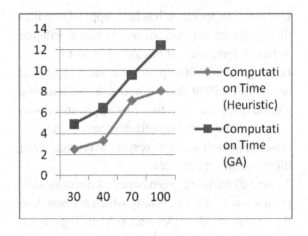

Figure 14. Proportion of drivers satisfaction based on GA and Heuristic

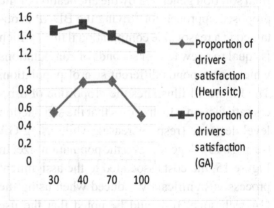

Figure 15. Minimum cost based on GA and Heuristic

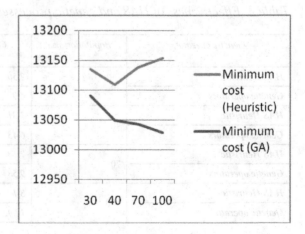

CASE STUDY

Our solution approach has been applied to a medium size company that covers 35 cities distributed around 2 base sites. It has 200 drivers that must be assigned to 3500 trips corresponding to a total amount of 13000 driving hours. Some learning can be obtained from this first application with respect to the computer effectiveness of the proposed approach and with respect to the quality of the solution set obtained.

Some charts are given here that illustrates the results issued from our study which are based on the two methods: HAS and GA. In Figure 13, we see that the computing time for all population size considered here, (30, 40, 70 and 100), is greater in HAS method than in using GA. The final solution generated by the application of the proposed approach for solving the BDAP is obtained in a reasonable computational time, which is equal to a few tens of seconds of computations which corresponds different size of population. The Figure 14 illustrates the proportion of drivers satisfaction which shows that the satisfaction level decrease (resp. increase), while using GA (resp. HAS), depending on the population size. In Figure 15, the cost associated to the assignment process is significantly reduced when using the GA technique. It should be noted that the use of genetic operators (crossover, mutation, and

inversion) contributes to improve intermediate solutions assignment (Table 3).

Experiments show the non-dominated sorting solutions belonging to set « DSND » that corresponds to the final population of solutions (Figure 16). This population size considered here has 100 candidates. We note that the maximum overall degree of satisfaction of drivers is almost 90%. This satisfaction incorporates an increased cost estimated around 115 extra hours. At the same time, the best solution in minimum cost is evaluated to around 28 hours of extra hours while the overall level of satisfaction is generated from this assignment solution is estimated to 40%. So, the use of our approach contributes in helping the decision makers to choose the most appropriate solution which fit with the political strategy followed by the RPTC.

CONCLUSION

In this communication, one of the main operations decision problem faced by RPTC has been dealt with using Mathematical Programming and Computational Intelligence methods. The proposed approach for solving the BDAP does not produce an exact solution in pure mathematical terms but appears to be quite adapted to give a real support to decision making. This approach

Table 3. Effectiveness of HAS and genetic processes for the BDAP

Genetic operators	Population size	Computing Time	Proportion of drivers satisfaction	Minimum cost
HAS Heuristic	30	2.50	0.53	13135
Genetic operators		4.9	1.44	13090
HAS Heuristic	40	3.31	0.86	13109
Genetic operators		6.43	1.5	13049
HAS Heuristic	70	7.14	0.87	13138
Genetic operators		9.58	1.38	13043
HAS Heuristic	100	8.1	0.49	13154
Genetic operators		12.43	1.24	13028

Figure 16. Non-dominated sorting solutions

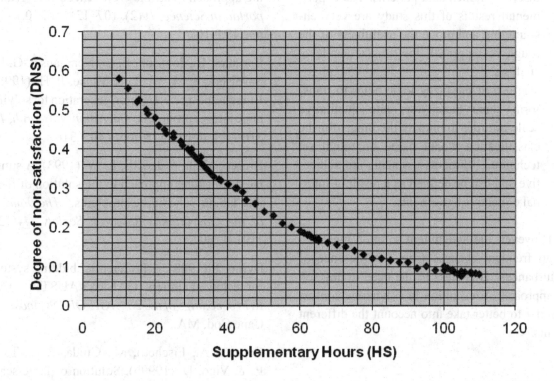

has some advantages highlighting two key factors in assessing the quality of an assignment solution: the cost of operations associated with the solution and the degree of satisfaction achieved for drivers. It provides, through a comprehensive process, an improved approximation of the set of non-dominated solutions attached to this bi-criterion decision problem. Moreover, this approach has dealt with the technique of rough sets in order to evaluate qualitatively the satisfaction of drivers through a predictive classification model. The significant benefits of the positive contribution of this approach are:

- The integration of a satisfaction index for drivers in the search of a solution for the BDAP;

- The search for a solution is finally based on minimizing the operating cost of the drivers;

- A set of assignment solutions is found at the end of the resolution process rather than a single solution. A subjective assessment by the decision makers of the RPTC can retain an appropriate final solution;

- The computation time required to find a satisfactory solution is very reasonable;

- The solving approach proposed here gives confidence to decision makers because it is clear;

- The use of the technique of GA has high robustness and easy implementation.

- The use of rough set approach is very helpful to extract satisfaction rules because it can be used to discover dependencies in data in order to predict future behavior and discover hidden data patterns. The experimental results of this study are very encouraging and prove the usefulness of the rough set approach for satisfaction level analysis.

- The use of the fuzzy logic helps to transform some parameters into qualitative scale in order to prepare the inputs for the classification model based on rough set technique. These assessments are subjective and can be defined as a result of internal collective agreements;

However, the assignment solution must undergo frequent changes from the presence of disturbances. This leads to the need to develop an approach of resolution in a dynamic context in order to better take into account the different events.

REFERENCES

Aickelin, U., & Dowsland, K. (2003). An indirect genetic algorithm for a nurse scheduling problem. *Journal of Computers and Operations Research, 31.*

Antosik, J. L. (1978). Automatic monthly crew assignment: A new approach. In *Proceedings of the 18th AGIFORS Symposium,* Vancouver, BC, Canada.

Awad, M., & Latifur, K. (2009). *Design and implementation of data mining tools.* Boca Raton, FL: Auerbach Publications. doi:10.1201/9781420045918

Balas, E., & Ho, A. (1980). Set covering algorithms using cutting planes, heuristics and sub-gradient optimization: A computational study. *Mathematical Programming Study, 12,* 37–60.

Ball, M., & Roberts, A. (1985). A graph partitioning approach to airline crew scheduling. *Transportation Science, 19*(2), 107–126. doi:10.1287/trsc.19.2.107

Barnhart, C., Johnson, E. L., Nemhauser, G. L., Savelsbergh, M. W. P., & Vance, P. H. (1998). Branch-and-price: Column generation for solving huge integer programs. *Operations Research, 46,* 316–329. doi:10.1287/opre.46.3.316

Brusco, M. J., & Jacobs, L. W. (1993). A simulated annealing approach to the solution of flexible labour scheduling problems. *The Journal of the Operational Research Society, 44*(12), 1191–1200.

Byrne, J. (1988). A Preferential bidding system for technical aircrew (QANTAS AUSTRALIA). In *Proceedings of the 28th AGIFORS Symposium,* Cape Cod, MA.

Caprara, A., Fischetti, M., Guida, P. L., Toth, P., & Vigo, D. (1999b). Solution of large-scale railway crew planning problems: The Italian experience . In Wilson, N. (Ed.), *Computer-aided transit scheduling* (pp. 1–18). Berlin, Germany: Springer-Verlag.

Caprara, A., Fischetti, M., & Toth, P. (1999a). A heuristic algorithm for the set covering problem. *Operations Research, 47,* 730–743. doi:10.1287/opre.47.5.730

Carraresi, P., Girardi, L., & Nonato, M. (1995). Network models, lagrangean relaxation and subgradients bundle approach in crew scheduling problems. In *Proceedings of the Sixth International Workshop on Computer-Aided Transit* (pp. 188-212).

Clement, R., & Wren, A. (1995). Greedy genetic algorithms, optimizing mutations and bus driver scheduling . In Wilson, N. (Ed.), *Computer-aided transit scheduling* (pp. 213–235). Berlin, Germany: Springer-Verlag.

Crainic, T. G., & Rousseau, J. M. (1987). The column generation principle and the airline crew scheduling problem. *INFOR, 25,* 136–151.

Curtis, S., Smith, B. M., & Wren, A. (1999). Forming bus driver schedules using constraint programming. In *Proceedings of the International Conference on the Practical Applications of Constraint Logic Programming* (pp. 239-254).

Curtis, S., Smith, B. M., & Wren, A. (2000). Constructing driver schedules using iterative repair. In *Proceedings of the 2nd International Conference on the Practical Applications of Constraint Technologies and Logic Programming.*

Dantzig, G. B., & Wolfe, P. (1960). Decomposition principles for linear programming. *Operations Research, 8*(1), 101–111. doi:10.1287/opre.8.1.101

Deb, K. (2001). *Multiobjective optimization using evolutionary algorithms.* Chichester, UK: John Wiley & Sons.

Deb, K., Pratap, A., Agarwal, S., & Meyarivan, T. (2002). A fast and elitist multi-objective genetic algorithm: NSGA-II. *IEEE Transactions on Evolutionary Computation, 6*(2), 181–197. doi:10.1109/4235.996017

Desaulniers, G., Desrosiers, J., Dumas, Y., Marc, S., Rioux, B., Solomon, M. M., & Soumis, F. (1997). Crew pairing at Air France. *European Journal of Operational Research, 97*(2), 245–259. doi:10.1016/S0377-2217(96)00195-6

Desaulniers, G., Desrosiers, J., Lasry, A., & Solomon, M. M. (1999). Crew pairing for a regional carrier . In Wilson, N. (Ed.), *Computer-aided transit scheduling* (pp. 19–41). Berlin, Germany: Springer-Verlag.

Desrochers, M., Gilbert, J., Sauve, M., & Soumis, F. (1992). CREW-OPT: Subproblem modeling in a column generation approach to urban crew scheduling . In Wilson, N. (Ed.), *Computer-aided transit scheduling.* Berlin, Germany: Springer-Verlag.

Desrochers, M., & Soumis, F. (1989). A column generation approach to the urban transit crew scheduling problem. *Transportation Science, 23*(1), 1–13. doi:10.1287/trsc.23.1.1

El Moudani, W., & Mora-Camino, F. (2000). A fuzzy solution approach for the roster planning problem. In *Proceedings of the 9th IEEE International Conference on Fuzzy Systems,* San Antonio, TX.

Elias, S. E. G. (1966). *A mathematical model for optimizing the assignment of man and machine in public transport run cutting* ([). Morgantown, WV: West Virginia University, Engineering Experiment Station.]. *Research Bulletin (Sun Chiwawitthaya Thang Thale Phuket),* 81.

Fischetti, M., Lodi, A., Martello, S., & Toth, P. (2001). A polyhedral approach to simplified crew and vehicle scheduling problems. *Management Science, 47*(6), 833–850. doi:10.1287/mnsc.47.6.833.9810

Fischetti, M., Martello, S., & Toth, P. (1989). The fixed job schedule problem with working-time constraints. *Operations Research, 37*(3), 395–403. doi:10.1287/opre.37.3.395

Fores, S., Proll, L. G., & Wren, A. (1999). An improved ILP system for driver scheduling . In Wilson, N. (Ed.), *Computer-aided transit scheduling* (pp. 43–62). Berlin, Germany: Springer-Verlag.

Forsyth, P., & Wren, A. (1997). *An ant system for bus driver scheduling* (Tech. Rep. No. 97.25). Leeds, UK: University of Leeds, School of Computer Studies.

Freling, R., Lentink, R. M., & Odijk, M. A. (2001). Scheduling train crews: A case study for the Dutch railways . In Voss, S., & Daduna, J. (Eds.), *Computer-aided scheduling of public transport* (pp. 153–165). Berlin, Germany: Springer-Verlag. doi:10.1007/978-3-642-56423-9_9

Glanert, W. (1984). A timetable approach to the assignment of pilots to rotations - Lufthansa. In *Proceedings of the 24th AGIFORS Symposium.*

Hoffman, K. L., & Padberg, M. (1993). Solving airline crew scheduling problems by branch-and-cut. *Management Science, 39*(6), 657–682. doi:10.1287/mnsc.39.6.657

Holland, J. H. (1992). *Adaptation in natural and artificial systems.* Cambridge, MA: MIT Press.

Horn, J., Nafploitis, N., & Goldberg, D. E. (1994). A niched Pareto genetic algorithm for multiobjective optimization. In *Proceedings of the First IEEE Conference on Evolutionary Computation*, Orlando, FL (pp. 82-87).

Hu, X., Lin, T. Y., & Jianchao, J. (2004). A new rough sets model based on database systems. *Fundamenta Informaticae*, 1-18.

Huang, T., Kecman, V., & Kopriva, I. (2006). *Kernel based algorithms for mining huge data sets.* Berlin, Germany: Springer-Verlag.

Jensen, R., & Shen, Q. (2004). Fuzzy-rough attribute reduction with application to web categorization. *Fuzzy Sets and Systems, 141*(3), 469–485. doi:10.1016/S0165-0114(03)00021-6

Kroon, L., & Fischetti, M. (2001). Crew scheduling for Netherlands railways "destination-customer.". In Voss, S., & Daduna, J. (Eds.), *Computer-aided scheduling of public transport* (pp. 181–201). Berlin, Germany: Springer-Verlag. doi:10.1007/978-3-642-56423-9_11

Kwan, A. S. K., Kwan, R. S. K., Parker, M. E., & Wren, A. (1999). Producing train driver schedules under differing operating strategies . In Wilson, N. (Ed.), *Computer-aided transit scheduling* (pp. 129–154). Berlin, Germany: Springer-Verlag.

Layfield, C. J., Smith, B. M., & Wren, A. (1999). Bus relief point selection using constraint programming. In *Proceedings of the 1st International Conference on the Practical Applications of Constraint Technologies and Logic Programming.*

Liping, Z. (2006). An heuristic method for analyzing driver scheduling problem. *IEEE Transactions on Systems, Man, and Cybernetics. Part A, Systems and Humans, 36*(3), 521–531. doi:10.1109/TSMCA.2005.853497

Marchettini, F. (1980). Automatic monthly cabin crew rostering procedure. In *Proceedings of the 20th AGIFORS Symposium*, New Delhi, India.

Marsten, R. E., Muller, M. R., & Killion, C. L. (1979). Crew planning at flying tiger: A successful application of integer programming. *Management Science, 25*(12), 1175–1183. doi:10.1287/mnsc.25.12.1175

Martello, S., & Toth, P. (1986). A heuristic approach to the bus driver scheduling problem. *European Journal of Operational Research, 24*(1), 106–117. doi:10.1016/0377-2217(86)90016-0

Mingozzi, A., Boschetti, M. A., Ricciardelli, S., & Bianco, L. (1999). A set partitioning approach to the crew scheduling problem. *Operations Research, 47*(6), 873–888. doi:10.1287/opre.47.6.873

Orlowska, E. (1998). *Incomplete information: Rough set analysis.* Heidelberg, Germany: Physica-Verlag.

Pawlak, Z. (1982). Rough sets. *International Journal of Computer and Information Sciences, 11*(5), 341–356. doi:10.1007/BF01001956

Pawlak, Z. (1991). *Rough sets: Theoretical aspects and reasoning about data.* Boston, MA: Kluwer Academic.

Peters, J. F., & Skowron, A. (2004). *Transactions on rough sets 1.* Berlin, Germany: Springer-Verlag.

Rubin, J. (1973). A technique for the solution of massive set covering problems, with application to airline crew scheduling. *Transportation Science, 7*(1), 34–48. doi:10.1287/trsc.7.1.34

Ryan, D. M., & Foster, B. A. (1981). An integer programming approach to scheduling. *Computer Scheduling of Public Transport*, 269-280.

Sarra, D. (1988). The automatic assignment model (SATURN - ALITALIA). In *Proceedings of the 28th AGIFORS Symposium*, Cape Cod, MA.

Sengoku, H., & Yoshihara, I. (1998). A fast TPS solver using GA on JAVA. In *Proceedings of the 3rd International Symposium on Artificial Life and Robotics* (pp. 283-288).

Smith, B. M., Layfield, C. J., & Wren, A. (2000). A constraint programming preprocessor for a bus driver scheduling system. In Freuder, E., & Wallace, R. (Eds.), *Constraint programming and large scale discrete optimization* (*Vol. 57*). Providence, RI: American Mathematical Society.

Smith, B. M., & Wren, A. (1988). A bus crew scheduling system using a set covering formulation. *Transportation Research*, *22*(2), 97–108. doi:10.1016/0191-2607(88)90022-2

Srinivas, N., & Deb, K. (1995). Multiobjective function optimization using nondominated sorting genetic algorithms. *Evolutionary Computation*, *2*(3), 221–248. doi:10.1162/evco.1994.2.3.221

Steinzen, I., Gintner, V., Suhl, L., & Kliewer, N. (2010). A time-space network approach for the integrated vehicle-and crew-scheduling problem with multiple depots. *Journal Transportation Science*, *44*(3).

Vercellis, C. (2009). *Business intelligence: Data mining and optimization for decision making*. New York, NY: John Wiley & Sons.

Ward, R. E., & Deibel, L. E. (1972, November). *The advancement of computerized assignment of transit operators to vehicles through programming techniques*. Paper presented at the Joint National Meeting of the Operations Research Society of America, Atlantic City, NJ.

Wren, A., Kwan, R. S. K., & Parker, M. E. (1994). Scheduling of rail driver duties. *Computers in railways–IV, 2*, 81-89.

Wren, A., & Rousseau, J. M. (1995). Bus driver scheduling: An overview . In Wilson, N. (Ed.), *Computer-aided transit scheduling* (pp. 174–187). Berlin, Germany: Springer-Verlag.

Wren, A., & Wren, D. O. (1995). A genetic algorithm for public transport driver scheduling. *Computers & Operations Research*, 22.

Yoshihara, I., & Sengoku, H. (2000, December 17-20). Scheduling bus driver's services based on genetic algorithm. In *Proceedings of the International Conference on Artificial Intelligence in Science and Technology*, Hobart, Tasmania (pp. 62-67).

Zhong, N., & Skowron, A. (2001). A rough set-based knowledge discovery process. *International Journal of Applied Mathematics and Computer Science*, *11*(3), 603–619.

Zimmermann, H. J. (1996). *Fuzzy set theory and its applications*. Boston, MA: Kluwer Academic.

Zitzler, E., Deb, K., & Thiele, L. (2000). Comparison of multiobjective evolutionary algorithms: Empirical results. *Evolutionary Computation*, *8*(2), 173–195. doi:10.1162/106365600568202

This work was previously published in the International Journal of Applied Metaheuristic Computing, Volume 2, Issue 2, edited by Peng-Yeng Yin, pp. 21-50, copyright 2011 by IGI Publishing (an imprint of IGI Global).

Chapter 14

A New Approach to Associative Classification Based on Binary Multi-Objective Particle Swarm Optimization

Madhabananda Das
KIIT University, India

Satchidananda Dehuri
Fakir Mohan University, India

Rahul Roy
KIIT University, India

Sung-Bae Cho
Yonsei University, Korea

ABSTRACT

Associative classification rule mining (ACRM) methods operate by association rule mining (ARM) to obtain classification rules from a previously classified data. In ACRM, classifiers are designed through two phases: rule extraction and rule selection. In this paper, the ACRM problem is treated as a multi-objective problem rather than a single objective one. As the problem is a discrete combinatorial optimization problem, it was necessary to develop a binary multi-objective particle swarm optimization (BMOPSO) to optimize the measure like coverage and confidence of association rule mining (ARM) to extract classification rules in rule extraction phase. In rule selection phase, a small number of rules are targeted from the extracted rules by BMOPSO to design an accurate and compact classifier which can maximize the accuracy of the rule sets and minimize their complexity simultaneously. Experiments are conducted on some of the University of California, Irvine (UCI) repository datasets. The comparative result of the proposed method with other standard classifiers confirms that the new proposed approach can be a suitable method for classification.

DOI: 10.4018/978-1-4666-2145-9.ch014

INTRODUCTION

Association and classification rule mining are two most promising data mining techniques for designing a classifier to solve the classification task in various domains of medical science (Giannopoulou, 2008, Huap et al., 2009), marketing management (Groth, 1997), telecommunication (Sasisekharan, Seshadri, & Weiss, 1996), disaster management (Shao & Fu, 2008), intrusion detection (Pei et al., 2004), life science (Wong & Li, 2006), insurance, biometric (Gutierrez et al., 2002), etc. The objective of association rule mining, in its basic form, extracts a set of the individual descriptive rules from the database with minimum support and confidence and the objective of classification rule mining is to design a compact and efficient predictive rule sets from the extracted rules. The integration of association rule mining with the classification rule mining has helped in overcoming some of the basic problems of classification rule mining which are illustrated below:

1. The integrated framework helps in better understandability of the classification problem. The standard classification system generates compact rules by using dominant biases and heuristics which sometimes result in generation of rules that may not be in agreement with the user, leaving behind many understandable rules undiscovered. With this framework, all the understandable rules are discovered using association rule mining and the compact classifier is designed from these rules thereby increasing the better understandability of the rules used by the classifier.
2. Another issue of rule discovery problem is the generation of interesting and useful rules. In quest of generating a compact classifier, the classification rule mining algorithm may leave behind many interesting and useful rules.
3. Classification rule requires the entire database to be loaded to the main memory for rule discovery which is not the case with associative classification rule mining.

This integrated framework for designing a classifier is referred to as associative classification rule mining or simply associative classification (Liu et al., 1998) (de la Iglesia, Reynolds, & Rayward-Smith, 2005; de la Iglesia, Richards, Philpott, & Rayward-Smith, 2006). Associative classification rule mining usually consists of two phases: rule extraction and rule selection. In the rule extraction phase, a large number of classification rules are extracted from a data set using an association rule mining technique that satisfies the pre-specified threshold values of the minimum support and confidence. In rule selection phase, a part of extracted rules are selected or targeted to design an accurate and compact classifier. The accuracy of the designed classifier usually depends on the specification of the minimum support and confidence. In the context of classification rule mining, their tuning has been discussed in the literatures (Coenen, Leng, & Zhang, 2005; Bayardo & Agrawal, 1999).

The initial framework for associative classification rule mining was proposed by Liu et al. (1998). Thereafter, many researchers have used the two phase associative rule mining for designing classifiers. Most of the two phase associative algorithms differ in the rule selection phase. All the statistical based associative classification rule mining concentrate on pruning of the rules using various heuristic algorithms in rule selection phase. These heuristic approaches prune the rules in order to make classifier compact without taking into account the interestingness of the rule, thereby compromising with the classification accuracy. However, the actual problem lied with the rule generation phase, which uses the exhaustive techniques like *Apriori* algorithms for generation of rules. For this reason, huge set of rules are extracted-that includes many uninteresting and conflicting rules.

After Brayado et al. (1999) work about the optimality of Pareto optimal rules with coverage and confidence, it has motivated many researches to incorporate meta-heuristic approaches in rule generation phase to generate Pareto optimal rules. The first pioneering work of using multi-objective meta-heuristic was done de la Iglesia et al. (2005) for the rule extraction. In 2005, Ishibuchi et al. has extended the work to rules selection phase and they showed the relation between Pareto optimal rules and Pareto optimal rulesets. Ishibuchi et al. (2005) showed that the Pareto optimal rules can be reduced to design a compact classifier without affecting the classification accuracy. They used multi-objective genetic algorithm which included expensive genetic operators like selection, crossover, mutation, etc., which increases the computation time. Also the binary string, used for representing the rules is divided into two parts: the first part used the gray code representation for numeric attributes and the second part represents the categorical attributes. This different representation of the attribute increases the string length making the representation bulky.

These shortcomings of the multi-objective genetic algorithm motivated us to adopt alike meta-heuristic in the two phases of associative classification algorithm which can increase the efficiency of the classification process and also reduce the computation time.

In this paper, we present, a binary multi-objective particle swarm optimization based associative classification rule mining techniques. In our approach, we adopt the particle swarm optimization based meta-heuristic to solve the multi-objective problem in both the phases of associative classification rule mining because of their relatively simple position updating equation and information sharing mechanism that can help in reducing the computation time.

Though this binary multi-objective meta-heuristics is not the first of its kind; however,

unlike the other meta-heuristic approaches, as mentioned in literature, we present a compact representation of the solution. Here we don't use different representation for encoding the numeric and categorical attribute in a solution. Rather, we transform the numeric attributes to discrete attribute and use the same representation for encoding both numeric and categorical attributes.

The relative simplicity, competitive performance and the information sharing mechanism of PSO as a single objective optimizer have made it a natural candidate to be extended for multi-objective optimization. A transfer of PSO to the multi-objective domain can be a natural progression with some intelligent modifications in the basic PSO algorithm. Changing a PSO to a multi-objective PSO (MOPSO) (Fieldsend & Singh, 2002; Coello, Pulido, & Lechuga, 2004) requires a redefinition of what a guide is, in order to obtain a front of optimal solutions (Pareto front) (Coello & Lechuga, 2002; Dehuri & Cho, 2009). In MOPSO, the Pareto-optimal solutions are used to determine the guide for each particle. A number of different studies have been published on Pareto approach based multi-objective PSO (MOPSO) (Fieldsend & Singh, 2002; Coello, Pulido, & Lechuga, 2004; Dehuri & Cho, 2009). Each of these studies implements MOPSO in a different fashion. In continuous MOPSO, the velocity is considered as the rate of change position but the same concept can be applied in case of binary search space. In case of binary search space, the position of a particle can be updated only flipping the bits. Thus in case of binary MOSPO, the velocity should represent the change in bits. Kennedy et al. (1997) attempted a binary PSO in 1997. Surprisingly, despite the considerable volume of research in MOPSO, there have been very little/no attempt made towards development of binary multi-objective PSO for solving the discrete optimization problem. Henceforth, our research effort in this direction can add one more dimension to the MOSPO domain.

In associative classification rule mining, the antecedent conditions of the rules having continuous values needs to be discretized (Han & Kamber, 2006). Hence during the rule extraction phase, we encode the generated rules in 0 or 1 which represent the occurrence of an antecedent condition in the rule. Also in rule selection phase, the occurrence of rule in a rule set can be well represented by binary value 0 or 1. These factors motivated us to develop a binary multi-objective particle swarm for associative classification rule mining. To the best of the knowledge of the authors, there is no such BMOPSO algorithm in the specialized literature for designing a classifier based on association rule mining.

The rest of the paper is organized as follows: Section 2 provides a brief literature review of the various meta-heuristics algorithms proposed for associative classification rule mining. Section 3 presents the description of associative classification rule mining and its various phases. In Section 4, a brief description of the binary MOPSO algorithm is provided. Section 5 discusses the experimental setups and the results of the experiment. Finally, a conclusion and future research direction is provided in Section 6.

RELATED WORK

Agrawal et al. (1996) first introduced the ARM for discovering association rules between items in large market basket datasets. Since then, ARM is playing a vital role either independently or coupled with some other methods (e.g., classification) in the area of data mining research. Bayardo and Agrawal (1999) used the concept of association rule mining for generating classification rules (confidence>90%). Kamal et al. (1997) also used the same concept to discover rules that describe individual class. However, none of them used association miner for design of a classifier from the rules. As this paper is focusing on associative

classification rule mining so the discussion of independent role of ARM is beyond the scope.

The framework that integrates association rule mining with classification rule mining was first proposed by Liu et al. (1998). Li et al. (2001) used the concept of class association rules to construct a class distribution associated FP tree for mining large database. For a classifier based on rules besides the accuracy of the classifier, its size is another vital aspect. In pursuit of high classification accuracy, many classifiers do not take into account their sizes, and contains numerous both essential and insignificant rules. Hence, Liu et al. (2009), has proposed a fast post-processing approach to remove insignificant rules generated from associative classification method. Chen et al. (2006), has proposed a new classifier based on extended association rule mining technique. In rule generation phase, many candidate item sets are filtered out and resulting a much smaller set of rules. The rules generated by classification association rule mining systems will be influenced by the choice of association rule mining parameters (typically support and confidence threshold). Coenen and Leng (2007) have examined the effect of the parameters on the accuracy of the classifier. Liu et al. (2011), has presented an efficient post processing method to prune redundant rules by virtue of the property of Galois connection, which inherently constrains rules with respect to objects. Li and Chen (2008) have proposed a multi-criteria model to induce optimal classification rules with better rates of accuracy, support and compactness. Finding "persistent rules" by independent application of association and classification algorithms to the same data to discover common or similar rules is another improvement in this direction (Rajasethupathy, Scime, Rajasethupathy, & Murray, 2009). A new associative classification technique, Ranked Multi-level Rule (RMR) algorithm was introduced by Thabtah and Cowling (2007), which generate rules with multiple levels. Li and Topor (2002) has presented an efficient algorithm for mining the optimal class association rule set using an upward closure property of pruning weak rules before they are actually generated. Most associative classification algorithms adopt the exhausted search method presented in Apriori algorithm to discover rules and require multiple passes over the database. Thabtah et al. (2006), has presented a solution called multi-class classification based on association rules to overcome the limitations.

Bayardo and Agrawal (1999) showed that Pareto optimal rules generated with support and confidence are the optimal set of rules. This motivated the researchers to consider rule mining as a multi-objective problem. de la Iglesia et al. (2005) proposed the Pareto based evolutionary multi-objective algorithm (EMO) for extraction of rules. They also used the concept of partial classification proposed by Kamal et al. (1997) for proposing a Pareto based rule discovery for each class. de la Iglesia et al. (2006) used the association rule mining for generation of associative rules for classification but did not provide detail of designing a compact classifier from extracted rules. Ishibuchi et al. in 2005 proposed EMO based approach for designing a classifier by selecting rules from the extracted associative rules without drastically degrading the accuracy and showed the relation between pareto optimal rules and pareto optimal rule sets. However, there is no description of the procedure of extraction of rules from the dataset. They concentrated more on the rule selection phase.

There are other bio-inspired meta-heuristics applied to classification rule mining as single/multi-objective problem. Holden and Freitas (2007) applied a hybrid PSO approach for classification rule mining. Their algorithm built each rule with the aim of optimizing that rule's quality individually, without interaction with other rules. Atlas and Akin (2009) have proposed a chaotic MOPSO based approach for classification rule mining. Carvalho and Pozo (2008) have proposed Pareto based rule generation technique using MOPSO approach for unordered data set. How-

ever, as per the best knowledge of the authors, there are no references on associative classification rule mining using MOPSO in the literature.

There are also references in the literature on application of fuzzy logic for extraction and selection of fuzzy if-then rules using support and confidence (Ishibuchi, Nozaki, Yamamoto, & Tanaka, 1995; Kaya, 2006). Many researchers have also applied other evolution based approach for associative classification mining. GAssist, XCS are some of the other approaches widely applied for classification rule mining. Bacardit and Butz (2007), has provide a comparison on XCS and GAssist based classifier. Though there is not much differences between both the classifier systems based on their performance but have pointed on the different problems of both the classifier systems.

ASSOCIATIVE CLASSIFICATION RULE MINING

Let us consider that we have a dataset with N tuples. From these N tuples, m ($m < N$) tuples are selected as training set denoted by P and remaining $N-m$ tuples are used as test set, denoted by T. The m training patterns containing M classes in n-dimensional continuous pattern space are represented as $x_p = (x_{p1}, x_{p2}, x_{p3}, \ldots, x_{pn})$, where $p = 1, 2, 3, \ldots, m$ and x_{pi} represents a value of ith attribute in p^{th} training pattern. The rule used for the classification problem is represented as:

Rule R_q: If x_1 is A_{q1} and x_2 is A_{q2} and . . . and x_n is A_{qn} then Class Cq, (1)

where R_q is the label for the qth rule. $x = (x_1, x_2, x_3, \ldots, x_n)$ is an n-dimensional pattern vector. The antecedent condition of a rule "$A_q => C_q$" is represented by $A_q = (A_{q1}, A_{q2}, A_{q3}, \ldots, A_{qn})$ where A_{qi} is the antecedent interval for i^{th} attribute, C_q is the class label. Here "x_i is A_{qi}" in (1) means that the "$x_i \in A_{qi}$". It should be noted the rule of the form (1) can have less than n antecedent conditions.

Recall that the associative classification rule mining is divided into two phases: *Rule generation* and *Rule selection*. We describe each of the phases one after another.

Rule Generation Phase

In associative classification rule mining (Agrawal, Mannila, Srikant, Toivonen, & Verkamo, 1996; Liu, Hsu, & Ma, 1998; Coenen & Leng, 2007), an association rule mining technique such as Apriori (Agrawal, Mannila, Srikant, Toivonen, & Verkamo, 1996) algorithm is used for extraction of the rules which satisfy the threshold of minimum support and confidence, respectively.

In association rule mining, the generated rules are evaluated using two most commonly used metrics, namely support and confidence. Let us represent the support and confidence of a rule "$A_q => C_q$" by $SUPP(A_q => C_q)$ and $CONF(A_q => C_q)$, respectively. The $SUPP(A_q => C_q)$ implies that the number of patterns compatible with both A_q and C_q. More formally, support of rule is defined as:

$$SUPP\left(A_q \Rightarrow C_q\right) = \frac{SUPP\left(A_q \cup C_q\right)}{|D|}. \qquad (2)$$

where $|D|$ is the cardinality of the data set D (i.e. $|D|=m$). Similarly, we can define the support for the antecedent condition A_q and class C_q. The confidence of $A_q \Rightarrow C_q$ is defined as:

$$CONF\left(A_q \Rightarrow C_q\right) = \frac{SUPP\left(A_q \Rightarrow C_q\right)}{SUPP\left(A_q\right)}. \qquad (3)$$

In partial classification (Liu, Hsu, & Ma, 1998; de la Iglesia, Reynolds, & Rayward-Smith, 2005; de la Iglesia, Richards, Philpott, & Rayward-Smith, 2006), the coverage is used instead of support for evaluation of rule. Let us denote coverage of a rule "$A_q => C_q$" as $COV(A_q => C_q)$. The

coverage metrics specifies how many number of patterns compatible with both antecedent conditions A_q and the class level C_q out of total number of patterns of class C_q. Mathematically,

$$COV\left(A_q \Rightarrow C_q\right) = \frac{SUPP\left(A_q \Rightarrow C_q\right)}{SUPP\left(C_q\right)}. \qquad (4)$$

Since the consequent class C_q is fixed for partial classification (i.e., the denominator is constant, the coverage can be considered same as that of support of a rule.

Rule Selection Phase

Let us consider a set of rules S extracted during the rule extraction phase. This candidate set of rules S become the input for rule selection phase. From these S rules, K rules set $(K < S)$ are selected to form the classifier. The selected rules must satisfy the three objectives stated below:

- The number of correctly classified training patterns by K should be maximum.
- The number of selected rules in K should be minimum.
- The total number of antecedent conditions in K should be minimum.

The third objective is viewed as the minimization of the total rule length since the number of antecedent condition of each rule is often referred to as the rule length.

When a new pattern x_p is to be classified by K-rule classifier, we choose a single winner rule with the maximum point coverage from K rules whose antecedent conditions is compatible with x_p. The consequent class of the winner rule is assigned to x_p.

The block diagram of the associative classification rule mining is shown in Figure 1.

Figure 1. Block Diagram for the associative classification rule mining

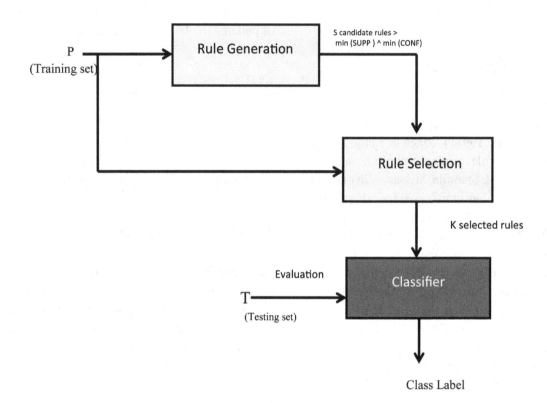

BMOPSO FOR ASSOCIATIVE CLASSIFICATION RULE MINING

In this Section, we describe our developed binary multi-objective particle swarm optimization (BMOPSO) applied for associative classification rule mining to extract Pareto optimal rules and Pareto optimal rulesets. The BMOPSO algorithm differs from continuous MOPSO algorithm in its particle representation and position update equation. As the guide selection strategy and repository maintenance strategy depends on objective space which is continuous in nature, we can adopt any of the well suited guide selection, depending on the movement strategy of the particle and repository Updation strategy from the specialized literature of continuous MOPSO.

We consider the associative classification rule mining as partial classification problem i.e., the extraction of rules is done w.r.t each class. The data sets used for designing the classifier are discretized by first equally dividing the difference between maximum value & minimum value of each predicate attribute into L divisions and then each predicate attribute is now transformed to a discrete value l where $l \in [1, L]$.

Binary MOPSO for Rule Extraction

In rule extraction phase, we apply the BMOPSO to obtain the Pareto optimal rules with respect to coverage and confidence.

The details of the BMOPSO for the rule extraction phase are given:

Figure 2. Particle representations of a rule in rule extraction phase

	l_1	l_2	l_3	l_4			l_L
A_1	0	1	0	1		...	1
A_2	1	1	0	0			0
A_3	0	1	1	0			1
					.		
					.		
					.		
A_n	0	1	1	1			0

Particle Representation

The antecedent condition of a rule R_q is encoded as a particle of the swarm. Based on the occurrence of an antecedent condition into the rule, it is represented by 0 or 1. Thus, a particle is represented as a matrix of $n \times l$ where n is the number of antecedent condition, and l is number of values of the antecedent condition. The block diagram of rule encoded as particle is shown in Figure 2.

This particle now actually represents the antecedent part of a rule having following antecedent conditions:

$$(A_1 = [2 \text{ V } 4 \text{ V}...\text{V L}]) \wedge (A_2 = [1\text{V } 2 \text{ V}...]) \wedge (A_3 = [2 \text{ V } 3 \text{ V}...\text{V L}]) \wedge ... \wedge (A_n = [2 \text{ V } 3 \text{ V}...]),$$

where $\wedge$ represents AND operation and V represents OR operation.

The particles are generated randomly. The generation of each bit of the particle is independent of each other i.e., *Prob* $[A_i=0] = Prob[A_i=1] = \frac{1}{2}$, where *Prob* represents the probability. The particle together constitutes the swarm and the size of the swarm is specified by user.

This encoding strategy of solution creates a compact representation of the particle unlike the one used in de la Iglesia, Reynolds, and Rayward-Smith (2005) and Ishibuchi and Nojima (2005).

Fitness Evaluation

Each particle of the swarm is evaluated with following two objectives:

Maximize {COV (R_q), CONF (R_q)},

such that

COV(R_q) > minCOV,

CONF(R_q) > minCONF,

where *COV(R_q)* and *CONF (R_q)* represents the coverage and confidence of rule R_q given by (4) and (3) respectively. The *minCOV* and *minCONF* are the threshold values for coverage and confidence respectively.

The objective of each particle is to maximize both coverage and confidence of a rule simultaneously subject to the constraint that the coverage and confidence of a rule is greater than their pre-specified threshold values. These constraints define the lower boundary for the feasible region of the solution space thereby reducing the search space. The threshold value for coverage and confidence are user-defined.

The fitness evaluation is carried out by evaluating coverage and confidence of each particle in the swarm in the following manner. The discrete value(s) of first predicate attribute (A_1) of the particle is matched with all the tuples of the input dataset one by one. Only those tuples from the dataset are extracted to form the subset Table S_1 whose A_1 matches with A_1 of the particle. Next the 2nd attribute A_2 of the said particle is matched with A_2 of all the tuples of S_1 one by one. Only those tuples from the discretized subset Table S_1 are extracted to form subset Table S_2 whose A_2 matches with A_2 of the particles. This process continues for all A_i where i=1...L. After A_L is matched, subset Table S_L is formed. Since partial classification is being considered, the coverage and confidence of a particle for a particular class is evaluated using Equations 4 and 3 and evaluate whether it violate the constraint.

The random initialization of the particle can result in a particle with all the antecedent condition set to zero, or finding of a set of patterns in which the antecedent conditions become compatible but none of consequent becomes compatible with specified class level. In this case, coverage and confidence of the particle are set to zero which makes the particle dominated by other particles with higher coverage and confidence in the domination test. Thus, the chance of such particle entering to the Pareto optimal set is ruled out.

Particle Movement

The particle new position is determined by using the velocity of the particle. The change in velocity is determined by the difference in particles present position with the personal best position and global best position respectively. The particle's velocity update equation is defined as

$$v_{t+1} = w \cdot v_t + rand_1() \ \theta_1 \ (pbest - x_t) + rand_2() \ \theta_2 \ (gbest - x_t) \qquad (5)$$

where v_t denotes the velocity of the particle in time t, w is the inertia, θ_1 represents the cognitive coefficient and θ_2 represents the interaction coefficient, $rand_1()$ and $rand_2()$ are the random numbers in the range [0,1]. *pbest* represents the personal best and *gbest* represents the global best position of a particle. w is the weight factor that controls the impact of previous history of velocity.

These values are determined empirically during the initial phases of the experiment. However, the inertia weight w, cognitive component θ_1 and social component θ_2 can be determined adaptively. Shi and Eberhart (1999) found a significant improvement in performance of PSO by linearly decreasing the weights over generation. This time varying inertial weights (TVIW) can be used in BMOPSO to adaptively determine the inertia weight using the Equation 6.

$$w = w_2 + \left(\frac{cycle_{max} - cycle}{cycle_{max}} \right) \times (w_1 - w_2) \qquad (6)$$

where w_1 and w_2 are the maximum and minimum inertia weight. $cycle_{max}$ is the maximum iteration (maximum generation) and $cycle$ represent the present iteration (generation).

Then, Ratnaweera, Halgamuge, and Watson (2004) introduced a time varying acceleration co-efficient (TVAC), which reduces the cognitive component, c_1 and increases the social component, c_2 of acceleration co-efficient with time. With a large value of c_1 and a small value of c_2 at the beginning, particles are allowed to move around the search space, instead of moving toward pbest. A small value of c_1 and a large value of c_2 allow the particles converge to the global optima in the latter part of the optimization. The TVAC is given in Equations 7 and 8 can be used in the BMOPSO to determine these factor adaptively.

$$c_1 = \left(c_{1i} - c_{1f}\right) \times \left(\frac{cycle_{\max} - cycle}{cycle_{\max}}\right) + c_{1f} \quad (7)$$

$$c_2 = \left(c_{2i} - c_{2f}\right) \times \left(\frac{cycle_{\max} - cycle}{cycle_{\max}}\right) + c_{2f} \quad (8)$$

The particle moves in the search space under the influence of two components i.e., the global best position of the swarm and its own personal best position. Under the influence of the guides, the particle moves to explore the least crowded area of the Pareto fonts (Nguyen & Kachitvichyanukul, 2010).

In case of continuous search space, the change in velocity can be directly applied to update the particle's position, i.e.,

$$x_{t+1} = x_t + v_{t+1} \quad (9)$$

However, our solution space is binary and thus the change in position is determined the rate of flipping of bits in the solution encoded in the particle. Thus the velocity cannot be applied directly to update the particle position. So, we adopt the notion of discrete PSO proposed by Kennedy and Eberhart (1997), where velocity is treated as probability of flipping of bits. Thus the equation for updating the position is given as:

$$x_{t+1} = \begin{cases} 1 & rand < W \\ 0 & otherwise \end{cases}, \quad (10)$$

where

$$W = 1/[1 + exp(-v_{t+1})]. \quad (11)$$

Equation 11 is used to normalize the velocity to lie in the range of [0, 1], thereby restricting the particle fly out of the search space.

The use of velocity as a probability for updating the particle position translates the continuous MOPSO to binary MOPSO.

Guide Selection Strategy

As the particle moves in the search space under the influence of its best global and local guide, so we need to select a best particles as the leader to guide other particles towards the optimal solutions. However, in multi-objective domain, we search for a set of optimal solutions. So a single guide will drive all the particles to a single optimal solution which may optimize one objective value compromising with the other. So, the guide selection strategy is important for multi-objective problem as it influences the movement of the particles in the search space. Thus we need to select guide for each particle so that the particle better explore the search space in search of a set of optimal solution that optimizes all objectives simultaneously. Hence, we need to store the indifferent solutions which represent the information of the Pareto front, which can be used as guide for particles in the swarm.

The selection of guide is done based on the objective values. Though the solution space is binary but the objective space is continuous. Hence, we can adopt the guide selection strategies of the continuous MOPSO for our BMOPSO algorithm.

Figure 3. Global leader selection using the k-medoid clustering approach

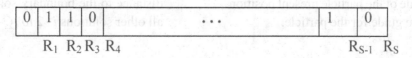

Among the three movement strategies proposed by Nugyen and Kachitvichyanukul (2010), our algorithm adopts the first movement strategy of selecting a global guide for exploring the least crowded area. The key idea of this movement strategy is to guide the swarm towards least crowded region of the Pareto front. Unlike the crowding strategy mention by Nugyen and Kachitvichyanukul (2010), here we find the least crowded area by using the *k*-medoid clustering technique (Dehuri & Cho, 2009). We use the *k*-medoid clustering technique to find the least crowded area as it provides better diversification of the solution which is one of the major drawback of crowding distance technique. In this method, *k* clusters are created on the objective space using the *k*-medoid clustering technique. '*k*' is the user defined parameter. From these *k* clusters, a least dense cluster is selected and the medoid of the cluster is used as the guide for the particle. This process is repeated for the entire swarm to explore the less explored region of the objective space.

The local guide for the particle is selected from the local memory of each particle which stores the non-dominated position obtained till the tth generation. We apply the sigma method (Mostaghim & Teich, 2003) to find the best local guide for the each particle. In this method, σ value is assigned to each non-dominated position stored in the local memory and the particle. The σ value is calculated using Equation 12:

$$\sigma = \frac{\left(K_2 f_1\right)^2 - \left(K_1 f_2\right)^2}{\left(K_2 f_1\right)^2 + \left(K_1 f_2\right)^2},$$ (12)

where K_1 and K_2 are the maximum objective values of the particle position in the swarm. The particle position in memory which has sigma value closest to the sigma value of the particle present position is selected as the guide for the particle.

Repository Maintenance Strategy

We use two repositories for storing the global and local guides. The size of these repositories cannot be allowed to grow too large as the search of the guides becomes complex which increases the CPU time and also the space requirement. Hence, we need to use some strategies to maintain these archives so that we can retain the best subset of the Pareto optimal solutions and also satisfy the space and time constraint.

The local memory is maintained using only the non-domination test so that it retains all the non-dominated position visited by the particle in every generation. A particle present position enters the local memory iff it is not dominated by previously stored non- dominated position of the particle. Also, if any of the previous stored particle position is dominated by the present position, it is removed from the local memory.

The repository of global leader grows faster than the repository of the local memory. So, along with the non-domination test, we need to adopt a secondary strategy for maintaining the external repository. After the non–domination test, if the size of repository exceeds the maximum size, we use the crowding distance strategy proposed in (Raquel, Prospero, & Naval, 2005) for maintaining the external repository. First, we calculate the crowding distance of the particles stored in external repository based on their fitness value. The crowding distance is calculated in the following manner:

- **Step 1:** Set the $Q=|REP|$ and assign the crowding distance $d_i=0$.
- **Step 2:** For each objective function $m=1, 2,...,M$, sort the set in worse order of f_m.
- **Step 3:** For $m=1, 2,.......,M$, assign a large distance to the boundary solutions and for all other solutions $j=2$ *to (Q-1)*, assign

$$d_{I_j^m} = d_{I_j^m} + \frac{f_m^{(I_{j+1}^m)} - f_m^{(I_{j-1}^m)}}{f_m^{\max} - f_m^{\min}} \qquad (13)$$

Then, we sort the repository particles in descending order depending on their crowding distances. After sorting, number of particles equal to the maximum size of the repository is retained and the remaining particles are deleted from the external repository.

The BMOPSO algorithm for the rule selection is shown below.

Algorithm: BMOPSO for Rule Extraction

- **Step 1:** Initialize the swarm *SW*:
  - Randomly generate $SWARM_{max}$ binary matrix (particles) each of size $n \times l$ as an initial swarm *SW* where $SWARM_{max}$ is a user-defined parameter called the population size.
- **Step 2:** Initialize the velocity of each particle:
  - For i=1 to $SWARM_{max}$,
    - $v_t[i]=0$ /*initializing each velocity with a string of 0's*/
- **Step 3:** Initialize the pbest of each particle:
  - For i=1 to SWARMmax,
  PBEST[i] = $SW_t[i]$
- **Step 4:** Evaluate the fitness of each particle /*compute coverage & confidence*/
- **Step 5:** Store the positions of the particles that represent non-dominated vectors in the repository *REP*.
- **Step 6:** WHILE maximum number of cycles has not been reached DO
  - If the number of particles in the repository REP is greater than the specified cluster size, compute the gbest for each particle in the REP by applying K-medoid clustering technique, medoid (k, REP) on two objective

criterions i.e. coverage & confidence. The medoid of the least dense cluster is selected as the gbest for a particle. Else randomly select a particle from the REP as a guide for the particle
  - Compute the velocity of each particle using the following expression bit wise:
  For *k=1 to n*
    - For *l=1 to L*
    - $v_{t+1}[i][k][l] = v_t[i][k][l] + rand() \theta_1 (pbest[i][k][l] - SW_t[i][k][l]) + rand() \theta_2 (gbest[i][k][l] - SW_t[i][k][l])$
    - /* *rand()* takes the values in the range (0...1)*/
  - Update the new positions of the particles $SW_{t+1}[i]$ bit wise:
  Calculate the threshold value
    - w=1/[1+exp(-$v_{t+1}[i][k][l]$)]
    If (*rand () < w*) then
    - $SW_{t+1}[i][k][l]=1$
    - else
    - $SW_{t+1}[i][k][l]=0$
    - End for
  - End for
  - Evaluate the fitness of each of the new particles in *SW*
  - Update the pbest of each particle
  - Update the contents of reposition *REP* by inserting all the currently non-dominated particles into the repository. Any dominated locations from the repository are eliminated in the process. Since the size of the repository is limited, whenever it gets full, a secondary criterion for retention called crowding sort technique is applied. We sort the particle fitness in descending order of their crowding distance and select |REP| elements to maintain the repository. The descended ordering of fitness values helps in

Figure 4. Representation of a particle for rule selection

finding less densely spaced solution thereby maintaining the diversity.

- END WHILE

```
K-medoid (k, REP)
{
Arbitrarily choose k particles as the
initial medoid;
REPEAT
Assign each remaining particle to the
cluster with the nearest medoid;
Randomly select a non-medoid parti-
cle, REP[R];
Compute the total cost, CT, of swap-
ping REP[j] with REP[R];
If CT < 0 then
Swap REP[j] with REP[R] to form the
new set of k medoid;
UNTIL no change;
}
```

Binary MOPSO for Rule Selection

In rule selection phase, we obtain Pareto optimal rule sets which form the compact and efficient classifier. The particle movement, the *gbest* selection and the repository maintenance strategy in rule selection phase is same as the one used in rule extraction phase. However, the particle representation and the fitness evaluation differ from the rule extraction phase. These are discussed in following subsections.

Particle Representation

In rule selection phase, each particle represents a rule set. The particle is represented as binary vector of size S. The inclusion/exclusion of a rule in the rule set is represented by 0 or 1. The particle representation is shown in Figure 4.

Fitness Evaluation

The fitness of a particle in rule selection phase is evaluated based on three objectives:

Maximize $f_1(S)$ where $f_1(S)$ is the number of correctly classified training patterns by S.

Minimize $f_2(S)$ where $f_2(S)$ is the number of selected rules in S.

Minimize $f_3(S)$ where $f_3(S)$ is the total number of antecedent condition over selected rules in S.

The first objective is maximized while the second and third objectives are minimized. The third objective can be viewed as the minimization of the total rule length since the number of antecedent condition of each rule is often referred to as the rule length.

EXPERIMENTAL STUDY

This section describes the experimental setup used to carry out the experiment and the results obtained from the experiments.

Dataset

- **Breast Cancer Wisconsin Data Set:** This breast cancer databases was obtained from the University of Wisconsin Hospitals, Madison from Dr. William H. Wolberg. The dataset has 9 attributes, 2 class labels and 699 instances. The class labels are malignant and benign. There are 16 instances with missing attributes. After removing the instances with missing attributes, 65.5% instances belong to benign class and 34.1% of the instance predicts the class malignant.

- **Balance Scale:** This dataset is generated to model the psychological experimental results. Each example is classified as having the balance scale tip to the right, left, or balanced. There are 4 numeric attributes, namely, the left weight, the left distance, the right weight, and the right distance and three class labels, namely, *left, right, or balanced*. 46.08% instances of the dataset belongs to the class label *left*, 7.84% instances of the dataset belongs to class label *balanced* and 46.08% instances belongs to class label *right*.

- **Wine Dataset:** The wine dataset is results of a chemical analysis of wine grown in the same region in Italy but derived from three different cultivars. The analysis determined the quantities of 13 constituents found in each of the three type of wine. The dataset has 13 attributes which are continuous in nature. The three classes have 59%, 71% and 48% instances respectively.

- **Glass Dataset:** The study of Glass dataset for classification is mainly motivated by the criminological investigation. The dataset has 10 attributes which are continuous in nature. These attributes together identifies 7 different types of glass. There are 214 instances in the dataset with no missing values. The types of glass which rep-

resent class attributes have the following distributions:
  - 163 Window glass (building windows and vehicle windows)
  - 87 float processed
  - 70 building windows
  - 17 vehicle windows
  - 76 non-float processed
  - 76 building windows
  - 0 vehicle windows
  - 51 Non-window glass
  - 13 containers
  - 9 tableware
  - 29 headlamps

- **Clevend Heart Dataset:** The Clevend heart dataset is a mixed dataset for identification of Clevend heard diseases. The data set has 303 instances out which 6 instances have missing attributes. There are 14 attributes out if which 8 are symbolic and 6 are numeric. 54.45% instances predict the occurrence of diseases and 45.54% instances predicts the non-occurrence of the disease.

Experimental Setup

In this associative classification rule mining, the concept of partial classification is applied to extract the rules. The Experiment is carried out with five datasets given in Table 1. The datasets are obtained from UCI repository (UCI Machine Learning Repository, 2010). We did not use incomplete patterns with missing values. All attributes were handled as real numbers.

The domain of each attribute was divided into (multiple) equal intervals by considering the equal width histogram technique.

We apply 4-fold cross validation in the both rule extraction and rule selection phase by considering a 75% training data and 25% testing data. We report the average result over its 4 runs. We examined 4x4 combinations of the following four specification of each threshold for each of the five data sets in Table 1:

Minimum confidence (minCONF): 50%, 60%, 70%, 80%

Minimum coverage (minCOV):

1%, 4%, 8%, 10%

These threshold values are manually set for each experiment by the user.

All the extracted rules for each combination of the two threshold values are used as candidate rules in the rule selection phase. The rules are extracted considering all attributes of the dataset.

The parameters used in the algorithm for both the phases to implement the different datasets are shown in Tables 2 and 3.

The parameters are determined empirically by performing some initial tuning of the algorithm. These parameters changes with the implementation of dataset. So, the algorithm requires tuning before implementation of each dataset.

We provide the comparisons of our algorithm with three other state-of-the art algorithm namely C4.5, K-nearest neighbors (K-NN) and Multi-layer perceptron (MLP).These algorithms are implemented with Tanagra, an open source data mining tool. We also compare the BMOPSO based classifier with NSGA-II based classifier. The NSGA-II based association rules mining technique is implemented by using the concepts provided in Ishibuchi and Nojima (2005) and de la Iglesia, Reynolds and Rayward-Smith (2005).

Experimental Results

We applied the multi-objective BMOPSO algorithm to Wisconsin breast cancer dataset in following manner: first we divided the dataset into 4 non-overlapping sets each containing 171 patterns. For each round of the 4-fold cross validation, 3 sets are used as training set and the last set is used as the test set. In rule extraction phase, we apply the BMOPSO algorithm on the training set to find the Pareto optimal solutions with maximum

Table 1. Data sets used in computational experiments

Data Set	Attributes	Patterns	Classes
Breast cancer Wisconsin	9	683*	2
Glass	10	214	6
Clevend Heart	13	297*	2
Balance Scale	4	626	3
Wine	13	178	3

*Incomplete patterns with missing values are not included

Table 2. Parameter setting for rule extraction phase

Data Set	Population	Iteration	W	Θ_1	Θ_2
Breast Cancer Wisconsin	200	100	0.4	1.76	2.5
Wine	200	100	1	1.76	2.5
Glass	200	100	1	1	1
Balance Scale	200	150	1	1.7	2.5
Clevend Heart	200	150	1	1.7	2.5

Table 3. Parameter setting for rule selection phase

Data Set	Population	Iteration	W	Θ_1	Θ_2
Breast Cancer Wisconsin	100	50	1	1.8	2.5
Wine	200	100	1	1.8	2.5
Glass	100	100	1	1.8	2.5
Balance Scale	200	150	1	1.8	2.5
Clevend Heart	100	50	1	1.8	2.5

Figure 5. Experimental results of BMOPSO on Breast Cancer Wisconsin Dataset a. Predictive accuracy before rule selection b. Predictive accuracy after rule selection c. Rule length before rule selection d. Rule length after rule selection e. Rule set size before rule selection f. Rule set size after rule selection

a. Predictive accuracy before rule selection

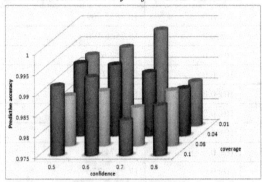

b. Predictive accuracy after rule selection

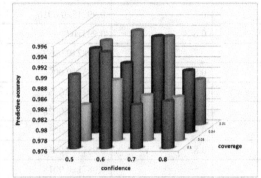

c. Rule length before rule selection

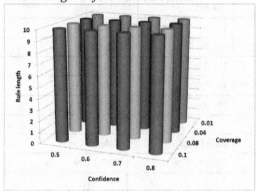

d. Rule length after rule selection

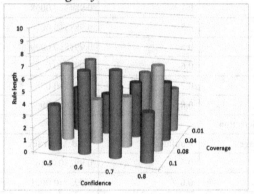

e. Rule set size before rule selection

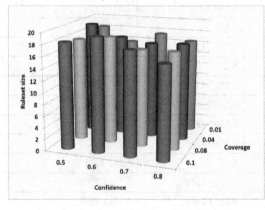

f. Rule set size after rule selection

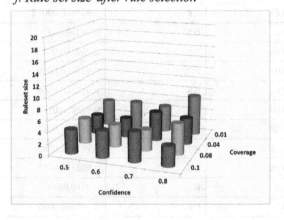

coverage and confidence satisfying the constraint $COV(R_q) > minCOV$ and $CONF(R_q) > minCONF$ respectively. Then, the rules corresponding to the Pareto optimal solutions are found. This process is repeated for 4 folds. After 4 folds, the predic-

tive accuracy of the candidate rules is evaluated. The predictive accuracy is found by dividing the number of test tuples classified over 4 fold divided by total number of tuples in the dataset. The extracted rules are used as candidate rules for the rule

Table 4. (a). Computational results of rule extraction phase with standard deviation (b). Computational results of rule selection phase with standard deviation

Rule Extraction				
Coverage	**Confidence**	**Pred accuracy**	**Rule Length**	**Ruleset Size**
0.01	0.5	0.99215±0.0016	10± 0.0	18.25±3.269
0.01	0.6	0.994±0.0025	10± 0.0	15.75±6.190
0.01	0.7	0.9982±0.0032	10± 0.0	17.5±2.6925
0.01	0.8	0.9857±0.0094	10± 0.0	16.5±3.354
0.04	0.5	0.9925±0.003	10± 0.0	19.5±1.12
0.04	0.6	0.9921±0.0028	10± 0.0	19.25±1.920
0.04	0.7	0.9903±0.0031	10± 0.0	16.75±1.7854
0.04	0.8	0.9864±0.0061	10± 0.0	17±3.7417
0.08	0.5	0.9871±0.0018	10± 0.0	17.75±2.770
0.08	0.6	0.9882±0.0037	10± 0.0	18.5±3.2016
0.08	0.7	0.9843±0.0058	10± 0.0	16.75±2.2776
0.08	0.8	0.9883±0.0016	10± 0.0	16.75±2.6810
0.1	0.5	0.9918±0.0029	10± 0.0	18.25±1.2990
0.1	0.6	0.9939±0.0027	10± 0.0	19.5±2.0616
0.1	0.7	0.9835±0.0084	10± 0.0	17.75±3.4911
0.1	0.8	0.9886±0.0004	10± 0.0	15.64±1.4911
Rule Selection				
Coverage	**Confidence**	**Pred Accuracy**	**Rule Length**	**Ruleset Size**
0.01	0.5	0.992±0.0016	5.75±2.4875	4.75±1.089
0.01	0.6	0.9939±0.0021	3.5±0.5	5±1.2247
0.01	0.7	0.9928±0.0032	5±1.4790	4.75±2.9155
0.01	0.8	0.9846±0.00017	3.8±0.7009	7.25±2.776
0.04	0.5	0.992±0.0032	3.5±0.5	3.5±1.0753
0.04	0.6	0.9892±0.0063	4.75±3.1125	3.5±1.1180
0.04	0.7	0.9943±0.0020	5±2.9155	5.25±1.2990
0.04	0.8	0.9878±0.0066	3.25±0.4330	4.75±1.0897
0.08	0.5	0.9828±0.0105	6.5±2.2913	4.5±0.8660
0.08	0.6	0.9875±0.0041	3.75±0.4330	4±0.7071
0.08	0.7	0.9846±0.0062	4.25±0.4330	3.75±0.4330
0.08	0.8	0.9843±0.0105	7±3	5.5±1.8025
0.1	0.5	0.99±0.0030	3.75±0.4330	4.25±1.2990
0.1	0.6	0.9943±0.0027	6.75±3.2692	4.75±1.0897
0.1	0.7	0.9843±0.0105	7±3	5.25±1.7854
0.1	0.8	0.9846±0.0100	4±2	3.5±1.2484

selection phase. In rule selection phase, we obtain non-dominating rulesets for each fold from which we select the rule set with maximum number of tuples classified. In case of tie, we consider the number rules as the first criteria and rule length as the second criteria for tie break. We reckon the

Table 5. Results of BMOPSO and NSGA-II obtained after rule extraction phase

Dataset	Predictive accuracy		Rulesets size		Rule Length	
	BMOPSO	NSGA-II	BMOPSO	NSGA-II	BMOPSO	NSGA-II
Breast Cancer Wisconsin	0.9886±0.0100	0.9801±0.0034	15.64±1.4911	15±2.345	10.0±0.0	10.0±0.0
Wine	0.9958±0.0104	0.9899±0.0142	34.53±1.652	40±2.12	10.0±0.0	10.0±0.0
Glass	0.8976±0.0021	0.8533±0.0045	17±3	15.98±2.87	10.0±0.0	10.0±0.0
Clevend Heart	0.9776±0.00123	0.9654±0.0032	55.86±1.98	54.87±3.95	10.0±0.0	10.0±0.0
Balancescale	0.9148±0.0030	0.9150±0.0076	43.12±3.87	57.76±4.21	10.0±0.0	10.0±0.0

predictive accuracy of classifier in similar manner as done in rule extraction phase. The average result over 4 runs is shown in Figure 5. We repeat the entire process for each pair of specified threshold coverage and confidence respectively.

From the results, we can clearly perceive that there is little or no decline in predictive accuracy. However, there is a great reduction in number of rules and rule length from rule extraction to rule selection phase. The tabulated results of the figure are shown in Table 4 (a) and Table 4 (b).

The implementation of the other datasets has confirmed the same result. We compared the predictive accuracy, number of rules and rule length obtained in each phase (i.e., rule extraction and rule selection) of BMOPSO based classifier with NSGA-II based classifier proposed in de la Iglesia, Reynolds and Rayward-Smith (2005) and Ishibuchi and Nojima (2005). The results are tabulated in Table 5 and 6.

The results reported in table are obtained with min COV of 0.1 and minCONF of 0.8. The results show that the BMOPSO based classifier has better predictive accuracy in both the phases of rule mining and also generates a compact classifier as compared to the NSGA-II based rule mining algorithm.

We have compared the predictive accuracy of different classification algorithms with our proposed algorithm. The predictive accuracy of all the algorithms on different datasets is shown in Table 7. The results reported in table are obtained with *minCOV* of 0.1 and *minCONF* of 0.8. The result clearly shows better performance of the algorithm in almost all data sets.

To confirm our proposition, we evaluate our classifier with other classifiers using the Friedman test (Demsar, 2006). The Friedman test is non-parametric equivalent of the repeated measures ANOVA. In this test we assign ranks to each classifiers based on their predictive accuracy on the

Table 6. Results of BMOPSO and NSGA-II obtained after rule selection phase

Dataset	Predictive accuracy		Rulesets size		Rule Length	
	BMOPSO	NSGA-II	BMOPSO	NSGA-II	BMOPSO	NSGA-II
Breast Cancer Wisconsin	0.9846±0.0100	0.9501±0.0121	3.5±1.2484	12.07±1.9	4±2	6±1
Wine	0.9758±0.0135	0.9587±0.0020	3.65±1.43	5.65±1.234	8.43±4.7	8.76±3.2
Glass	0.8376±0.0039	0.7804±0.0005	10.54±2.54	14.32±2.54	7.65±1.07	8±1
Clevend Heart	0.9277±0.00178	0.8706±0.0109	5±2	8.76±3.87	5.12±3.432	6.43±5.6
Balancescale	0.9148±0.0314	0.8687±0.0165	8±2.56	13.08±3.987	2.68±.1.56	3.13±1.98

Table 7. Accuracy of different algorithms on the 5 datasets with their standard deviation

Data set	BMOPSO	NSGA-II	C 4.5	K-NN	MLP
Breast Cancer Wisconsin	0.9846±0.0100	0.9501±0.0121	0.9456±0.0058	0.9630±0.0058	0.9626±0.0027
Wine	0.9758±0.0135	0.9587±0.0020	0.5645±0.03125	0.6719±0.3493	0.6875±0.05845
Glass	0.8376±0.0039	0.7804±0.0005	0.6533±0.178	0.68395±0.5242	0.6026±0.02011
Clevend Heart	0.9277±0.00178	0.8706±0.0109	0.7650±0.0128	0.8116±0.0067	0.8000±0.0158
Balancescale	0.9148±0.0314	0.8687±0.0165	0.7765±0.0075	0.8490±0.0039	0.9054±0.00547

datasets. Then we obtain the Friedman statistics given by Equation 15.

$$F_F = \frac{(N-1)\chi_F^2}{N(K-1) - \chi_F^2} \qquad (14)$$

$$\chi_F^2 = \frac{12N}{k(k+1)}\left[\sum_j R_j^2 - \frac{k(k+1)^2}{4}\right] \qquad (15)$$

where R_j^2 is the average rank, N is the numbers of datasets and k is the number of classifiers. Using the Friedman statistic F_F, we obtain a better statistic which is distributed according to F- distribution with k-1 and $(k$-1$)(N$-1$)$ degree of freedom.

In our case, the ranks are shown in Table 8. Using the average rank and Equation 15, we obtain X_F^2 as 16.8 and F_F as 21. The critical F distribution value for (5-1) and (5-1)(5-1) =16 degree of freedom for $\alpha_{0.005}$=5.64 and $\alpha_{0.01}$=4.77

which are smaller than F_F value hence we can reject the null hypothesis. We further proceed with the post hoc test to find which classifier is the best. For this, we calculate critical difference (*C.D*) given by Equation 16.

$$C.D = q_\alpha \sqrt{\frac{k(k+1)}{6N}} \qquad (16)$$

At $q_{0.05}$, the C.D is 2.728. The difference between the best average rank and the worst average rank of the classifiers (i.e., between the BMOPSO and C 4.5) is greater than the C.D value. This helps us conclude that BMOPSO algorithm has better performance than the C4.5 algorithms. However, no conclusion can be drawn for the other algorithms as the difference is less than C.D value. At $q_{0.10}$, C.D is 2.459. The difference between the average rank of BMOPSO and MLP is 2.6 which are greater than the CD value. Thus MLP performs worse than BMOPSO. However

Table 8. Ranks assigned to different classifier based on their standard deviations

Data set	BMOPSO	NSGA-II	C 4.5	k-NN	MLP
Breast Cancer Wisconsin	0.9846±0.0100(1)	0.9501±0.0121 (2)	0.9456±0.0058 (5)	0.9630±0.0058 (3)	0.9626±0.0027 (4)
Wine	0.9758±0.0135 (1)	0.9587±0.0020(2)	0.5645±0.03125 (5)	0.6719±0.3493 (4)	0.6875±0.05845(3)
Glass	0.8376±0.0039 (1)	0.7804±0.0005(2)	0.6533±0.178 (4)	0.68395±0.5242(3)	0.6026±0.02011(5)
Clevend Heart	0.9277±0.00178(1)	0.8706±0.0109(2)	0.7650±0.0128 (5)	0.8116±0.0067 (3)	0.8000±0.0158 (4)
Balancescale	0.9148±0.0314 (1)	0.8687±0.0165(3)	0.7765±0.0075 (5)	0.8490±0.0039 (4)	0.9054±0.00547(2)
Avg Rank	1.0	2.2	4.8	3.4	3.6

the average rank difference between BMOPSO and KNN approaches the C.D value (3.4-1 =2.4≈2.459). But Nemenyi test don't conclude anything regarding the performance of NSGA-II based classifier and BMOPSO based classifier. So we proceed with Bonferroni-Dunn test. This test is same as Nemenyi's test expect for the critical values of $\alpha/(k-1)$. In this test, the critical value $q_{0.05}$ and $q_{0.10}$ for 5 classifiers are 2.498 and 2.241 respectively. So the C.D value are $q_\alpha\sqrt{\dfrac{5.6}{6.5}} = q_\alpha$ i.e., 2.498 and 2.241. The difference between the average ranks of the two meta-heuristic based classifiers is 1.2, which is less than the C.D. Thus we can say that there is no significance difference in performance of the two meta-heuristic based classifiers.

CONCLUSION

In this article, we have presented a BMOPSO based associative rule mining classifier where the rule mining is carried out in two steps: rule extraction and rule selection. We have presented a compact representation of rules for the rule extraction phase. We have also shown that the rule selection procedure helps in building an efficient and compact classifier without reducing the predictive accuracy of the classifier. Also we have evaluated the performance of the BMOPSO classifier with the other classifiers and have seen that its performance is better than the other classifiers. However, we have compared only the machine learning based classifiers and NSGA-II based classifiers. Hence as a future scope we need to evaluate its performance with other bio-inspired algorithms. Also we have considered only few datasets. There are other UCI datasets on which its performance needs to be evaluated.

REFERENCES

Agrawal, R., Mannila, H., Srikant, R., Toivonen, H., & Verkamo, I. A. (1996). Fast discovery of association rules . In Fayyad, U. M., Piatetsky-Shapiro, G., Smyth, P., & Uthurusamy, R. (Eds.), *Advances in knowledge discovery and data mining* (pp. 369–376). Menlo Park, CA: AAAI.

Akin, E., & Alatas, B. (2009). Multi-objective rule mining using a chaotic particle swarm optimization algorithm. *Knowledge-Based Systems*, *22*(6), 455–460. doi:10.1016/j.knosys.2009.06.004

Bacardit, J., & Butz, M. V. (2007). Data mining in learning classifier systems: comparing XCS with GAssist. In T. Kovacs, X. Llorà, K. Takadama, P. L. Lanzi, W. Stolzmann, & S. W. Wilson (Eds.), *Proceedings of the International Conference on Learning Classifier Systems* (LNCS 4399, pp. 282-290).

Bayardo, R. J., Jr., & Agrawal, R. (1999). Mining the most interesting rules. In *Proceedings of the 5th ACM SIGKDD International Conference on Knowledge Discovery and Data Mining* (pp. 145-153). New York, NY: ACM Press.

Carvalho, A. B., & Pozo, A. (2008). Non-ordered data mining rules through multi-objective particle swarm optimization: Dealing with numeric and discrete attributes. In *Proceedings of the International Conference on Hybrid Intelligent Systems* (pp. 495-500). Washington, DC: IEEE Computer Society.

Chen, G., Liu, H., Yu, L., Wei, Q., & Zhang, X. (2006). A new approach to classification based on association rule mining. *Decision Support Systems*, *42*, 674–689. doi:10.1016/j.dss.2005.03.005

Coello, C. A., & Lechuga, M. (2002). MOPSO: A proposal for multiple–objective particle swarm optimization. In *Proceedings of the 9th IEEE World Congress on Computational Intelligence*, Honolulu, HI (pp. 1051-1056). Washington, DC: IEEE Computer Society.

Coello, C. A., Pulido, G. T., & Lechuga, M. S. (2004). Handling multiple objectives with particle swarm optimization. *IEEE Transactions on Evolutionary Computation*, 8(3), 256–279. doi:10.1109/TEVC.2004.826067

Coenen, F., & Leng, P. (2007). The effect of threshold values on association rule based classification accuracy. *Data & Knowledge Engineering*, 345–689. doi:10.1016/j.datak.2006.02.005

Coenen, F., Leng, P., & Zhang, L. (2005). Threshold tuning for improved classification association rule mining. In T. B. Ho, D. Cheung, & H. Liu (Eds.), *Proceedings of the 9th Pacific-Asia Conference on Advances in Knowledge Discovery and Data Mining* (LNCS 3518, pp. 216-225).

de la Iglesia, B., Reynolds, A., & Rayward-Smith, V. J. (2005). Developments on a multi-objective metaheuristic (MOMH) algorithm for finding interesting sets of classification rules. In *Proceedings of the Conference on Evolutionary Multi Criterion Optimization* (pp. 826-840).

de la Iglesia, B., Richards, G., Philpott, M. S., & Rayward-Smith, V. J. (2006). The application and effectiveness of a multi-objective metaheuristic algorithm for partial classification. *European Journal of Operational Research*, 169, 898–917. doi:10.1016/j.ejor.2004.08.025

Dehuri, S., & Cho, S.-B. (2009). Multi-criterion pareto based particle swarm optimized polynomial neural network for classification: A review and state-of-the-art. *Journal of Computer Science Review*, 3, 19–40. doi:10.1016/j.cosrev.2008.11.002

Demsar, J. (2006). Statistical comparisons of classifiers over multiple datasets. *Journal of Machine Learning Research*, 7, 1–30.

Fieldsend, J. E., & Singh, S. (2002). A multi-objective algorithm based upon particle swarm optimization, an efficient data structure and turbulence. In *Proceedings of the Workshop on Computational Intelligence*, Brimingham, UK (pp. 37-44).

Giannopoulou, E. G. (2008). *Data mining in medical and biological research*. Rijeka, Croatia: INTECH.

Groth, R. (1997). *Data mining: A hands-on approach for business professionals*. Upper Saddle River, NJ: Prentice Hall.

Gutierrez, F. J., Lerma-Rascon, M. M., Salgado-Garza, L. R., & Cantu, F. J. (2002). Biometrics and data mining: Comparison of data mining-based keystroke dynamics methods for identity verification. In C. A. Coello Coello, A. de Albornoz, L. E. Sucar, & O. C. Battistutti (Eds.), *Proceedings of the Second Mexican International Conference on Advances in Artificial Intelligence* (LNCS 2313, pp. 460-469).

Han, J., & Kamber, M. (2006). *Data mining concepts and techniques*. San Francisco, CA: Morgan Kaufmann.

Holden, N. F. (2007). A hybrid PSO/ACO algorithm for discovering classification rules in data mining. In *Proceedings of the GECCO Conference Companion on Genetic and Evolutionary Computation*, London, UK (pp. 2745-2750).

Huap, S. H., Wulsin, L. R., Li, H., & Guo, J. (2009). Dimensionality reduction for knowledge discovery in medical claims database: Application to antidepressant medication utilization study. *Computer Methods and Programs in Biomedicine*, 93(2), 115–123. doi:10.1016/j.cmpb.2008.08.002

Ishibuchi, H., Nakashima, T., & Murata, T. (109-133). Three-objective genetics-based machine learning for linguistic rule extraction. *Information Sciences*, 136.

Ishibuchi, H., & Namba, S. (2004). Evolutionary multiobjective knowledge extraction for high dimentional pattern classification problems. In X. Yao, E. K. Burke, J. A. Lozano, J. Smith, J. J. Merelo-Guervós, J. A. Bullinaria et al. (Eds.), *Proceedings of the 8th International Conference on Parallel Problem Solving from Nature* (LNCS 3242, pp. 1123-1132).

Ishibuchi, H., & Nojima, Y. (2005). Accuracy-complexity tradeoff analysis by multiobjective rule selection. In *Proceedings of the Workshop on Computational Intelligence in Data Mining* (pp. 39-48).

Ishibuchi, H., Nozaki, K., Yamamoto, N., & Tanaka, H. (1995). Selecting fuzzy if-then rules for classification problems using genetic algorithms. *IEEE Transactions on Fuzzy Systems, 3*, 260–270. doi:10.1109/91.413232

Kamal, A., Manganaris, S., & Srikant, R. (1997). Partial classification using association rules. In *Proceedings of the AAAI Conference on Knowledge Discovery in Databases* (pp. 115-118).

Kaya, M. (2006). Multi-objective genetic algorithm based approaches for mining optimized fuzzy association rules. *Soft Computing, 10*, 578–586. doi:10.1007/s00500-005-0509-5

Kennedy, J., & Eberhart, R. (1997). A discrete binary version of the particle swarm algorithm. *IEEE Transactions on Systems, Man, and Cybernetics, 5*, 4104–4108.

Li, H.-L., & Chen, M.-H. (2008). Induction of multiple criteria optimal classification rules for biological and medical data. *Computers in Biology and Medicine, 38*, 42–52. doi:10.1016/j.compbiomed.2007.07.006

Li, J. S., & Topor, R. (2002). Mining the optimal class association rule set. *Knowledge-Based Systems, 15*, 399–405. doi:10.1016/S0950-7051(02)00024-2

Li, W. H., & Pei, J. (2001). Accurate and efficient classification based on multiple class-association rules. In *Proceedings of the 1st IEEE International Conference on Data Mining* (pp. 369-376). Washington, DC: IEEE Computer Society.

Liu, B., Hsu, W., & Ma, Y. (1998). Integrating classification and association rule mining. In *Proceedings of the 4th International Conference on Knowledge Discovery and Data Mining* (pp. 80-86).

Liu, H., Liu, L., & Zhang, H. (2011). A fast pruning redundant rule method using galois connection. *Applied Soft Computing, 11*, 130–137. doi:10.1016/j.asoc.2009.11.004

Liu, H., Sun, J., & Zhang, H. (2009). Post processing of associative classification rules using closed sets. *Expert Systems with Applications, 36*, 6659–6667. doi:10.1016/j.eswa.2008.08.046

Morishita, S. (1998). On classification and regression. In *Proceedings of the First International Conference on Discovery Science* (pp. 40-57).

Mostaghim, S., & Teich, J. (2003). Strategies for finding good local guides in multiobjective particle swarm optimization. *IEEE Swarm Intelligence Symposium*, 26-33.

Nguyen, S., & Kachitvichyanukul, V. (2010). Movement strategies for multi-objective particle swarm optimization. *International Journal of Applied Metaheuristic Computing, 1*(3), 59–79. doi:10.4018/jamc.2010070105

Pei, J., Upadhyaya, S. J., Farooq, F., & Govidaraju, V. (2004). Data mining for intrusion detection: Techniques, applications, and systems. In *Proceedings of the 20th International Conference on Data Engineering* (p. 877). Washington, DC: IEEE Computer Society.

Rajasethupathy, K., Scime, A., Rajasethupathy, K. S., & Murray, G. R. (2009). Finding "persistent rules": Combining association and classification results. *Expert Systems with Applications, 36,* 6019–6024. doi:10.1016/j.eswa.2008.06.090

Raquel, C., Prospero, C., & Naval, J. (2005). An effective use of crowding distance in multiobjective particle swarm optimization. In *Proceedings of the Conference on Genetic and Evolutionary Computation,* Washington, DC (pp. 257-264). New York, NY: ACM Press.

Ratnaweera, A., Halgamuge, S. K., & Watson, H. C. (2004). Self-organizing hirarchical particle swarm optimizer with time varying acceleration coefficients. *IEEE Transactions on Evolutionary Computation, 8*(3), 240–255. doi:10.1109/TEVC.2004.826071

Sasisekharan, R., Seshadri, V., & Weiss, S. M. (1996). Data mining and forecasting in large-scale telecommunication networks. *IEEE Expert Intelligent Systems and their Applications, 11*(1), 37-43.

Shao, L. S., & Fu, G. X. (2008). Disaster prediction of coal mine gas based on data mining. *Journal of Coal Science and Engineering, 14*(3), 458–463. doi:10.1007/s12404-008-0099-9

Shi, Y., & Eberhart, R. (1999). Empirical study of paticle swarm otpimization. In *Proceedings of the IEEE World Congress on Evolutionary Computation* (pp. 6-9). Washington, DC: IEEE Computer Society.

Thabtah, F., Cowling, P. I., & Hammoud, S. (2006). Improving rule sorting, predictive accuracy, and training time in associative classification. *Expert Systems with Applications, 31,* 414–426. doi:10.1016/j.eswa.2005.09.039

Thabtah, F. A., & Cowling, P. I. (2007). A greedy classification algorithm based on association rule. *Applied Soft Computing, 7,* 1102–1111. doi:10.1016/j.asoc.2006.10.008

UCI. (2010). *Machine learning repository.* Retrieved from http://archive.ics.uci.edu/ml

Wong, S., & Li, C. S. (Eds.). (2006). *Life science data mining.* Singapore: World Scientific.

This work was previously published in the International Journal of Applied Metaheuristic Computing, Volume 2, Issue 2, edited by Peng-Yeng Yin, pp. 51-73, copyright 2011 by IGI Publishing (an imprint of IGI Global).

Chapter 15
Extraction of Target User Group from Web Usage Data Using Evolutionary Biclustering Approach

R. Rathipriya
Periyar University, India

K. Thangavel
Periyar University, India

J. Bagyamani
Government Arts College, India

ABSTRACT

Data mining extracts hidden information from a database that the user did not know existed. Biclustering is one of the data mining technique which helps marketing user to target marketing campaigns more accurately and to align campaigns more closely with the needs, wants, and attitudes of customers and prospects. The biclustering results can be tuned to find users' browsing patterns relevant to current business problems. This paper presents a new application of biclustering to web usage data using a combination of heuristics and meta-heuristics algorithms. Two-way K-means clustering is used to generate the seeds from preprocessed web usage data, Greedy Heuristic is used iteratively to refine a set of seeds, which is fast but often yield local optimal solutions. In this paper, Genetic Algorithm is used as a global optimizer that can be coupled with greedy method to identify the global optimal target user groups based on their coherent browsing pattern. The performance of the proposed work is evaluated by conducting experiment on the msnbc, a clickstream dataset from UCI repository. Results show that the proposed work performs well in extracting optimal target users groups from the web usage data which can be used for focalized marketing campaigns.

DOI: 10.4018/978-1-4666-2145-9.ch015

INTRODUCTION

Customer/User segmentation has been a critical element of the marketing and is one of the most important strategic concepts contributed by the marketing discipline to business firms (Shina et al., 2004). User Segmentation is the process of developing effective schemes for categorizing and organizing meaningful groups of customers. A user segment is a group of prospects or customers who are selected from a database based on characteristics they possess or exhibit. It also allows company to differentially treat consumers in different segments. User Profiling is the process of analyzing the customers of each segment in order to generalize, describe or name this set of customers based on common characteristics. It is the process of understanding and labeling a set of users. It provides valuable information about users/customers so marketers can furnish stronger, more targeted offers and each user segment is the target group of users.

One-to-one marketing is the ideal marketing strategy, in which every marketing campaign or product is optimally targeted for each individual customer; but this is not always possible. Therefore, segmentation is required to distinguish similar users and put them together in a segment/group. Doubtlessly using segmentation to understand user's needs is much easier, faster and more economical than uniquely investing to understand them particularly (Jonker et al., 2004).

With proper market segmentation, enterprises can arrange the right web pages, services and resources to each target user group and build a close relationship with them. Market segmentation has consequently been regarded as one of the most critical elements in achieving successful modern marketing and customer relationship management.

Click stream data is a sequence of Uniform Resource Locators (URLs) browsed by the user within a particular period of time. To discover group of users with similar behavior and motivation for visiting the particular website can be found by clustering. Traditional clustering (Lee et al., 2008) is used to segment the web users or web pages in to groups based on the existing similarities. When a clustering method is used for grouping users, it typically partitions users according to their similarity of browsing behavior under all pages. But, in the most cases web users behave similarly only on a subset of pages and their behavior is not similar over the rest of the pages. Therefore, traditional clustering methods fail to identify such user groups.

To overcome this problem, concept of Biclustering was introduced. Biclustering (Bleuler et al., 2004; Chakraborty et al., 2005; Madeira et al., 2004) was first introduced by Hartigan (1972). Biclustering is the simultaneous clustering of rows and columns of the data matrix. In literature, biclustering algorithms are widely applied to the gene expression data. In this paper, it is used to mine clickstream data in order to extract target usage groups. These groups are analyzed to determine user's behavior which is an important element in the E-Commerce applications.

The rest of the paper is organized as follows. In Section 2, some of the existing work related to the biclustering approaches and user segmentations are discussed. Methods and materials required for biclustering approach are described in Section 3. Section 4 focuses on the proposed Biclustering approach using Genetic Algorithm. Analysis of experimental results is discussed in Section 5. Section 6 concludes the paper.

RELATED WORK

Tsai et al. (2004) developed a market segmentation methodology based on product specific variables such as purchased items and the associative monetary expenses from the transactional history of customers to address the unreliable results of segmentation based on general variables like customer demographics. Shina et al. (2004) used three clustering methods such as K-Means,

Self-Organizing Map, and fuzzy K-Means for segmentation to find properly graded stock market brokerage commission rates based on transactional data. Punj and Stewart (1983), suggest two approaches for clustering. The first was hierarchical clustering, to determine the number of clusters, and the second was nonhierarchical clustering, for fine-tuning the results. In which, hierarchical cluster analysis was extremely time consuming, it is rarely used in practice. Jonker et al. (2004) used the Genetic Algorithm (GA) to determine the optimized marketing strategy to integrate customer segmentation and customer targeting. In Abraham (2003), intelligent-miner" (i-Miner) is introduced and optimized concurrent architecture of a fuzzy clustering algorithm is used to discover web data clusters and a fuzzy inference system to analyze the trends of the web site visitors.

Koutsonikola et al. (2009) proposed a biclustering approach for web data, which identifies groups of related web users and pages using spectral clustering method on both row and column dimensions. Araya et al. (2004) proposed methodology for target group identification from web usage data which improved the customer relationship management e.g., financial services. Xu et al. (2010) presented bipartite spectral clustering for co-clustering of web users and pages and the impact of using various clustering algorithms is also investigated in that paper. In web mining, there is no related work that has been applied specific biclustering algorithms for discovering the coherent browsing patterns. Zong et al. (2010) introduces Coclustering approach to the web logs in semantic space. In Bleuler et al. (2004) and Chakraborty et al. (2005), a genetic algorithm based biclustering algorithm has been presented. The main focus of these algorithms was to evolve high volume biclusters but they may fail to discover highly coherent "interesting" biclusters.

Among the various biclustering methods proposed in the literature (Madeira et al., 2004), most approaches are based on greedy heuristics that iteratively refine a set of biclusters. These algo-

rithms can be considered as local search methods which are fast but often yield suboptimal results. To escape from the local optimal problem, greedy biclustering methods and evolutionary approach namely Genetic Algorithm (GA) are combined in the proposed work. It extracts global optimal target group based on their browsing patterns. The results show that GA outperforms the greedy procedure by identifying optimal coherent user behavioral patterns. These patterns are very useful in the decision making for target marketing. Target Marketing systems analyze patterns of users' browsing interest and to provide personalized services which match user's interest in most business domains, benefiting both the user and the merchant.

METHODS AND MATERIALS

Preprocessing

Clickstream data pattern is one of the web usage data formats that have been taken for analyzing the user's behavior. It is converted into web user access matrix A by using (1) in which rows represent users and columns represent pages of web sites. Let A(U, P) be an 'n x m' user access matrix where U be a set of users and P be a set of pages of a web site. It is used to describe the relationship between web pages and users who access these web pages. The element a_{ij} of A(U,P) represents frequency of the user u_i of U visit the page p_j of P during a given period of time.

$$a_{ij} = \begin{cases} \text{Hits}\left(u_i,\ p_j\right), & \text{if } p_j \text{ is visited by } u_i \\ 0, & \text{otherwise} \end{cases} \quad (1)$$

where Hits(u_i, p_j) is the count/frequency of the user u_i accesses the page p_j during a given period of time.

Average Correlation Value

Average Correlation Value (ACV) (Bagyamani et al., 2010), is a coherence measure which is used to measure the degree of coherence of the biclusters. This measure has advantage of calculating similarity depending on the pattern but not on the absolute magnitude of the spatial vector. Matrix $B = (b_{ij})$ has the ACV which is defined by the following function,

$$ACV(B) = \max\left\{\frac{\sum\limits_{i=1}^{n}\sum\limits_{j=1}^{n}\left|r\_row_{ij}\right|-n}{n^2-n}, \frac{\sum\limits_{k=1}^{m}\sum\limits_{l=1}^{m}\left|r\_col_{kl}\right|-m}{m^2-m}\right\}$$

(2)

r_row_{ij} is the correlation between row i and row j, r_col_{kl} and is the correlation between column k and column l. A high ACV suggests high similarities among the users or pages.

Greedy Heuristic Algorithm

A greedy algorithm repeatedly executes a search procedure which tries to maximize the bicluster based on examining local conditions, with the hope that the outcome will lead to a desired outcome for the global problem (Bleuler et al., 2004).

Algorithm: Greedy Heuristic Algorithm

- Start with initial bicluster.
- For every iteration
  - Add/ remove the element (user/page) to/from the bicluster which maximize the objective function.
- End for

Binary Representation of Biclusters

Each enlarged and refined bicluster is encoded as a binary string as shown in Figure 1. The length of the string is the number of rows plus the number of columns of the user access matrix A (U, P). A bit is set to one when the corresponding user or page is included in the bicluster. These binary encoded biclusters are used as initial population for genetic algorithm.

Volume of Bicluster

Volume of bicluster B (U, P) is defined as the number of elements in bicluster and denoted as VOL (B (U, P)).

$$VOL(B(U,P)) = |U| \times |P|$$

(3)

where, |U| is the number of users in the B and |P| is the number of pages in B.

Degree of Overlapping

The degree of overlapping (Das et al., 2008) among all bicluster is defined as follows:

$$R = \frac{1}{|U| * |P|}\sum_{i=1}^{|U|}\sum_{j=1}^{|P|}T_{ij}$$

$$where$$

(3)

$$Tij = \frac{1}{(N-1)} * \left(\sum_{k=1}^{N}W_k(a_{ij}) - 1\right)$$

where N is the total number of biclusters, |U| represents the total number of users, and |P| represents the total number of pages in the data matrix A.

Figure 1. A binary encoded chromosome representing a bicluster

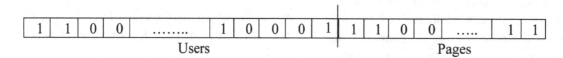

The value of $w_k(a_{ij})$ is either 0 or 1. If the element $a_{ij} \in A$ is present in the kth bicluster, then $w_k(a_{ij})$ = 1, otherwise 0. Hence, the R index represents the degree of overlapping among the biclusters. The range of R index is $0 \leq R \leq 1$.

Extraction of Target User Groups

Biclustering approach based on Genetic Algorithm is used to build user profile model by using inputs from a database to predict user behavior. After a model has been created based on historical data, it can then be applied to new data in order to make predictions about unseen behavior. The proposed algorithm is used to identify the optimal coherent biclusters in terms of volume and quality. Basically, the proposed work is divided into two phases namely seed generation phase and optimal target group extraction phase. In seed generation phase, K-Means clustering algorithm is applied on the both dimensions of the user access matrix to generate the initial biclusters called seeds. In second phase, greedy method and Genetic Algorithm (GA) are used. Seeds are enlarged and refined using greedy method. GA is used as global optimizer to identify the global optimal biclusters. Initial procedure for GA is done using Two-way K-Means clustering and Greedy K-Means biclustering (Rathipriya et al., 2011).

Generally, the frequency of visiting the web pages by the users of a web site may rise or fall synchronously according to their browsing interest. Though the magnitude of their interest levels may not be close, but the pattern they exhibit can be very much similar. Therefore, a new correlation based fitness function is designed in this paper which captures pattern similarity among the web users and pages of a web site. Our proposed biclustering framework using GA is interested in finding such coherent patterns of bicluster of users and with a general understanding of users' browsing interest. This method makes significant contribution in the field of web mining, E-Commerce applications and etc. One-to-one

relation between web users and pages of a web site is not appropriate because web users are not strictly interested in one category of web pages. Therefore, the proposed algorithm is designed to discover the overlapping coherent biclusters from clickstream data patterns. These overlapped coherent biclusters have high degree of correlation among subset of users and subset of related pages of a web site. These optimal biclusters are treated as target user groups.

Evolutionary Biclustering Algorithm for Target User Group

Evolutionary algorithms such as GA model natural processes, such as selection, recombination, mutation, migration, locality and neighborhood. Figure 2 shows the structure of a simple evolutionary algorithm. At the beginning of the computation a number of individuals (the population) are initialized. Usually, GA is initialized with the population of random solutions. In order to avoid random interference, biclusters obtained from greedy search procedure are used to initialize GA. This will result in faster convergence compared to random initialization. Maintaining diversity in the population is another advantage of initializing with these biclusters.

The objective function is then evaluated for these individuals. If the optimization criteria are not met the creation of a new generation starts. Individuals are selected according to their fitness for the production of offspring. Parents are recombined to produce offspring. All offspring will be mutated with a certain probability. The fitness of the offspring is then computed. The offspring are inserted into the population replacing the parents, producing a new generation. This cycle is performed until the optimization criteria are reached.

P is the set of biclusters in each population, *mp* is the probability of mutation, *cp* is the fraction of the population to be replaced by crossover in each population, N is the number of biclusters

Figure 2. Structure of evolutionary algorithm

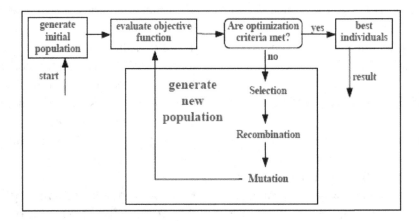

in each population. The biclustering framework using genetic algorithm is given below. Parameter setting for Evolutionary Biclustering Algorithm is given in Table 1.

Fitness Function

The main aim of the proposed work is to extract high volume coherent biclusters from web usage data. The following fitness function F (I, J) is used to extract global optimal coherent bicluster.

$$
F\ (I,\ J)\ =\ \begin{cases} |I| * |J|, & \text{if } 1-\ \text{ACV}\left(\text{bicluster}\right) \le \delta \\ 0, & \text{Otherwise} \end{cases} \quad (4)
$$

where |I| and |J| are number of rows and columns of bicluster and δ is set to 0.1

Table 1. Configuration parameters for proposed work

Crossover Probability (cp)	0.7
Mutation Probability (mp)	0.01
Population Size (N)	120
Generation or Maximum Number of Iteration	100
Initial ACV Threshold	0.9526

Algorithm: Evolutionary Biclustering Algorithm for Target User Group

- **Input:** User Access Matrix A
- **Output:** Set of optimal Target User Groups
- Step 1.// Initial Bicluster Formation using Two-way K-Means Clustering
  - Compute k_u user clusters and k_p page clusters from preprocessed click-stream data.
- Step 2. Combine k_u and k_p clusters to form $k_u \times k_p$ biclusters called seeds.
- Step 3.// Seed Enlargement and Refinement using Greedy Heuristic
  - For each seed do
  - nodes= list of users and pages which is not in the seed
  - For each element in nodes do
    - Add the element to the seed which maximize the ACV of the seed
  - End (for)
  - nodes1 = list of users and pages in the seed.
  - For each element in nodes1 do
    - Remove the element from the seed which maximize the ACV of the seed
  - End (for)

- End (for)
- Step 4. Set enlarged and refined seed as initial population P.
- Step 5.// Extraction of optimal target users groups using GA
  - ◦ Binary encoding of Bicluster
- Step 6. Evaluate the fitness of individuals in P
- Step 7. For i =1 to max_generation
  - ◦ Selection of the best individuals from P
  - ◦ Recombine the selected individuals and produce new offsprings using crossover and mutate operator
  - ◦ Insert the best offsprings in to P
  - ◦ Evaluate the fitness of P
  - ◦ End(For)
- Step 10. Return the set of optimal biclusters as target user groups.

In GAs based biclustering, an initial population is the set of binary encoded biclusters. At each generation, fitness of each individual is evaluated in the evaluation step and recombined with others on the basis of its fitness. The expected number of times an individual is selected for recombination is proportional to its fitness relative to the rest of the population. New individuals are created using two main genetic recombination operators known as crossover and mutation. Crossover operates by selecting a random location in the genetic string of the parents (crossover point) and concatenating the initial part of one parent with the final part of the second parent to create a new child. A second child is simultaneously generated using the remaining parts of the two parents.

Mutation is provided to occasional disturbances in the crossover operation by inverting one or more genetic elements during the reproduction process. This operation insures diversity in the genetic strings over long periods of time and prevents stagnation in the convergence of the optimization technique. In addition to fitness, generation crossover rate and mutation rate, the size

of the population, encoding and selection strategy are called configuration parameters which should be specified. While the termination condition is not met, which might be number of generations or a specific fitness threshold, the processes of selection, recombination, mutations and fitness calculations are done.

EXPERIMENTAL RESULTS AND ANALYSIS

The experiments are conducted on the well-known benchmark clickstream dataset called msnbc dataset which was collected from MSNBC.com[1] portal. This dataset was taken from UCI repository and the original data are preprocessed using equation 1. There are 989,818 users and only 17 distinct items, because these items are recorded at the level of URL category, not at page level, which greatly reduces the dimensionality.

The length of the clickstream record starts from 1 to 64. Average number of visits per user is 5.7. Intuitively, very small and very large number of URL category visited may not provide any useful information about the user's behavior. Thus, the length of the record having less than 5 is considered as a short and record length greater than 15 is considered as a long. During data filtering process, short and long records are removed from the dataset.

Table 2 shows the characteristic of the seeds which are obtained by Two-way K-Means clustering and Greedy Heuristic. To avoid random interference, very tightly correlated biclusters obtained using greedy method is used as initial population for GA. Moreover, it results in quick convergence and provides more number of potential biclusters. These biclusters have high ACV and high volume which is obvious from Table 3. This approach shows excellent performance at finding high degree of overlapped coherent biclusters from web data.

Table 2. Characteristic of seeds

	Two-way K-Means Clustering	K-Means + Greedy Heuristic
No. of Seeds	120	120
Average ACV	0.4711	0.9103
Average Volume	695.9	1699.8
Overlapping Degree	0	0.0192

Table 3. Performance of biclustering using GA

Initial Population	Average Volume	Average ACV	Overlapping Degree
Random Initialization	8373.6	0.9412	0.1196
K-Means + Greedy Heuristic	12517.7	0.9829	0.2056

From the Table 3, it has been observed that GA in combination with the local search method such as Greedy Heuristic yields the largest biclusters compared to the results of the GA with random initialization of population.

Initialization procedure for GA using K-Means and Greedy method extracts largest volume biclusters from web usage data than random initialization procedure which is graphically represented in the Figure 3. In this work, a correlation based fitness function is used to capture browsing pattern similarity among users or pages. Moreover, users in the global optimal biclusters are highly correlated whose ACV is close to 1. From the results, it can be concluded that greedy bicluster-

ing is combined with GA extracts high volume overlapping coherent biclusters with high ACV. Comparison of different biclustering methods is represented graphically in Figure 4.

From Table 4, it has been observed that K-Means + Greedy Heuristics+ GA biclustering performs well than biclustering using Greedy Heuristic biclustering alone for extracting the global optimal user groups from preprocessed web usage data such as clickstream data. Target user groups are constructed from global optimal biclusters and it is defined as a set of pairs of page and weight of the page. Target user group is also called as an aggregated usage profile. Aggregated usage profile is the set of all pages whose weight is greater than and equal to min_weight threshold.

Figure 3. Comparison for various initalization procedure for GA

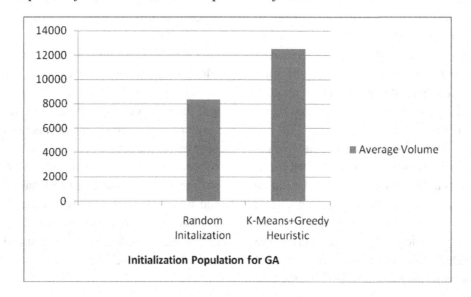

Figure 4. Comparison for biclustering methods

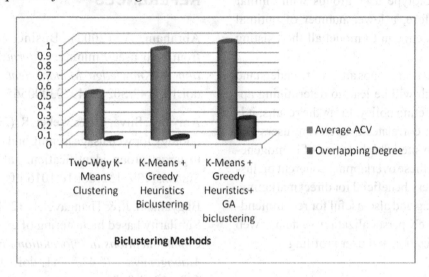

Aggregated Usage Profile of Bicluster B(u',p')
= { p, w(p) | p ∈ p', w(p) ≥ min_weight }

where $p' = \{p_1, p_2, \ldots, p_n\}$, a set of *n* page- in B and each page uniquely represented by its associated URL and the w(p) is the mean value of the frequency visit count of the page p in the bicluster. The advantage of these biclustering approaches is that they build prototypes that may be directly used as a basis for personalization.

Table 4. Comparison of average volume and ACV of biclusters

	Average Volume	Average ACV	Over-lapping Degree
Two-Way K-Means Clustering	695.9	0.4711	0
K-Means + Greedy Heuristics Biclustering	1699.8	0.9103	0.0192
K-Means + Greedy Heuristics+ GA biclustering	12517.7	0.9829	0.2056

Interpretation of Biclusters

The interpretation of biclustering results is also used by the company for focalized marketing campaigns to improve their performance of the business. Each bicluster represents the user segment from clickstream data based on their browsing interest. The aggregate target user groups are tabulated in Table 5, which shows the users interested web pages. It is also observed that setting a higher number of biclusters, leads

Table 5. List of aggregated target user groups

Target User Group ID	Page ID	List of Pages
1	1, 4, 7, 15,16	Frontpage, local, misc, travel, msn-news
2	8, 9, 13,14	weather, health, summary, bbs
3	2, 3, 5, 6, 10, 11, 12, 14, 17	News, tech, opinion, on-air, living, business, sports, bbs, msn-sports
4	2,3,5, 12, 14	News, tech, opinion, health, sports, bbs
5	1,4, 6, 7, 10, 11	Frontpage, local, on-air, misc, living, business

to various prototype user groups with similar values. Therefore, a lower number of optimal biclusters was enough to model all the existing user segments.

These target user groups are used to understand the users which will be lead to determining appropriate marketing policy. From the results, it is obvious that it correlates the relevant users and pages of a web site in high degree of homogeneity. Analyzing these overlapping coherent biclusters could be very beneficial for direct marketing, target marketing and also useful for recommending system, web personalization systems, web usage categorization and user profiling.

CONCLUSION

The proposed algorithm is used to detect target groups of customers from the historic dataset using evolutionary biclustering algorithm. This algorithm identifies the optimal coherent user segments in terms of biclusters. It provides valuable information about the user for offering the right offer to the right user and also possible for personalization services. The key feature of the proposed work is the extraction of target groups based on their coherent browsing patterns. For this, a new coherence based similarity measure is introduced in the paper. It captures pattern based closeness of the users but not the closeness of values which has a wide range of E-Commerce applications like marketing, advertising, recommendation and etc. The proposed evolutionary biclustering approach extracts the target user groups from the preprocessed web data. Once the target user groups are known, promotions can be tailored to each users group's interests and attitudes. The results of this approach are used in targeted promotional campaign.

REFERENCES

Abraham, A. (2003). Business intelligence from web usage mining. *Journal of Information & Knowledge Management*, 2, 375–390. doi:10.1142/S0219649203000565

Araya, S., Silva, M., & Weber, R. (2004). A methodology for web usage mining and its application to target group identification. *Fuzzy Sets and Systems*, 139–152. doi:10.1016/j.fss.2004.03.011

Bagyamani, J., & Thangavel, K. (2010). SIMBIC: Similarity based biclustering of expression data. *Communications in Information Processing and Management*, 70, 437–441. doi:10.1007/978-3-642-12214-9_73

Bleuler, S., Prelic, A., & Zitzler, E. (2004). An EA framework for biclustering of gene expression data. In . *Proceedings of the Congress on Evolutionary Computation*, 1, 166–173.

Chakraborty, A., & Maka, H. (2005). Biclustering of gene expression data using genetic algorithm. In *Proceedings of the IEEE Symposium on Computational Intelligence in Bioinformatics and Computational Biology* (pp. 1-8).

Das, C., Maji, P., & Chattopadhyay, S. (2008). A novel biclustering algorithm for discovering value-coherent overlapping σ-Biclusters. *Advanced Computing and Communications*, 148-156.

Hartigan, J. A. (1972). Direct clustering of a data matrix. *Journal of the American Statistical Association*, 123–129. doi:10.2307/2284710

Jonker, J.-J., Piersma, N., & Van den Poel, D. (2004). Joint optimization of customer segmentation and marketing policy to maximize long-term profitability. *Expert Systems with Applications*, 27, 159–168. doi:10.1016/j.eswa.2004.01.010

Koutsonikola, V. A., & Vakali, A. (2009). A fuzzy bi-clustering approach to correlate web users and pages. International . *Journal of Knowledge and Web Intelligence, 1*(1-2), 3–23. doi:10.1504/IJKWI.2009.027923

Lee, C.-H., & Fu, Y.-H. (2008). Web usage mining based on clustering of browsing. In *Proceedings of the Eighth International Conference on Intelligent Systems Design and Applications* (Vol. 1, pp. 281-286).

Madeira, S. C., & Oliveira, A. L. (2004). Biclustering algorithms for biological data analysis: A survey. *IEEE/ACM Transactions on Computational Biology and Bioinformatics, 1*, 24–45. doi:10.1109/TCBB.2004.2

Punj, G., & Stewart, D. W. (1983). Cluster analysis in marketing research: Review and suggestions for application. *JMR, Journal of Marketing Research, 20*, 134–148. doi:10.2307/3151680

Rathipriya, R., Thangavel, K., & Bagyamani, J. (2011). Evolutionary biclustering of clickstream data. *International Journal Computer Science, 8*, 32–38.

Shina, H. W., & Sohnb, S. Y. (2004). Segmentation of stock trading customers according to potential value. *Expert Systems with Applications, 27*, 27–33. doi:10.1016/j.eswa.2003.12.002

Srivastava, J., Cooley, R., Deshpande, M., & Tan, P. N. (2000). Web usage mining: Discovery and applications of usage patterns from web data. *SIGKDD Explorations, 1*(2), 12–23. doi:10.1145/846183.846188

Tsai, C. Y., & Chiu, C. C. (2004). A purchase-based market segmentation methodology. *Expert Systems with Applications, 27*, 265–276. doi:10.1016/j.eswa.2004.02.005

Xu, G., Zong, Y., Dolog, P., & Zhang, Y. (2010). Co-clustering analysis of weblogs using bipartite spectral projection approach. In R. Setchi, I. Jordanov, R. J. Howlett, & L. C. Jain (Eds.), *Proceedings of the 14th International Conference on Knowledge-Based and Intelligent Information and Engineering Systems* (LNCS 6278, pp. 398-407).

Zong, Y., Xu, G., Dolog, P., & Zhang, Y. (2010). Co-clustering for weblogs in semantic space. In L. Chen, P. Triantafillou, & T. Suel (Eds.), *Proceedings of 11th International Conference on Web Information Systems Engineering* (LNCS 6488, pp. 120-127).

ENDNOTE

[1]   http://archive.ics.uci.edu/ml/datasets/

This work was previously published in the International Journal of Applied Metaheuristic Computing, Volume 2, Issue 3, edited by Peng-Yeng Yin, pp. 69-79, copyright 2011 by IGI Publishing (an imprint of IGI Global).

Chapter 16
Dynamic Assignment of Crew Reserve in Airlines

Walid Moudani
Lebanese University, Lebanon

Félix Mora-Camino
Ecole Nationale de l'Aviation Civile (ENAC–DGAC), France

ABSTRACT

The Crew Reserve Assignment Problem (CRAP) considers the assignment of the crew members to a set of reserve activities covering all the scheduled flights in order to ensure a continuous plan so that operations costs are minimized while its solution must meet hard constraints resulting from the safety regulations of Civil Aviation as well as from the airlines internal agreements. The problem considered in this study is of highest interest for airlines and may have important consequences on the service quality and on the economic return of the operations. A new mathematical formulation for the CRAP is proposed which takes into account the regulations and the internal agreements. While current solutions make use of Artificial Intelligence techniques run on main frame computers, a low cost approach is proposed to provide on-line efficient solutions to face perturbed operating conditions. The proposed solution method uses a dynamic programming approach for the duties scheduling problem and when applied to the case of a medium airline while providing efficient solutions, shows good potential acceptability by the operations staff. This optimization scheme can then be considered as the core of an on-line Decision Support System for crew reserve assignment operations management.

INTRODUCTION

The management of planning problems confronting logistics service providers frequently involves complex decisions. For more than four decades now the crew assignment problem (to flights, duties, reserve activities, etc.) of the airlines has captured the attention of the Management and Operations Research community since crew members management is extremely complicated and may generate many problems that hinder the smooth operation, influencing the total company

DOI: 10.4018/978-1-4666-2145-9.ch016

operations profit (Rosenberger, 2001). Therefore for airlines the efficient management of their crew staffs is considered to be a question of the highest economic relevance. In the process of managing airline operations, once the scheduling of flights is performed by the commercial department and once the nominal crew assignment is completed, the problem of scheduling the reserve activities is raised. The latter activity is, first, to scale the size of the crew reserve and then allocate the available crew on reserve activities, taking into account the various constraints. In the case considered here, we assume that the solutions of the crew scheduling problem (CSP) and of the size of reserve problem are already established; thus, we focus our interest only on the assignment of crew to reserve activities. The amount of necessary investments and the high operational costs supported by the air transportation sector as well as its hard competitive environment, has led airlines to pursue a permanent improvement in their management practice at the planning and operations levels. With the deregulation policies adopted by the air transportation administrations of many countries, this goal has gained a major momentum. To meet their daily commitments, airlines assign or reassign their available crew to the scheduled or unscheduled flights and reserve activities while satisfying regulations and other operational constraints. When these assignments are done in an ineffective way, they have to support substantial overcosts which can impair their survival as a firm. The crew reserve assignment problem is then of considerable importance for airlines; and consequently, since the pioneering work of (Simpson, 1969), there has been a lot of research done to improve its solution (Yoshihara & Sengoku, 2000; Sohoni, Johnson, & Bailey, 2004; Liping, 2006). From the point of view of complexity theory, this problem is considered to be a difficult one since it presents a combinatorial choice within time and space dimensions with complicating ingredients from the operational

constraints. It appears also that this problem is strongly related to the scheduling problem of crews duties (preventive medical visits, training, license renewal, etc.). Thus, a global solution for the crew reserve assignment problem is requested. In this communication a solution approach based on efficient enumerative methods is proposed to solve simultaneously these related problems.

In most airlines, the crew reserve assignment is established empirically. Sometimes even no crew reserve is planned, and operational problems are solved on time (Gaballa, 1979). The decisions are not optimal and can be very expensive. In fact, some operators spend little time on this task and have focused more on development of crew rostering. The development of crew reserve planning is often based on fixed ratios and the occurrence of specific events and periods (bridges between holidays, vacation periods, etc.). This leads to an availability of reserves approximately equal for each day of the month with the exception of those periods where the size of the reserve is increased. However, an effective crew reserve planning should allow airlines to cope with unexpected operational situations such as: failure of a crew (illness, delayed delivery, absence, late arrival of the crew at the airport, etc.), technical failure generating a change in aircraft type, strikes, and delayed flights (traffic congestion or bad weather) which prevents the crew scheduled for a flight to be available (Gaballa, 1979; Sohoni, Bailey, Martin, Carter, & Johnson, 2003; Sohoni, Johnson, & Bailey, 2004). The programming of crew reserve activities may affects the productivity of the airlines. An unused crew on reserve represents a loss of potential flight hours, but a higher loss is generated when no crew reserve is available and the call for a backup crew becomes unavoidable. The call for a crew who was given a leave is never desirable given the disruption it will produce in his personal life.

In our proposed approach, we introduce the CRAP as a mono-criterion decision problem where the criterion is representative of the additional crew operations cost. A set of hard and soft constraints are considered while solving the CRAP. This paper is organized as follows: an overview of research found in literature about this problem is presented; followed by a mathematical formulation for this problem and a description of the proposed solution approach since the classical mathematical solutions approach present high complexity. Our proposed solution approach deals with the CRAP based on cost minimization using Dynamic Programming technique in order to build an efficient workload. We illustrate our solution approach with a numerical example applied to a medium size charter airline followed by discussion and analysis of the obtained results. Finally, we conclude by an upraise of this new approach and point out related new ideas to be tackled in the future.

The Airlines Crew Reserve Assignment Problem

In general, the CRAP is treated as part of the crew scheduling problem which is in general performed once the schedule of the trips has been established for the next period (often a month) and once the available aircraft have been assigned to the scheduled flights. Airline CSP is an important problem for airlines which is a worldwide NP-hard combinatorial optimization problem with considerable economic significance (Barutt & Hull, 1990). The basic airline CSP concerns the daily assignment of the crew members to round trips for all the scheduled flights so that the total service time is minimized. Two classes of constraints are considered in order to produce the workload for the crew members over the planning period: hard constraints whose violation impair the security of the trip (crew qualifications, national regulations concerning duration of work and rest times, training requirements) and soft constraints (internal airline rules, agreements with unions regarding

the crews working and remuneration conditions, holidays and declared claims by the crew members) which are relevant to build the crew members schedule but whose relaxation should lead to lower cost solutions (Barnhart, Cohn, Johnson, Klabjan, Nemhauser, & Vance, 2003). While some of these soft constraints are common to most airlines, others are only relevant for specific classes of airlines and some few are particular to a given airline. The primary objective sought by airlines at this level of decision making is to minimize the crew members operations costs, so in most research studies, the CSP has been formulated as a mono-criterion cost minimization problem. It is known that mathematical programming formulations of this problem are generally too complex to be exactly solved for real-size applications. A sub optimal but widely accepted approach to tackle more efficiently the CSP consists in decomposing it in two sub-problems of lower difficulty. In the first sub-problem, the airlines are involved in the construction of an efficient set of rotations (a rotation is a sequence of flights which starts and ends at the same airline base), called crew pairings (CPP), while meeting all relevant legal regulations. In the second sub-problem, the airlines consider the assignment of the crew members to the generated set of rotations so that an effective workload is obtained for each crew member (Curtis, Smith, & Wren, 1999; Moudani & Mora-Camino, 2000; Moudani, Cosenza, Coligny, & Mora-Camino, 2001). In fact, crew cost is the second greatest operational expense for airlines, exceeded only by the cost of fuel consumption (Andersson, Housos, Kohl, & Wedelin, 1997). As an example, in Anbil, Gelman, Patty, and Tanga (1991), it is reported that American Airlines spent 1.3 billion dollars on crew costs in 1991. Therefore, a low cost solution to the CPP can save airlines millions of dollars per year. Almost every major airline is using a system that automates CSP. However, computational tests, conducted with actual data, led to the conclusion that many of the solutions provided by these systems needed significant improvement (Desaulniers, Desrosiers, Lasry, & Solomon, 1999).

As mentioned above, the crew reserve scheduling allows airlines to cope with unexpected operational situations such as: failure of a crew (illness, delayed delivery, absence, late arrival of the crew at the airport, etc.), technical failure generating a change in aircraft type, strikes, and delayed flights (traffic congestion or bad weather) which prevents the crew for a scheduled flight to be available. Most of these researches didn't take into consideration the crew reserve assignment within the global CSP while treating separately several crucial duties (the crew reserve, the crew preventive medical visits, the crew training, the renewal of license, etc.). In this paper, we focus our interest in building dynamically the crew reserve schedule while taking into consideration the assignment of the crucial duties.

Literature Overview of the Crew Assignment Problem

The literature about the CSP is very rich and can be divided in two main classes, namely theoretical and application-based papers. The first class of theoretical papers deals with formulations and algorithms for simplified CSPs with only a few constraints. In the second class those algorithms are applied to solve problems which arise from practical applications. Many studies have tried to solve the crew assignment problem either using optimization methods or complex type of heuristic methods. In the past, a variety of approaches using exact methods and efficient heuristics have already been proposed for solving airline CSP. Also there were a large number of contributions on various extensions to the basic problem. Unfortunately, the exact numerical solution of the associated large scale combinatorial optimization problem is very difficult to obtain. Early rules of thumb have been quickly overrun by the size of the practical problems encountered (hundreds or

thousands of crew to be assigned to at least as many duties) and by the complexity of the set of constraints to be satisfied, leading very often to poor performance solutions. More recently, with the enhancement of computer performances, optimization approaches have been proposed to solve this problem: mathematical programming methods (large scale linear programming and integer programming techniques) (Curtis, Smith, & Wren, 1999; Sohoni, Johnson, & Bailey, 2004), artificial intelligence methods (logical programming, simulated annealing, neural networks, fuzzy logic and genetic algorithms) as well as heuristic approaches and their respective combinations (Brusco, 1993; Yoshihara & Sengoku, 2000; Aickelin & Dowsland, 2003). Many studies refer to the CRAP which is a static decision problem, based on a monthly table of flights, and devoted exclusively to the minimization of airlines operations costs. The set of papers gathered from literature dealt with the CSP faced by bus transport companies, railways companies (Caprara, Fischetti, Guida, Toth, & Vigo, 1999; Freling, Lentink, & Odijk, 2001; Kroon & Fischetti, 2001) and airlines (Desaulniers, Desrosiers, Lasry, & Solomon, 1999). Thus, our presentation for the related works in literature covers the airlines sector while illustrating the main related research activities. A set of various approaches have been proposed to tackle with the CSP. These approaches which are detailed and discussed here can be classified. Firstly, approaches for solving set covering and/or set partitioning problems. They are based on classical techniques of optimization in graphs. Secondly, the column generation method which consists in generating a subset of rotations since the number of possible flights is extremely high and cannot be treated globally; and subsequently, generating eligible workloads for each crew member. Thirdly, evolutionary algorithms and heuristic methods that generate solutions directly hoped to be effective while overcoming the computer constraints (CPU time, memory space).

Set Covering and Partitioning Problems

Rubin (1973) and Crainic and Rousseau (1987) are among the first having proposed to deal with the problem of building the airlines CSP as a set covering/partitioning problem. The choice of a formalism corresponding to this approach led in general to the definition of two sub-problems: the generation of the constraint matrix where each column corresponds to an individual acceptable workload; and, the optimization of this matrix by using an appropriate technique. Numerous studies have subsequently been developed based on this two steps solution approach. Some proposed to generate a subset of feasible workloads by finding disjoint paths in a directed graph where each path corresponded to a feasible workload. Then, a coupling method was used to improve the quality of the initial workloads. This process was supposed to be continued until no improved solutions could be found. A different solution approach, (Freling, Lentink, & Odijk, 2001; Sohoni, Johnson, & Bailey, 2004; Qi, Bard, & Yu, 2004; Zeghal & Minoux, 2005), aimed to consider the problem in a global approach rather than local and was based on a heuristic to generate a large number of rotations instead of just covering some scheduled flights. Then the primal simplex method was used to solve the set partitioning problem through linear relaxation while the dual variables were used to assess the trips with a negative marginal cost, thereby reducing the overall cost associated to the problem. This process was to be repeated until no new rotations could be included the constraint matrix. Ryan and Foster (1981) developed such heuristic to find the integer solution since the method of branch and bound presents enormous difficulties. Less remotely Hoffman and Padberg (1993) presented a solution approach using the branch and cut method to solve set partitioning problems of large dimensions. This method provided a gain of 0.5% in terms of operating cost associated with the construction

of pairings with respect to the method of Rubin (1973). This technique uses a heuristic method to find a solution with integer values where cuts (or constraints) are added to reduce the search space in the polyhedron of solutions. Smith and Wren (1988) formulated the driver scheduling problem as an integer linear programming (ILP) model but they developed a set of heuristics which reduced the problem size. The ILP model was solved by first relaxing the 0-1 conditions, and then using a branch and bound process which took advantage of acquired knowledge about crew scheduling. Schaefer et al. (2005) developed better approximate solution method for airline crew scheduling under uncertainty due to disruptions where they provided a lower bound on the cost of an optimal crew schedule in operations.

Column Generation Approach

Optimizing the construction of workloads with a formulation of set covering/partitioning problems requests in theory to enumerate all eligible workloads and then solve the problem where each column corresponds to a decision variable. The resolution of this problem becomes intractable considering computer time when it reaches a large size. Hence the importance of the method of column generation which, unlike the traditional approach of set covering/partitioning, is able to solve problems with a huge number of variables, by repeatedly generating the matrix of constraints (Crainic & Rousseau, 1987; Desaulniers, Desrosiers, Lasry, & Solomon, 1999). Using this technique leads to a decomposition of the main problem into: firstly, a master problem for solving a problem of set covering or set partitioning, and secondly, a sub-problem for constructing the columns associated with the rotations. The master problem involves solving a relaxed linear program associated with a branch and bound method to find an integer solution. The sub-problem allows the construction of columns associated with rotations at minimum marginal cost. It may be formulated as a constrained shortest path research problem in a graph. If it is no more possible to generate negative marginal cost columns, the current solution is optimal. This technique allows the user to choose between obtaining an optimal solution or to get a feasible solution at an acceptable cost.

The general idea, introduced by Dantzig and Wolfe (1960), was to solve a sequence of reduced problems, where each reduced problem contains only a small portion of the set of variables (columns). After a reduced problem is solved, a new set of columns is obtained by using dual information of the solution. The column generation algorithm converges once it has established that the optimal solution based on the current set of columns cannot be improved upon by adding more columns. The set of columns can be enumerated or generated heuristically. Then the optimal solution of the reduced problem is the optimal solution of the overall problem. If (some of) the decisions variables values need to be integer, different algorithms, exact or approached, can be used. Branch-and-bound algorithms are most often used to solve the resulting (mixed) integer programming problem (Barnhart, Johnson, Nemhauser, Savelsbergh, & Vance, 1998). Hoffman and Padberg (1993) described a branch-and-cut algorithm where cutting planes were generated based on the underlying structure of the polytope. In Caprara, Toth, and Vigo (1998), Lagrangian heuristics and variable fixing techniques have been used to find integer solutions. Mingozzi, Boschetti, Ricciardelli, and Bianco (1999) suggested an approach to compute a dual solution of the LP-relaxation combined with Lagrangian relaxation (Carraresi, Girardi, & Nonato, 1995). Afterwards, the dual solution can be used to reduce the number of primal variables such that the resulting problem is solved by a branch and bound algorithm. Furthermore, the different procedures use lower bounds of previous procedures and improve them. In some of these procedures Lagrangian relaxation and column generation are used. A completely different formulation has been introduced by Fischetti,

Lodi, Martello, and Toth (2001), namely a single commodity flow formulation. Numerical experiments carried out have shown that the best results, in computing time, are obtained by integrating multiple columns in the matrix of constraints. This reduces the number of calls to the column generation process that generates a waste of time much larger than a few pivoting operations in the simplex method.

Heuristics Methods

The earliest work on CSP, four decades ago, developed exhaustive tree search approaches which failed to produce good solutions even for small problems. A set of different heuristics are proposed to produce schedules quite quickly. Reasonable results were obtained in some practical and lightly constrained situations. Thus, heuristic approaches have provided very early an alternative to overcome the constraints associated with CSP. These methods are often intelligent since they are inspired by the knowledge gained by practitioners who have developed empirically simplified methods solving quite satisfactorily the problem. Among these methods we retained the day by day method, the pilot by pilot method, a combination of both and a method considering attractiveness of rotations and staff seniority:

The "day by day" method is proposed, for the case of airlines, in which the assignment of the crew is performed, chronologically, based on the day unit time (Sarra, 1988). They proposed to solve the pre-allocation of activities such as training, medical and flight crew desires. These activities are pre-assigned to different crew and become hard constraints during the process of crew scheduling. This method leads to daily assignments more and more degraded and can generate a lot of uncovered rotations. Using this method makes it difficult to predict the consequences of previous decisions on the difficult problems to solve in the coming days. This can be seen especially in peak periods (summer and holidays).

The "pilot by pilot" method is proposed to perform the assignment by considering the crew members one after the other (Byrne, 1988). In the basic approach, we select arbitrary a crew and assigning to him a monthly workload. This process continues until all crew is assigned a monthly workload. The disadvantage of this method is relative to the unequal treatment which is subject to different crew, as the first crew selected will be favored over the latter. Here too, at the end of the assignment process, in order to cover the residual rotations, it will be necessary to assign overtime, and/or to over-size the crew.

The combination of the "day by day" and "pilot by pilot" methods: This solution approach aims to combine the two previous methods proposed in literature (Antosik, 1978) by assigning the scheduled rotations to qualified pilots while minimizing the disproportionate workload (flight time, number of working days, number of rotations desired by pilots and assigned to them, number of rotations unwanted by pilots and which do not affect them) between different crew members. First, the method "pilot by pilot" is applied where each crew is assigned as many rotations as possible. The first objective here is to cover all rotations scheduled by pilots. Once this step is completed, the method "day by day" is used to reduce the disproportionate workloads and other parameters (working hours, the number of working days and the degree of satisfaction) related to the workload of crew.

A heuristic method based on the attractiveness of rotations and staff seniority has been proposed by Glanert (1984). It consists in constructing sequentially the monthly workload for each crew member, ranked in descending order of seniority. Each crew is assigned a satisfactorily workload by giving higher priority to his preferred activities (a particular rotation, a rest period, an attractive destination, etc.). Then, a validation of the workload set is performed to check if the residual rotations can be covered. This process continues until all the rotations are covered. An improvement of this approach based not only the seniority of the crew but also on the attractiveness of rotations has been proposed by Giafferri, Hamon, and Lengline (1982). This approach has some drawbacks since it may lead to unbalanced distribution of workloads.

However, some heuristic approaches have been developed in order to perform the assignment of pilots to reserve activities. One approach developed on behalf of a North American airline is detailed hereafter (Ananda, Nirup, & Gang, 1996). Here the crew members are classified in a workload ascending order. In this company, the crew is paid a Minimum Guaranteed (MG) wage in addition to their premiums. The company must make maximum use of the crew while trying to stay below the MG equivalent workload level. This leads to accumulate the number of flown hours for each crew throughout the assignment period (Figure 1). In the example, this leads to consider that pilot "A" has priority on pilot "B"

Figure 1. Distribution of workload related to minimum guaranteed wage

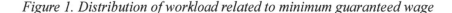
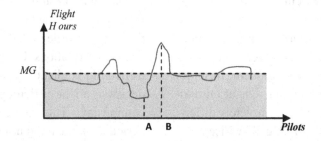

to be assigned reserve activities. Thus, crew members can be assigned a priority class according to their workload. In Figure 2 three classes of priority are considered: Pilots Large Priority (PLP), Pilots Medium Priority (PMP), and Pilots lOw Priority (POP).

Besides considering the pre-assigned workload, it is up to each company to make adaptations to the crew member classification method. The company can consider additional criteria such as the number of weekends assigned to the crew members in the month, the number of pilots absence due to illness, favorable conditions for crew members (combined working periods), the number of reserve activity carried on by the crew members during the assignment period, and the number of days leave taken by each crew member. This list is not exhaustive, but gives an overview of the different criteria that can be considered to perform crew member classification. Note that this classification process can be subject to negotiation within the airlines. All this can result in the adoption of very different classification processes from one airline to another.

Evolutionary Methods

A pure evolutionary algorithm has been proposed in the literature to tackle with this problem while overcoming the complexity structure and calculation. Genetic Algorithm (GA) is a powerful computerized heuristic search and optimization method (Holland, 1976; Goldberg, 1989; Mitchell,

1996). Curtis, Smith, and Wren (1999) have presented a genetic algorithms approach which has been successfully used to overcome some limitations of the established integer linear programming solution techniques. The proposed approach starts by generating all probable potential shifts according to well developed heuristics. This study was originally performed to solve urban bus operations, although it was also successfully applied to rural and short interurban services. In Majumdar and Bhunia (2010), an airline CSP with interval valued time parameters has been proposed considering the service time (including rest time) of each crew as interval. In this case, the problem (with interval objective) has been formulated as an assignment problem using interval arithmetic and existing recently developed complete definitions due to Majumdar and Bhunia (2006) of interval order relations with respect to the pessimistic decision makers' preference. To solve this interval valued CSP, two different methods have been proposed: (1) an elitist genetic algorithm (EGA) with interval valued fitness function and (2) EGA approach after converting it into a multi-objective assignment problem with crisp objectives considering both the centre and width values of the corresponding intervals. The results of the proposed methods have been compared with the help of an example and to study the effect of changes of various genetic parameters on the performances of both the methods, sensitivity analyses have been done.

Levine (1996) developed a heuristic method like hybrid genetic algorithm consisting of a

Figure 2. Classification of crew members for reserve activities based on their workload

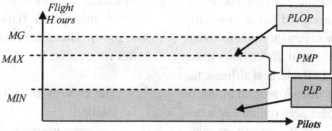

steady-state genetic algorithm and a local search heuristic to solve the same problem. Ozdemir and Mohan (2001) also proposed genetic algorithm for CSP in airlines. A bi-criterion approach for the airline crew rostering problem was proposed by Moudani, Cosenza, and Mora-Camino (2001) and Moudani and Mora-Camino (2011). In this approach, the solution is associated with acceptable satisfaction levels for the crew staff. Curtis, Smith, and Wren (2000) have investigated an approach using constraint programming, and subsequently produced promising results on relatively small problems with a mixture of neural networks method and repair heuristics. Forsyth and Wren (1997) reports on the early stages of the development of an ant colony system for driver scheduling. Lucic (1999) has proposed a solution approach for CSP in airlines by using a Simulated Annealing (SA) approach: first heuristics methods are used in order to generate an acceptable initial solution and then a SA method is supposed to improve the initial solution.

ANALYSIS OF THE RESERVE ACTIVITIES ASSIGNMENT PROBLEM

Problem Description of the CRAP

In this section, we introduce the CRAP faced by airlines which are either passengers and/or freight carriers with regular and/or irregular operations. The crew assignment to reserve activities is combinatorial by nature since it allocates discrete resources (crew) to discrete operations (reserve activities) and this could be enough to qualify it as complex, however, since the set of reserve activities to which crew are assigned can change daily and may be known only several hours before start time, its temporal and spatial dimensions make it even more complex. In the largest airlines, the CRAP is solved using integer programming or artificial intelligence packages run on dedicated large computer devices. To the operations man-

agers they appear as "black boxes" which realize in an automatic way local trade-offs between the different alternatives building up a global solution. Since computing times can be long, no or little sensitivity and robustness analysis of the results are performed. In smaller airlines, the assignment of the crew reserve is made manually and is based on experience and practical knowledge. However, whatever the size of the airline and the selected solution approach, in practical ways, crew reserve assignment is achieved through two stages: a planning stage where the assignment is performed periodically supposing nominal operations and a regulation stage where amendments to the assignment plan are performed to cope with significant unexpected events. These perturbations are mainly composed of cancelled or delayed flights which can be caused by equipment breakdowns, corrective maintenance operations, adverse meteorological conditions, congestion effects at airports and delayed availability of crew members.

Airlines Reserve Activities, Related Costs, and Global Constraints

The human resources management unit of an airline establishes in advance (a month ahead for instance) the schedule of the crew for the next operations period which includes some days off during which they are not supposed to be solicited by the airline. The planning of these effective holidays follows some rules specified in the airlines regulations and in the company's accreditation with the association of airline pilots or their union. During the remaining time period, the crew must stay at disposal when required for operational and reserve activities (Weir, 2002; Sohoni, Johnson, & Bailey, 2004; Qi, Bard, & Yu, 2004). There are four different operational states for a crew:

1. The crew is in service and assigned to a rotation;
2. The crew is assigned to a reserve activity;

3. The crew is on duty but not assigned to a rotation, he is assigned either to a training task, or to administrative duties or trade union or reserve activity;

4. The crew is on leave (for one or more consecutive days). Usually the day can be subdivided into two reserve periods (e.g., from 0h to 12h and from 12h to 24h). To each of these reserve periods and for each base of operations a variable number of crew can be assigned.

The costs associated to airlines crew operations are already detailed in Holloway (1997). The assignment of crew to reserve activities must take into account this immobilization cost since this reserve activity generates an additional workload for crew which must be remunerated by the airline. This is specially the case when the assignment of reserve activities generates overtime periods. In general, most crew do not show particular interest for this kind of activity because the hour rate is significantly lower than that of flight hours and more than that, this is considered to be a tedious activity. Thus, the occurrence for each crew of this activity during an operational period is usually limited.

Regarding the global constraint for the CRAP, the national and international regulations as well as the collective agreements within airlines generate a set of constraints that may be either hard or soft for the CRAP. These constraints consider diverse aspects such as the duration of the legal working time and crew license (Sohoni, Bailey, Martin, Carter, & Johnson, 2003; Qi, Bard, & Yu, 2004). The CRAP must satisfy several constraints such as:

*   A sufficient number of crew is assigned to each reserve activity in order to cover as much as possible all incidents related to the flights operated during this period;
*   This reserve activity must also cover a number of unscheduled flights (open trips);
*   The assignments to reserve activity should not disrupt the schedule of crew rostering;

*   The assignment should comply with regulations and collective agreements which limit the number of reserve activities provided for each crew during an operational assignment period.

The Size of Crew Reserve

The choice of the number of pilots which will be assigned to reserve activities during an operational period is turned difficult mainly by the multiplicity of factors to take into account: available statistics about reserve activity assignment and its connection with adverse or exceptional weather conditions and with peak periods (end of the month, weekend, holidays, etc.). Moreover, the end of an operational period presents always a high request for reserve crew when the regulatory limits in regard to the workload of crew members are very often approached. The beginning of the period may also correspond to a peak of reserves considering the assignment changes necessary to insure continuity between the prior period and the current period. Moreover, the assignment of crew to reserve activities can be increased significantly during weekends and holidays periods with the occurrence of non-scheduled flights. In general, the determination of the necessary levels of reserve activities has been performed using statistical (Mitchell, 1977; Dillon & Kontogiorgis, 1999) and econometric approaches (Holloway, 1997). Nevertheless it should be noted that when adopting an optimization approach we get a particularly difficult problem since it has both a stochastic dimension (occurrence of disturbances) and a combinatorial dimension (the allocation of pilots to reserve activities). The number of reserve pilots should be sufficient to prevent the disruption of flight schedules leading to delayed or canceled flights and generating additional costs for the airline related with the accommodation of passengers in hotels, the payment of compensations to passengers, or refund of tickets, etc. On the other hand, if the number of crew members in reserve is excessive, this can generate direct over costs (compensation reserve periods) and indirect over costs since the crew are turned unavailable during the rest period following each regulatory activity of reserve. Haydon (1973) proposed a probabilistic approach for sizing reserve crew based on historical data following a binomial distribution. This leads to define some extra reserve crew using probability calculations

Modeling the Assignment of Crew Members to Reserve Activities

Mathematical Formulation

In literature, the optimization approach for solving the assignment problems adopted considers unanimously a cost criterion. But the social history of this sector in recent decades has shown that everything was not just a question of economic costs but on the contrary, some elements which are difficult to be quantified could easily disrupt the solutions and lead to spectacular poor economic performance. The crew reserve assignment has been formulated as a zero-one integer mathematical programming problem where the crew reserve cost is the criterion to be minimized under a finite set of hard constraints (Crainic & Rousseau, 1987; Ryan, 1992). The classic formulation of the crew reserve assignment corresponds to a partition problem manipulating sets of binary decisions variables. The mathematical formulation is given as follows:

$$\text{Minimize} \sum_{i=1}^{n} \left(FH_i + \sum_{r=1}^{R} \delta_r x_{ir} \right) \quad (1)$$

where δ_r is the equivalence in flight hours for the reserve activity "r", x_{ir} is a binary variable taking the value "1" if the crew "i" is assigned to the reserve activity "r", 0 otherwise. FH_i is the number of flight hours performed by the crew "i" ($i \in \{0, 1, \cdots, n\}$) where n is the number of crew

members in consideration during the assignment period. R represents the total number of reserve activities.

The Constraints:

Among the constraints associated with the problem of assigning crew to reserve activities, we state mainly:

Constraint 1:

$$\sum_{r=1}^{R} x_{ir} \leq NBRES, \ \forall \quad i = 1, \cdots, n \quad (2)$$

This constraint ensures that the maximum number of reserve activities as each crew is allowed to be performed during a service period shall not exceed a given threshold $NBRES$.

Constraint 2:

$$\bar{x}_{ir} \in X_{ir} = \left\{ \{0\}, \{0, 1\} \right\},$$

$$\forall \quad i = 1 \cdots n; r = 1 \cdots R \quad (3)$$

There are two cases for making the decision to assign a crew to a reserve activity:

- Either the pilot « i » is unavailable to perform the reserve activity « r »; in this case, the decision variable x_{ir} must be equal to « 0 »: $X_{ir} = \{0\}$;
- Either he is available to perform the reserve activity « r »; in this case, the value of the decision variable x_{ir} is equal to 0 or 1 depending on the interest associated with this choice: $X_{ir} = \{0, 1\}$.

Constraint 3:
The following constraint takes into account the availability of crew members for reserve activities. This constraint is formulated by:

$$\overline{y}_i^T .A. \overline{x}_i = \sum_{r=1}^{R} \left(\sum_{k=1}^{m} y_{ik} . a_{kr} \right) . x_{ir} = 0 \qquad (4)$$

where the variables y_{ik} are binary variables that is equal to « 1 » if the crew « i » is assigned to rotation « k », « 0 » otherwise. The nth rotation corresponds to other activities of the pilot and his rest period. The parameters a_{kr} are such as:

$$\overline{z}_i^T .\overline{x}_i = 0, \forall \, i = 1, \cdots, n \text{ where } z_i^T = \left(\overline{y}_i^T A \right)$$
with $A = (a_{kr})_{k=1...m, r=1...R}$ and $\overline{y}_i^T = (y_{ik})_{k=1...m}, \forall \, i = 1 \cdots n$

Constraint 4:

This constraint ensures the distribution of reserve activities while complying with regulations and internal collective agreements. Here, we are assured that two reserve activities assigned to a pilot shall be separated from each other. It is written as:

$$x_{ir} + \sum_{s \in O_r} x_{is} \leq 1, \ \forall \, i = 1, \cdots, n \qquad (5)$$

where O_r is the set of reserve activities belonging near the reserve activity "r" assigned to the crew "i". That means no reserve activity belonging to a time interval around the reserve activity "r" should be assigned to the crew "i".

Constraint 5:

This constraint ensures that to each reserve activity is assigned a well defined number of qualified crew members. It is written as:

$$\sum_{i \in P_r} x_{ir} \geq L_r, \ \forall \, r \in R \quad (6)$$ where P_r is the set of qualified crew members able to carry on the flights belonging to reserve activity « r » and L_r is the requested number of crew members for reserve activity « r ».

Other constraints considered as not essential may be taken into account in this mathematical formulation. These constraints represent the claims of crew members that the airline is willing to consider, as well as other internal rules from the agreements made between the airline and union groups. The repeated non-satisfaction of these constraints leads to decrease the degree of satisfaction of each of the crew members. The following paragraph applies to define the concept. Some important comments can be done:

- Exact solutions for the resulting large scale combinatorial optimization problems are not available in an acceptable computing time and in general, airlines have produced their crew reserve assignment schedules using simplistic heuristics. However during the last decade, a number of major airlines have developed optimization-based techniques to solve the CRAP by developing improved heuristics based on column generation techniques and exact solution approaches based on constraints logic programming.

- Other relevant concerns can be added as soft constraints to the mathematical formulation of the CRAP. However, constraints relative to the crew members' satisfaction levels are difficult to add to the standard mathematical formulation of the CRAP since they depend in a complex way of the considered decision variables. Another difficulty is relative to the choice of individual versus collective crew satisfaction constraints. So it appears more appropriate to introduce this concern as such criterion to the overall assignment problem.

Analysis of the Crew Reserve Assignment Problem

Some observations can be done about the formulation of the assignment problem:

First, the combinatorial nature of the problem appears clearly through constraints (2) and (6): From constraint (6), the number of binary decisions variables is *n x m* and theoretically, there are 2^{nxm} possible solutions. Taking for instance a small sized problem with 100 reserve activities and 6 pilots, if the mean runtime to generate and evaluate each solution is 10^{-9} seconds, then, since $2^{600} > 10^{180}$, a pure enumerative approach should take more than 3×10^{163} years to get the solution. This is more than unacceptable. In fact, it can be shown that, according to computational complexity theory, this problem is NP-Hard (Garey & Johnson, 1979). Considering the formulation of constraints (4) and the definition of sets P_r and O_r, it is clear that a lot of preprocessing must be performed before getting a tractable formulation of the problem. Note also that every delay or reschedule of a flight can imply a need to recalculate these constraints (since overlapping of flights can happen) and these sets (a modification of the required minimum capacity can occur when the demand estimated for a flight is altered by a change in the time schedules).

Second, the above algebraic formulation is a static one and does not allow directly the consideration of several duties constraints such as: medical visit, office duties, training, trade union, etc. (they can be expressed easily through temporal logic expressions). These duties take a prominent part in air transportation operations and affect significantly the performance of the crew reserve assignment solutions. A way to introduce in the mathematical formulation of the problem the duties operations requiring the planned immobilization of crew at a given site, is to consider dummy flights to be run within imposed time windows (defined from imperative regulations edited by the national or international air transportation authorities) and whose positions depend on the past state of the crew (total number of rotations, medical state, training, total flight durations). So, the reactive nature of this decision problem, which is a factor of increased difficulty, is also clearly displayed here.

Third, we proceed to choose the pilots who likely to be assigned in reserve activities. Reducing the number of pilots is done here by eliminating those that are not available to be allocated to a reserve activity. Unavailable pilots may be: crew in a scheduled service period (rotation, administrative work or labor, training, etc.), crew in mandatory rest period, in a statutory leave, or in sickness period of crew member. In our study, we proceed at first to eliminate not available candidates. From this idea, it is important to create subsets of crew members who meet qualifications and geographic area. The second step is to compare the characteristics required for the reserve activity in question with the available crew members. This eliminates those whose qualifications are not required for this reserve activity.

A Solution Strategy for the Crew Reserve Assignment Problem

In this section is proposed a practical solution strategy which tackles the complexity of the above CRAP. This new solution approach developed here to address the problem of assigning crew members to reserve activities is a combinatorial problem. The choice of a separate criterion approach, coupled with the set of constraints defined above, allows considering an optimal resolution based on the principle of Dynamic Programming (DP) which is a proven and effective technique for the treatment of combinatorial problems (Nemhauser, 1966). The proposed solution strategy, named *DynCRAP*, considers the main duties appointments as hard constraints for the crew reserve assignment whose solution opens a set of opportunities to cope with lighter duties operations.

Several reasons have led to the adoption of this technique. The first, DP is recognized to be an efficient technique to treat combinatorial problems in a dynamic environment. This technique is appropriate here since the succession of reserve activities provides the problem with a dynamic nature while the separability structure of the constraints insures its applicability. The second is that DP is well adapted to cope with perturbations since it can be implemented recursively, and can provide a feedback solution to deviations from nominal conditions. In addition, the key point for using DP in an efficient way in this particular application seems to be the identification of each reserve activity **r** (real or dummy) with a set of stages of the search process to which is attached the set of states P_r. This choice is judicious since it reduces drastically the number of states at each stage (its upper limit is then equal to "**n**", the number of required crew in the airlines) and, because it is compatible with the on-line calculation of the sets P_r, avoiding data storage and processing difficulties. These characteristics make the method compatible with the use of small computing equipments. The proposed approach seems compatible with very different operational characteristics such as: the composition and size of the fleet, passengers and freight transportation, regular and non-regular operations, regional, domestic and international flights.

Principle and Implementation of the Resolution Scheme

Before applying the resolution scheme of dynamic programming to solve the CRAP, we must define two essential elements: the stages and the states associated with the assignment process. A stage is associated with the assignment of a single pilot to a reserve activity "*r*". Thus, to assign "*k*" pilots to a reserve activity, "*k*" stages should be considered ($k = 1 .. |L_r|$). The reserve activities are arranged in chronological order. To each state is associated an available crew for the considered stage and the number of reserve activities he has done till this point. Recall that the number of reserve activities carried out by a pilot during an assignment period is limited by a well-defined threshold, "*NBRES*". Let I_k^i be the set of states

associated with the pilot "*i*" in stage "*k*" of reserve activity "*r*" and which is defined as:

$$I_k^i = \{(i,0),(i,1),\cdots,(i,NBRES)\}, \ \forall \ i \in P_r \quad (7)$$

where P_r is the set of available crew members associated to the reserve activity "*r*" at stage "*k*". Let E_k be the set of states associated to the available crew member "*i*" at stage "*k*" of reserve activity "*r*":

$$E_k = \bigcup_{i=1}^{P_r} I_k^i \quad (8)$$

where L_r is the number of pilots that can be used for reserve activities during the considered assignment period.

Solving the CRAP using the DP technique leads to the following mathematical formulation. Let K is the number of stages to be considered during the assignment period; N is the number of available pilots; E_k is the set of states associated to stage "*k*"; $hs_k(e_{k-1}, x_k)$ represents the number of additional flight hours generated following the decision x_k that permits to move from state « e_{k-1} » to state « e_k »; $\sum_{k=1}^{K} hs_k(e_{k-1}, x_k)$ represents the sum of the additional flight hours associated to the sequence decisions $x = (x_1, x_2, \cdots x_k)$ that permits to move from the initial state « e_0 » to state « e_k »; and $F_k(e_{k-1}, x_k) = e_k$ represents the transition state.

The solution of this problem leads to find an optimal sequence $x = (x_1, x_2, \cdots x_K)$ that starts from the initial state " e_0 " up to state " e_K " while minimizing the following function:

$$AFF(k, e_k) = MIN \left\{ \sum_{k=1}^{K} hs_k(e_{k-1}, x_k) / x_k \in X_k; e_k = F_k(e_{k-1}, x_k) \quad \forall \ k = 1 \cdots K \right\} \quad (9)$$

Figure 3. Decomposition process of the dynamic programming in stages

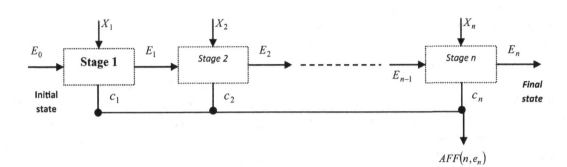

$$AFF(n, e_n)$$

The principle of optimality of dynamic programming (Figure 3), shows that whatever the decision in stage "k" that brings us from state $e_{k-1} \in E_{k-1}$ to state $e_k \in E_k$, the portion of the policy between "e_0" and "e_{k-1}" must be optimal. By applying this principle of optimality, we can calculate step by step $AFF(K, e_K)$ using the following recurrence equation:

$$AFF(k, e_k) = \underset{\{x_k \in X_k / e_k = F_k(e_{k-1}, x_k)\}}{MIN} \left\{ hs_k(e_{k-1}, x_k) + AFF(k-1, e_{k-1}) \right\}$$
(10)

with $AFF(0, e_0) = 0$

Advantage of this Approach and Interpretation of Complexity

The resolution scheme by DP has a drawback related to its time complexity which increases exponentially depending on the number of states considered at each stage. The resolution process will be overwhelmed by a number of states not useful. Thus, reducing the number of available crew members for a stage will lead to reduce this complexity. In this study, we propose to reduce the number of crew members likely to be assigned at every stage. This reduces the number of states to be considered and subsequently reduce the time complexity. Among the key elements associated with the DP, we note the following:

- The total number of paths estimated in the solution of problem has increased in a linear fashion, proportional only to the number of states at each stage. Therefore, the original combinatorial problem has been significantly reduced. The memory space required for the algorithm developed here to remember throughout the resolution process is of order $O(|E_k| \times N)$, where $|E_k|$ is the cardinality of the states in stage "k", N is the number of stages.

- In any stage, the optimal policy is given from one or more states of this stage towards an end point. The estimated computation time to explore the different possible solutions is considerably reduced. It is of order $O(|E_k|^2 \times N)$, where $|E_k|$ is the cardinality of the states in stage "k", N is the number of stages.

NUMERICAL APPLICATIONS

Application to the Case of a Charter Airline

Since airlines can present very different or coordinated operations with other airlines, the above general solution strategy must be customized.

The proposed solution strategy has been adapted to an international medium size charter airline operating on a non regular basis for different customers (travel agencies, customers associations, etc.). This charter airline, like many others, commits herself months ahead to realize unique transportation operations using its own crew or some hired additional crew. To maximize her profit two goals are pursued: the maximization of her revenue through the selling of the maximum number of cost-effective flights, this is the major goal of her marketing department, and the minimization of operations costs through the efficient assignment of the crew reserve on the scheduled flights, this is one of the main goals of the operations department. When the available crew is used near full capacity, which is the more desirable situation, an interaction must take place between these two departments, since it is the solution of the crew assignment problem (duties constraints included) which can tell if it is feasible and profitable for the airline to include additional flights in its planning and consequently, additional reserve activities. The considered charter airline operates from three bases on a medium range basis: Middle East, Europe, and part of North Africa. His permanent fleet is composed of 15 aged 120-150 seats jets with approximately the same operating costs. Since the charter airline sells flights and not seats, no payload concern is necessary and the capacity problem is not acute. Considering the structure of the operated network (passenger's airports and maintenance bases), the average of the flight duration is about two hours and half and since the effective planning horizon is three months, the total number of flights (m) considered at a time in the corresponding assignment problem is about two thousand.

The proposed solution has been applied for solving a problem of *100 pilots*. The assignment period in consideration presents around *400 reserve activities* where to each one is assigned a well defined number of pilots (in average 3 pilots are assigned to a reserve activity). The airline

considered in our study has considered that the minimum workload for each crew member is 60 hours of flying. Different situations (flights plans and a set of constraints) have been considered and the solutions obtained by the proposed method have appeared to be significantly superior to those obtained from lengthy manual procedures. Also, various sensitivity analysis have been conducted through the use of the evaluation function of the crew reserve assignment. Simulations and visual analysis will be used to validate the accuracy of the improved approach. In our case, we have considered two cases in order to evaluate the performance of the proposed solution strategy. The considered cases are: assignment of crew to reserve activities while ensuring at least one reserve activity for a crew member; assignment of crew to reserve activities without constraints. Our algorithm simulates these real scheduled activities by allowing the experts to define a number of practical soft constraints (holiday period, etc.) in order to be able to get the appropriate decisions.

Performance Evaluation

This section describes some characteristics of tests conducted using the solution approach adopted in order to generate dynamically the different optimal crew reserve assignment solutions. We proceed to evaluate the performance of the proposed approach by analyzing the responding time and some various sensitivity analyses that can be conducted through the use of some metrics measure. The solution method has been developed using Visual C++ programming language on a PC computer equipped with a Pentium processor. For the size of problems considered, the response time of the system is about 5 minutes. It presents an acceptable computing time which is compatible with on-line use in an operations management environment. A set of graphical displays have been developed to submit in a more comprehensible way the computed solutions to the operations staff.

Table 1. Computing time (in sec.) related to the size of crew members and the reserve activities

# of crew members	Number of reserve activities			
	50	100	200	400
10	10	23	46	90
20	32	55	99	173
50	89	130	275	410
75	112	179	452	740
100	160	340	631	1057

Efficiency in Computing Time

This section describes the efficiency of the proposed solution approach while considering different problem size. Concerning the response time consumed by the system which is stated in Table 1, it presents a much shorter computing time than with pre-existent Artificial Intelligence or Mathematical Programming methods and this response time is compatible with online use in an operational environment. The solutions obtained by the proposed method have appeared to be significantly superior to those obtained from lengthy manual procedures. Results show for each set of crew members the requested computing time under several sizes of reserve activities or positions (RP).

Analysis of the Solution Quality

Concerning the quality performance of the proposed approach, several experiments were reviewed in order to test and compare this performance while considering several cases. We report also some metrics measure to evaluate the quality of the proposed solution. Two examples are presented here to analyze the quality of the

Figure 4. (a) Distribution of the workload among the crew members after covering the reserve activities; (b) generating supplementary hours; (c) distribution of the reserve activities

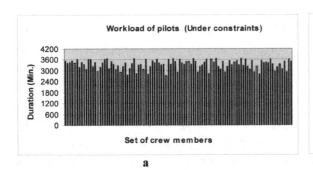

a

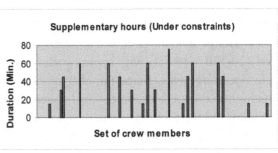

b

c

solution approach developed in our study. In the first example, we consider that each crew must perform at least one reserve activity during an assignment period. In the second example, we eliminate this constraint. These two cases considered here contribute to evaluate two different situations of the problem which affect directly the final quality of the solution.

First Case: Assignment Problem with a Constraint (At Least One Reserve Activity)

The introduction of a constraint concerning the assignment of pilots to at least a reserve activity during an assignment period can generate overtime. Adding a reserve activity to pilot having workload somehow closed to the Minimum Guarantee or who has already got overtime will generate additional overtime. In Figures 4a and 4b, the assignment solution to reserve activities shows that additional time equal to 12 flight hours has been generated. To be noted here that the minimum monthly workload is evaluated to 60 hours.

Figure 4c also presents the distribution of the number of reserve activities among the crew members since it is clear that each crew has been entrusted to carry out at least one reserve activity during an assignment period.

Second Case: Problem without Constraint

In this case, we eliminate the constraint related to the minimum number of reserve activities for each crew member. The elimination of this constraint has the effect of reducing the number of overtime

Figure 5. (a) Distribution of the workload after the assignment of the crew members to reserve activities; (b) generating supplementary hours; (c) distribution of the reserve activities

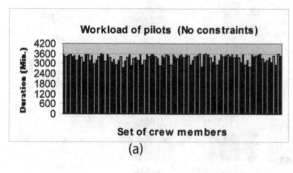

(a)

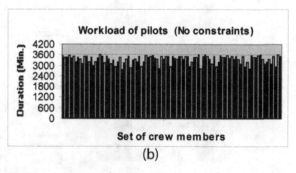

(b)

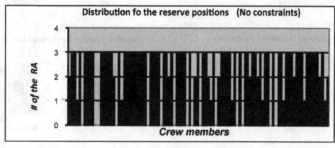

(c)

hours because the pilots have a lower workload becomes priority to perform the reserve activities without exceeding the maximum number of the reserve activities defined for an assignment period. In Figures 5a and 5b, we find that the number of additional hours is greatly reduced in this case.

Figure 5c also affects negatively the distribution of the number of reserve activities among the crew members. It shows unfair distribution of the reserve activities since the primary objective is to minimize the overtime associated to workload distribution among crew members during an assignment period.

Solution Analysis and Comparative Study

In order to achieve the performance evaluation of the DynCRAP, we have compared it with the some intelligence computational tools developed in the literature and which dealt with the CRAP such as: Simulated Annealing optimization (SimRSAR) and Genetic Algorithm (GenRSAR). Moreover, some parameters are introduced, in order to analyze the quality of the generated solutions, such as: the standard deviation of the distribution of the number of reserve activities and the percent-

Figure 6. Analysis statement related to several factors

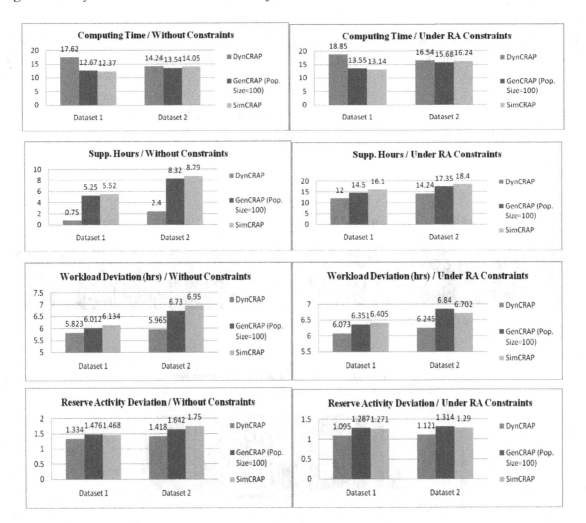

Table 2. Reported results based on the number of pilots without RA constraints

# pilots	DynCRAP				GenCRAP (Pop. Size=100)				SimCRAP			
	CT	SH	SDW	SDRA	CT	SH	SDW	SDRA	CT	SH	SDW	SDRA
Dataset 1	*17.62*	*0.75*	*5.823*	*1.334*	*12.67*	*5.25*	*6.012*	*1.476*	*12.20*	*5.52*	*6.134*	*1.468*
Dataset 2	*14.24*	*2.40*	*5.965*	*1.418*	*13.54*	*8.32*	*6.73*	*1.642*	*14.05*	*8.79*	*6.95*	*1.75*

age assessing the non satisfaction of pilots. This deviation of the distribution of reserve activities is expressed as follows:

$$\sigma_{dist} = \sqrt{\frac{\sum_{i=1}^{N}\left(NBRES - \sum_{k=1}^{K} x_{ik}\right)^2}{N}}$$

where $\sum_{k=1}^{K} x_{ik}$ is the number of reserve activities performed by the crew member "*i*" during the considered assignment period. *NBRES* is the maximal number of reserve activities that each crew member must not exceed during an assignment period and "*N*" is the number of crew members considered.

The results of this comparison, which are based on 2 different datasets, are reported in Tables 2 and 3 and Figure 6. In dataset 1 (resp. dataset 2),

the schedule period presents the lowest (resp. highest) perturbations and absences among the crew members. Also, in both datasets, we have considered 100 pilots to be assigned in order to cover 400 reserve activities. Each reserve activity session is considered as equivalent to 1.5 of flying hours. We focus in our study on four factors such as: Computing Time (CT) expressed in seconds, Supplementary Hours (SH) generated under several conditions, the Standard Deviation of the Workload (SDW), and finally the Standard Deviation of the Reserve Activity (SDRA). Each of these factors is reported under the considered intelligence computational tool.

DynCRAP approach is the best for any dataset since it is based on an optimistic method while the others are of type greedy heuristics. DynCRAP outperforms all the considered methods GenRSAR and SimRSAR for any datasets (Figure 6). The results reported show that the CT factor is the

Table 3. Reported results based on the number of pilots with At least 1 RA

# pilots	DynCRAP				GenCRAP (Pop. Size=100)				SimCRAP			
	CT	SH	SDW	SDRA	CT	SH	SDW	SDRA	CT	SH	SDW	SDRA
Dataset 1	*18.85*	*12*	*6.073*	*1.095*	*13.55*	*14.5*	*6.351*	*1.287*	*13.14*	*16.1*	*6.405*	*1.271*
Dataset 2	*16.54*	*14.24*	*6.245*	*1.121*	*15.68*	*17.35*	*6.840*	*1.314*	*16.24*	*18.40*	*6.702*	*1.290*

highest with DynCRAP method while it is not the case for the remaining three factors. The Supplementary Hours (SH) generated from assigning the crew to the set of reserve activities is the lowest while comparing it to the other methods. It shows also that the SDW and SDRA factors are reduced when using DynCRAP comparing to SimCRAP and GenCRAP methods.

Considering any dataset, we conclude that the SDRA, in the case of no constraint on RA, is higher than the one with constraint on RA. Moreover, the SDW is the lowest in the case of no constraint. While it increases in the case of constraint on RA. We can state also that we can avoid the increasing of SH by relaxing the constraint on RA.

In conclusion, DynCRAP shows promising and competitive performance compared with others computational intelligence tools in terms of solution qualities. The performance of GenCRAP and SimCRAP is comparable since there is no significant difference between them for any datasets. We note here that GenCRAP outperforms SimCRAP for any dataset.

CONCLUSION

In this communication one of the main operational problem faced by airlines, assigning crew members to reserve activities has been dealt with. The proposed approach is based on the technique of Dynamic Programming leading to produce an exact solution in mathematical terms and appears to be quite adapted to the operational context of the airlines and provides, through a comprehensive process for the decision-makers, improved legible and dynamic solutions. It allows airlines to explore the optimal sets of solutions that can drive the cost minimization of the airline and reduce the process complexity. This study constitutes a part of a large cooperation research project for the design of efficient Decision Support Systems for airlines operations with the ultimate objective to

improve their economical performances in their hard competitive environment. A case study considering a charter airline has been performed to validate a customization of the proposed solution approach. In perspectives, a Decision Support System should integrate several important decision problems involving airlines resources aspects (fleet assignment problem, maintenance scheduling problem, crew assignment to flights, yield management, flight programming, ground logistics, catering, etc.) to improve the efficiency of airlines operations, this is another important goal of this common research.

REFERENCES

Aickelin, U., & Dowsland, K. (2003). An indirect genetic algorithm for a nurse scheduling problem. *Computers & Operations Research*, 31.

Ananda, R., Nirup, K., & Gang, Y. (1996). System operations advisor: A real-time decision support system for managing airline operations at United Airlines. *Interfaces*, *26*(2), 50–58. doi:10.1287/inte.26.2.50

Anbil, R., Gelman, E., Patty, B., & Tanga, R. (1991). Recent advances in crew-pairing optimization at American Airlines. *Interfaces*, *21*(1), 62–74. doi:10.1287/inte.21.1.62

Andersson, E., Housos, E., Kohl, N., & Wedelin, D. (1997). Crew pairing optimization. In Yu, G. (Ed.), *Operations research in the airline industry*. Boston, MA: Kluwer Academic.

Antosik, J. L. (1978). Automatic monthly crew assignment: A new approach. In *Proceedings of the 18th Airline Group International Federation of Operational Research Societies Symposium*.

Barnhart, C., Cohn, A., Johnson, E., Klabjan, D., Nemhauser, G., & Vance, P. (2003). Airline crew scheduling. In Hall, R. (Ed.), *Handbook of transportation science* (pp. 493–521). Boston, MA: Kluwer Academic.

Barnhart, C., Johnson, E. L., Nemhauser, G. L., Savelsbergh, M. W. P., & Vance, P. H. (1998). Branch-and-price: Column generation for solving huge integer programs. *Operations Research, 46*, 316–329. doi:10.1287/opre.46.3.316

Barutt, J., & Hull, T. (1990). Airline crew scheduling: Supercomputers and algorithms. *SIAM News, 23*(6).

Brusco, M. J., & Jacobs, L. W. (1993). A simulated annealing approach to the solution of flexible labor scheduling problems. *The Journal of the Operational Research Society, 44*(12), 1191–1200.

Byrne, J. (1988). A preferential bidding system for technical aircrew (Qantas Australia). In *Proceedings of the 28th Airline Group International Federation of Operational Research Societies Symposium*.

Caprara, A., Fischetti, M., Guida, P. L., Toth, P., & Vigo, D. (1999). Solution of large-scale railway crew planning problems: The Italian experience. In Wilson, N. H. M. (Ed.), *Computer-aided transit scheduling* (pp. 1–18). Berlin, Germany: Springer-Verlag.

Caprara, A., Toth, P., & Vigo, D. (1998). Modeling and solving the crew rostering problem. *Operations Research, 46*, 820–830. doi:10.1287/opre.46.6.820

Carraresi, P., Girardi, L., & Nonato, M. (1995). Network models, Lagrangian relaxation and subgradients bundle approach in crew scheduling problems. In *Proceedings of the Sixth International Workshop on Computer-aided Transit Scheduling* (pp. 188-212).

Chanas, S., & Kuchta, D. (1996). Multiobjective programming in optimization of interval objective functions–A generalized approach. *European Journal of Operational Research, 94*(3), 594–598. doi:10.1016/0377-2217(95)00055-0

Crainic, T. G., & Rousseau, J. M. (1987). The column generation principle and the airline crew scheduling problem. *INFOR, 25*, 136–151.

Curtis, S., Smith, B. M., & Wren, A. (1999). Forming bus driver schedules using constraint programming. In *Proceedings of the International Conference on the Practical Applications of Constraint Logic Programming* (pp. 239-254).

Curtis, S., Smith, B. M., & Wren, A. (2000). Constructing driver schedules using iterative repair. In *Proceedings of the 2nd International Conference on the Practical Applications of Constraint Technologies and Logic Programming*.

Dantzig, G. B., & Wolfe, P. (1960). Decomposition principles for linear programming. *Operations Research, 8*, 101–111. doi:10.1287/opre.8.1.101

Desaulniers, G., Desrosiers, J., Lasry, A., & Solomon, M. M. (1999). Crew pairing for a regional carrier. In Wilson, N. H. M. (Ed.), *Computer-aided transit scheduling* (pp. 19–41). Berlin, Germany: Springer-Verlag.

Dillon, J. E., & Kontogiorgis, S. (1999). US Airways optimizes the scheduling of reserve flight crews. *Interfaces, 29*, 123–131. doi:10.1287/inte.29.5.123

Fischetti, M., Lodi, A., Martello, S., & Toth, P. (2001). A polyhedral approach to simplified crew and vehicle scheduling problems. *Management Science, 47*, 833–850. doi:10.1287/mnsc.47.6.833.9810

Forsyth, P., & Wren, A. (1997). *An ant system for bus driver scheduling* (Research Report No. 97.25). Leeds, UK: University of Leeds School of Computer Studies.

Freling, R., Lentink, R. M., & Odijk, M. A. (2001). Scheduling train crews: A case study for the Dutch railways. In Wilson, N. H. M. (Ed.), *Computer-aided scheduling of public transport* (pp. 153–165). Berlin, Germany: Springer-Verlag. doi:10.1007/978-3-642-56423-9_9

Gaballa, A. (1979). Planning callout reserve for aircraft delays. *Interfaces, 9*, 78–86. doi:10.1287/inte.9.2pt2.78

Garey, M. R., & Johnson, D. S. (1979). *Computers and intractability: A guide to the theory of NP-completeness*. New York, NY: Freeman.

Giaferri, C., Hamon, J. P., & Lengline, J. G. (1982). Automatic monthly assignment of medium-haul cabin crew. In. *Proceedings of the Airline Group International Federation of Operational Research Societies Symposium, 22*, 69–95.

Glanert, W. (1984). A timetable approach to the assignment of pilots to rotations - Lufthansa. In *Proceedings of the 24th Airline Group International Federation of Operational Research Societies Symposium.*

Goldberg, D. E. (1989). *Genetic algorithms in search, optimization and machine learning.* Reading, MA: Addison-Wesley.

Haydon, B. (1973). Aircrew standby policies for long haul carriers. In *Proceedings of the 13th Airline Group International Federation of Operational Research Societies Symposium.*

Hoffman, K. L., & Padberg, M. (1993). Solving airline crew scheduling problems by branch-and-cut. *Management Science, 39*, 657–682. doi:10.1287/mnsc.39.6.657

Holloway, S. (1997). *Straight and level: Practical airline economics*. London, UK: Ashgate.

Klabjan, D., Johnson, E. L., Nemhauser, G. L., Gelman, E., & Ramaswamy, S. (2001). Airline crew scheduling with regularity. *Transportation Science, 35*(4), 359–374. doi:10.1287/trsc.35.4.359.10437

Kroon, L., & Fischetti, M. (2001). Crew scheduling for Netherlands railways "Destination-Customer". In Wilson, N. H. M. (Ed.), *Computer-aided scheduling of public transport* (pp. 181–201). Berlin, Germany: Springer-Verlag. doi:10.1007/978-3-642-56423-9_11

Levine, D. (1996). Application of a hybrid genetic algorithm to airline crew-scheduling. *Computers & Operations Research, 23*, 547–558. doi:10.1016/0305-0548(95)00060-7

Liping, Z. (2006). A heuristic method for analyzing driver scheduling problem. *IEEE Transactions on Systems, Man, and Cybernetics. Part A, Systems and Humans, 36*(3).

Majumdar, J., & Bhunia, A. K. (2006). Elitist genetic algorithm approach for assignment problem. *AMO- Advanced Modeling and Optimization, 8*(2), 135-149.

Majumdar, J., & Bhunia, A. K. (2007). Elitist genetic algorithm for assignment problem with imprecise goal. *European Journal of Operational Research, 177*, 684–692. doi:10.1016/j.ejor.2005.11.034

Majumdar, J., & Bhunia, A. K. (2010). Solving airline crew-scheduling problem with imprecise service time using genetic algorithm. *AMO - Advanced Modeling and Optimization, 12*(2).

Marsten, R., & Shepardson, F. (1981). Exact solution of crew scheduling problems using the set partitioning model: Recent successful applications. *Networks, 11*, 165–177. doi:10.1002/net.3230110208

Mingozzi, A., Boschetti, M. A., Ricciardelli, S., & Bianco, L. (1999). A set partitioning approach to the crew scheduling problem. *Operations Research, 47*, 873–888. doi:10.1287/opre.47.6.873

Mitchell, J. S. (1977). Goal programming for scheduling reserve pilot availability. In *Proceedings of the Airline Group International Federation of Operational Research Societies Symposium.*

Mitchell, M. (1996). *An introduction to genetic algorithms.* Cambridge, MA: MIT Press.

Moudani, W., Cosenza, C. A., De-Coligny, M., & Mora-Camino, F. (2001). A bi-criterion approach for the airlines crew rostering problem. In *Proceedings of the First International Conference on Evolutionary Multi-Criterion Optimization,* Zurich, Switzerland.

Moudani, W., & Mora-Camino, F. (2000). A fuzzy solution approach for the rostering planning problem. In *Proceedings of the 9th IEEE International Conference on Fuzzy Systems,* San Antonio, TX.

Moudani, W., & Mora-Camino, F. (2011). Management of bus driver duties using data mining. *International Journal of Applied Metaheuristic Computing, 2*(2).

Nemhauser, G. L. (1966). *Introduction to dynamic programming.* New York, NY: John Wiley & Sons.

Ozdemir, H. T., & Mohan, C. K. (2001). Flight graph genetic algorithm for crew scheduling in airlines. *Information Sciences, 133*, 165–173. doi:10.1016/S0020-0255(01)00083-4

Qi, X., Bard, J., & Yu, G. (2004). Class scheduling for pilot training. *Operations Research, 52*(1), 148–162. doi:10.1287/opre.1030.0076

Rosenberger, J. M. (2001). *Topics in airline operations.* Unpublished doctoral dissertation, Georgia Institute of Technology, Atlanta, GA.

Rubin, J. (1973). A technique for the solution of massive set covering problems, with application to airline crew scheduling. *Transportation Science, 7*, 34–48. doi:10.1287/trsc.7.1.34

Ryan, D. M. (1992). A solution of massive generalized set partitioning problems in aircrew rostering. *The Journal of the Operational Research Society, 43*(5), 459–467.

Ryan, D. M., & Foster, B. A. (1981). An integer programming approach to scheduling. In *Proceedings of the International Conference on Computer Scheduling of Public Transport* (pp. 269-280).

Sarra, D. (1988). The automatic assignment model (Saturn - Alitalia). In *Proceedings of the 28th Airline Group International Federation of Operational Research Societies Symposium.*

Schaefer, A. J., Johnson, E. L., Kleywegt, A. J., & Nemhauser, G. L. (2005). Airline crew scheduling under uncertainty. *Transportation Science, 39*(3), 340–348. doi:10.1287/trsc.1040.0091

Simpson, R. W. (1969). *Scheduling and routing models for airline systems* (Tech. Rep. No. R-68-3). Cambridge, MA: MIT Press.

Smith, B. M., Layfield, C. J., & Wren, A. (2000). A constraint programming preprocessor for a bus driver scheduling system. In *Proceedings of the Workshop on Constraint Programming and Large Scale Discrete Optimization* (pp. 131-150).

Smith, B. M., & Wren, A. (1988). A bus crew scheduling system using a set covering formulation. *Transportation Research, 22A*, 97–108.

Sohoni, M., Bailey, G., Martin, K., Carter, H., & Johnson, E. (2003). Delta optimizes continuing-qualification-training schedules for pilots. *Interfaces, 33*, 57–70. doi:10.1287/inte.33.5.57.19253

Sohoni, M., Johnson, E., & Bailey, T. (2004). Long-range reserve crew manpower planning. *Management Science, 50*(6), 724–739. doi:10.1287/mnsc.1030.0141

Weir, J. D. (2002). *A three phase approach to solving the bidline generation problem with an emphasis on mitigating pilot fatigue through circadian rule enforcement.* Unpublished doctoral dissertation, Georgia Institute of Technology, Atlanta, GA.

Yoshihara, I., & Sengoku, H. (2000). Scheduling bus driver's services based on genetic algorithm. In *Proceedings of the International Conference on Artificial Intelligence in Science and Technology* (pp. 62-67).

Zeghal, F. M., & Minoux, M. (2006). Modeling and solving a crew assignment problem in air transportation. *European Journal of Operational Research, 175*(1), 187–209. doi:10.1016/j.ejor.2004.11.028

This work was previously published in the International Journal of Applied Metaheuristic Computing, Volume 2, Issue 3, edited by Peng-Yeng Yin, pp. 45-68, copyright 2011 by IGI Publishing (an imprint of IGI Global).

Chapter 17

DIMMA–Implemented Metaheuristics for Finding Shortest Hamiltonian Path between Iranian Cities Using Sequential DOE Approach for Parameters Tuning

Masoud Yaghini
Iran University of Science and Technology, Iran

Mohsen Momeni
Iran University of Science and Technology, Iran

Mohammadreza Sarmadi
Iran University of Science and Technology, Iran

ABSTRACT

A Hamiltonian path is a path in an undirected graph, which visits each node exactly once and returns to the starting node. Finding such paths in graphs is the Hamiltonian path problem, which is NP-complete. In this paper, for the first time, a comparative study on metaheuristic algorithms for finding the shortest Hamiltonian path for 1071 Iranian cities is conducted. These are the main cities of Iran based on social-economic characteristics. For solving this problem, four hybrid efficient and effective metaheuristics, consisting of simulated annealing, ant colony optimization, genetic algorithm, and tabu search algorithms, are combined with the local search methods. The algorithms' parameters are tuned by sequential design of experiments (DOE) approach, and the most appropriate values for the parameters are adjusted. To evaluate the proposed algorithms, the standard problems with different sizes are used. The performance of the proposed algorithms is analyzed by the quality of solution and CPU time measures. The results are compared based on efficiency and effectiveness of the algorithms.

DOI: 10.4018/978-1-4666-2145-9.ch017

INTRODUCTION

The shortest Hamiltonian path is a well-known and important combinatorial optimization problem. The goal of Traveling Salesman Problem (TSP) is to find the shortest path that visits each city in a given list exactly once and then returns to the starting city. The distances between n cities are stored in a distance matrix **D** with elements d_{ij} where $i, j = 1, 2, ..., n$ and the diagonal elements d_{ij} are zero (Hahsler & Hornik, 2007). In the 1920's, the mathematician and economist Menger publicized this problem among his colleagues in Vienna. In the 1930's the problem reappeared in the mathematical circles of Princeton (Applegate et al., 1998). In the 1940's, mathematician Flood publicized the name, TSP, within the mathematical community at mass (Lawler et al., 1985). Despite this simple problem statement, solving the TSP is difficult, since it belongs to the class of NP-complete problems (Johnson & Papadimitriou, 1985). The importance of the TSP arises besides from its theoretical appeal from the variety of its applications. Typical applications in operations research include vehicle routing, computer wiring, cutting wallpaper, job sequencing, and job scheduling. Recognizing methods to solve TSP and find the appropriate method is one of the important research areas (Hahsler & Hornik, 2009). In this paper, a comparative study is conducted between Simulated Annealing (SA) and Tabu Search (TS) algorithm as single-solution-based algorithms, Ant Colony Optimization (ACO), and Genetic Algorithm (GA) as population-based-algorithms to find the shortest Hamiltonian path for 1071 Iranian cities.

The proposed algorithms are implemented based on Design and Implementation Methodology for Metaheuristic Algorithms (DIMMA). DIMMA has three sequential phases that each of them has several steps (Figure 1). These phases are as follows: initiation, in which the problem

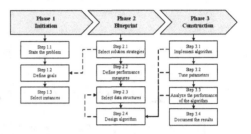

Figure 1. Steps in each phases of DIMMA. (Adapted from [Yaghini & Kazemzadeh], 2010)

in hand must be understood precisely, the goal of designing metaheuristic must be clearly defined, and the instances must be selected. The next phase is blueprint, the important goals of this phase are selecting solution strategy, defining performance measures, and designing algorithm for the solution strategy. The last phase is construction in which implementing the designed algorithm, parameters tuning (parameter setting), analyzing its performance, and finally documentation of results must be done (Yaghini & Kazemzadeh, 2010).

The contribution of our paper could be summarized as follows. In this paper, we proposed four effectiveness and efficient hybrid algorithms for solving TSP problem. High quality of the solution and short computation time show the efficiency and effectiveness of the algorithms. Tuning parameter has been done using DOE based on sequential methodology. Using metaheuristic algorithms with local search to finding the shortest Hamiltonian path between the Iranian cities is the practical contribution of this paper.

The paper is organized as follows. In section two, literature review is conducted. In the third section, the initiation phase of the proposed algorithms is presented. In the fourth section, blueprint, and in the fifth section construction phase of algorithms are discussed. In section six, the shortest Hamiltonian path between Iranian cities is derived by using the proposed algorithms. Conclusion is presented in section seven.

LITERATURE REVIEW

The TSP problem has different types, such as Probabilistic Traveling Salesman Problem (PTSP), Traveling Salesman Problem with Time Windows (TSPTW), Multicommodity Traveling Salesman Problem (MTSP), and Railway Traveling Salesman Problem (RTSP).

The PTSP is a topic of theoretical and practical importance in the study of stochastic network problems. It provides researchers with a modeling framework for exploring the stochastic effects in routing problems. Designing effective algorithms for stochastic routing problems is a difficult task. This is due to the element of uncertainty in the data, which increases the difficulty of finding an optimal solution in a large search space (Jaillet, 1988; Bertsimas, 1988).

The TSPTW involves finding the minimum cost tour in which all cities are visited exactly once within their requested time windows. This problem has a number of important practical applications including scheduling and routing (Savelsbergh, 1985).

The MTSP presents a more general cost structure, allowing for solutions that consider the quality of service to the customers, delivery priorities and delivery risk, among other possible objectives. In the MTSP, the salesman pays the traditional TSP fixed cost for each arc visited, and a variable cost for each of the commodities being transported across the network (Balas, 1979, 1985). The RTSP in which a salesman using the railway network to visit a certain number of stations to carry out business, starting and ending at the same station, and having as goal to minimize the overall time of the journey. It is related to the Generalized Asymmetric Traveling Salesman Problem (GATSP) (Sarubbi et al., 2008; Gutin, 2003; Noon & Bean, 1991). TSP solution methods can be divided into two groups including exact methods and approximate methods. Although exact algorithms exist for solving the TSP, like enumeration method, branch and bound and

dynamic programming but it has been proved that for large-size TSP, it is almost impossible to generate an optimal solution within a reasonable amount of time. Exact methods can solve only small instances to optimality. The approximate methods are classified into metaheuristic and heuristic techniques. Metaheuristics, instead of exact algorithms, are extensively used to solve such problems. For constructing and improving initial solutions in metaheuristics, the heuristic methods such as nearest neighbor, 2-opt, and 3-opt methods can be used.

Attempts to solve the TSP were futile until the mid-1950's when a method for solving the TSP is presented (Dantzig et al., 1954). They showed the effectiveness of their method by solving a 49-city instance. A branch and bound model developed with penalty tour building for solving TSP and transportation routing problems (Radharamanan & Chobi, 1986). A branch and cut algorithm is presented for the MTSP (Sarubbi et al., 2008). An effective procedure is presented that finds lower bounds for the TSP based on the 1-tree using a learning-based Lagrangian relaxation technique (Zamani & Lau, 2010).

Many metaheuristic methods are presented for solving TSP. A new crossover operator is used for the TSP problem, the greedy selection crossover operator, which is designed for path representation and performed at gene level. It can utilize local precedence and global precedence relationship between genes to perform intensive search among solution space to reproduce an improved offspring (Cheng & Mitsuo, 1994). An efficient genetic algorithm is used to solve the TSP with precedence constraints. The key concept of the proposed algorithm is a topological sort, which is defined as an ordering of vertices in a directed graph (Moon et al., 2002).

Bontoux solved the GTSP by using memetic algorithm. In this problem, the set of cities is divided into mutually exclusive clusters. The objective of the GTSP consists in visiting each cluster exactly once in a tour, while minimizing the sum of the routing costs (Moscato & Cotta, 2005; Bontoux et al., 2009). Balaprakash et al. (2010) customized two metaheuristics, an iterated local search algorithm and a Memetic algorithm to solve the PTSP. The three initial solution generators are proposed under a genetic algorithm framework for solving the PTSP. This paper used a set of numerical experiments based on heterogeneous and homogeneous PTSP instances to test the effectiveness and efficiency of the proposed algorithms (Liu, 2010).

Marinakis proposed a new hybrid algorithmic nature inspired approach based on Particle Swarm Optimization (PSO), Greedy Randomized Adaptive Search Procedure (GRASP), and Expanding Neighborhood Search (ENS) strategy for the solution of the PTSP (Marinakis & Marinaki, 2010). The ant colony optimization algorithm usually falls into local optimal solution and cannot select the path with high pheromone concentration quickly in solving the TSP. Tiankun proposed an improved MAX-MIN Ant System (MMAS) algorithm based on pheromone concentration re-initialization to overcome this problem (Tiankun et al., 2009). Zhang presented a paper that is a survey of natural computation for the traveling salesman problem (Zhang, 2009).

The work of Ghoseari and Sarhadi is similar to our paper. They tried to find the shortest Hamiltonian path for 360 Iranian cities using ant colony optimization (Ghoseari & Sarhadi, 2009). The weaknesses of their work are: (1) they did not use the standard problems to evaluate their algorithm, (2) the parameter tuning method was not an appropriate approach, and (3) the solving CPU time (5523 seconds) was very long.

In our paper, the number of cities is increase to 1071 cities, the design of experiments approach is used for parameter tuning, the standard problems with different dimensions are used to evaluate the proposed algorithms, and the solving CPU time is decreased dramatically.

INITIATION PHASE

In initiation phase of DIMMA methodology, the problem in hand must be understood precisely, the goal of designing metaheuristic must be clearly defined, and the instances must be selected. First, we state the TSP problem. Given a list of n cities and their pairwise distances (or costs), the problem is to find a minimum tour that visits each city exactly once. In other words, we want to find a minimum Hamiltonian tour between a set of cities. In the next step, the goal of solving TSP should be defined. Finding a minimum tour in acceptable time is the goal of this case study.

Next, the TSP instances are selected. We choose instances from TSPLIB (Reinelt, 1991) that is one of the popular sources for TSP instances (Table 1). Selected instances must have enough diversity in terms of size. Therefore, we select 12 instances from TSPLIB with different sizes.

Table 1. Problem instances for TSP from TSPLIB

#	Problem name	Number of cities	Global optima		#	Problem name	Number of cities	Global optima
1	ulysses16	16	6859		7	gr202	202	40160
2	ulysses22	22	7013		8	a280	280	2579
3	eil51	51	426		9	pcb442	442	50778
4	berlin52	52	7542		10	gr666	666	294358
5	kroA100	100	21282		11	pr1002	1002	259045
6	rd100	100	7910		12	u1060	1060	224094

BLUEPRINT PHASE

The next phase of DIMMA methodology is blueprint, the important goals of this phase are selecting solution strategy, defining performance measures, and designing algorithm for the solution strategy.

In the first step of blueprint phase, solution strategy should be selected. Because the TSP is an NP-hard problem (Garey & Johnson, 1979), we have selected the metaheuristic approach to solve it. We use simulated annealing, ant colony optimization, genetic algorithm, and tabu search as the solution methods for TSP.

Next, the performance measures for solving TSP with the proposed algorithms will be defined. The defined goal is to find a good solution in a reasonable time. Therefore, performance measures, which are used for TSP problem, are solution quality and CPU time. The global optima are known for the selected instances (Table 1). Therefore, for solution quality measure we use relative gap from global optima. In the following sections, the structure and design of the proposed algorithms are described.

The Proposed Simulated Annealing Algorithm

The simulated annealing algorithm has been introduced in 1983 by Kirkpatrick et al. (1983). This memoryless metaheuristic is base on Metropolis algorithm and prevents remaining in the local optimal with the transition towards low quality solution. The SA algorithm, which is a single-solution-based metaheuristic, has very high ability to be combined with a heuristic algorithm. The components of the proposed SA algorithm are solution representation, initial solution, neighbourhood structure, initial temperature, cooling schedule, and termination criteria.

To represent solution, an array that its length is the number of cities is used. In each element of this array, the number of a specific city is stored

(Figure 2). Finally, the sequence of cities that located in this array is indicating the constructed path. To create the initial solution, a nearest neighbor heuristic method is used. Using this algorithm leads to create appropriate initial solution and speed up the trend of algorithm toward the optimal solution.

One of the main components of the SA algorithm is neighborhood structure. Common methods such as inversion, insertion, and swap are not suitable for large-scale problems (Basu & Ghosh, 2008). Therefore, to solve such problems using local search methods is appropriate. In this way, for every move, several cities selected randomly and replacement with a determined number of nearest cities are reviewed. Ultimately, the best city is selected to substitute. This approach leads to better solution quality and reduce CPU time to reach the best solution (Janaki et al., 1996).

Initial temperature has a large influence on the performance of the proposed algorithm. High initial temperature leads to low performance and running time of algorithm is increased. Low initial temperature does not search solution space well and it cannot escape from local optimal. In parameters tuning, the appropriate amount of the initial temperature is adjusted.

In the proposed algorithm for cooling schedule, the linear and geometric methods are used (Equations 1, and 2).

Linear cooling function: $T_{i+1} = T_i - \beta$

$$(1)$$

Figure 2. Solution representation

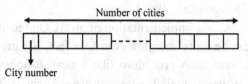

Number of cities

City number

Figure 2. Solution representation

Geometric cooling function: $T_{i+1} = T_i \times \beta$

$$(2)$$

where T_i represents the temperature at iteration i and β is a cooling rate (specified constant value). In parameters tuning section, with considering the quality of solution, an appropriate function is selected for reducing temperature. In the proposed algorithm, the number of iterations without improvement and the final temperature are considered as termination conditions. In the parameter setting, the type of termination condition is determined.

The Proposed Ant Colony Optimization

Dorigo proposed a common framework for the applications and algorithmic variants of a variety of ant algorithms. Algorithms that fit into the Ant Colony Optimization (ACO) metaheuristic framework called in the following ACO algorithms (Dorigo, 1992, Dorigo et al., 1996). The ACO is inspired by the foraging behavior of ant colonies. The ACO is a population-based algorithm, which is a powerful algorithm for solving combinatorial problems. One of the ACO algorithms is Ant Colony System (ACS) in which, pheromone evaporation is interleaved with tour construction (Dorigo & Gambardella, 1997a, 1997b). In this paper, ACS algorithm combined with local search method.

The components of the proposed ACO algorithm are solution representation, initial pheromone value, number of ants, q_0, α, β, the local evaporation coefficient ζ, public evaporation coefficient ρ, and the termination condition (Dorigo et al., 1996).

A first implementation of an ACO algorithm can be quite straightforward. In fact, if a greedy construction procedure like a nearest-neighbor heuristic is available, one can use as a construction graph. The same graph used by the construction procedure, and then it is only necessary to add pheromone trail variables to the construction graph and define the set of artificial ants. The parameter ρ is used to avoid unlimited accumulation of the pheromone trails and it enables the algorithm to forget bad decisions previously taken. The heuristic value is used by the ants' heuristic rule to make probabilistic decisions on how to move on the graph. α and β are two parameters which determine the relative influence of the pheromone trail and the heuristic information. With probability q_0 the ant makes the best possible move as indicated by the learned pheromone trails and the heuristic information (in this case, the ant is exploiting the learned knowledge); while with probability $(1-q_0)$ it performs a biased exploration of the arcs. In the proposed algorithm, only one ant (the best-so-far ant) is allowed to add pheromone after each iteration. In addition to the global pheromone trail updating rule, in this algorithm the ants use a local pheromone update rule that they apply immediately after having crossed an arc. In fact, the best trade-off between solution quality and computation time seems to be obtained when using a small number of ants between two and ten. In the proposed algorithm, after making the tour by the ants in a colony, the ant that has produced the best solution, update the pheromone on the possible arcs and increases the rate of pheromone. Once all the ants have terminated their ant solution construction procedure, a pheromone update phase is started in which pheromone trails are modified. Termination condition in the proposed algorithm is the number of iterations without improvements that is specified in parameters tuning section (Dorigo & Gambardella, 1997a, 1997b).

The Proposed Genetic Algorithm

Genetic algorithms are categorized as metaheuristics and are a particular class of evolutionary algorithms that use techniques inspired by evolutionary biology such as inheritance, mutation, selection, and crossover. Genetic algorithms paired with local search techniques are categorized as Memetic Algorithms (MA) (Moscato & Norman, 1992; Holland, 1975; Hart et al., 2005).

The components of the proposed GA algorithm are representation, population size, parent selection, recombination, replacement, and termination condition. To represent solution, an array that its length is the number of cities is used (Figure 2). To create the initial solution, a nearest neighbor heuristic method is used. The role of population is to hold possible solutions. However, it is verified that the time of convergence towards an absorption state becomes longer as the population size increases.

Tournament selection method considers fitness by basing selection probabilities on relative rather than absolute fitness. In this paper, tournament method with tournament size 2 ($k=2$) is used where each time we pick two members of the population at random (with replacement), and the one with the highest fitness value is selected. If amount of k is large, more of the fitter individuals control selection pressure and algorithm does not achieve appropriate solution. In order that the algorithm can find solutions better than those represented in the current population, it is required that they are transformed by the application of variation operators or search operators, the order crossover as one of recombination method is used in the proposed algorithm. The main steps of order crossover are as follows:

1. Choose an arbitrary part from the first parent.
2. Copy this part to the first child.
3. Copy the numbers that are not in the first part, to the first child, by starting right from cut point of the copied part, then using the order of the second parent, and wrapping around at the end.
4. Analogous for the second child, with parent roles reversed.

The mutation operators alter an individual to another form. In the proposed algorithm instead of simple swap (pick two cities at random and swap their positions), local search is applied. In local search, many swaps are surveyed and the best swap is selected. The role of replacement is to distinguish among individuals based on their solution quality. The elitism method is used for replacement, which guaranty that the good solutions do not remove. The termination condition is number of iteration without improvement that this value is set in parameters tuning section (Holland, 1975; Glover, 1986; Dreo et al., 2006).

The Proposed Tabu Search Algorithm

Tabu Search (TS) was first proposed by Glover in an article published in 1986 (Glover, 1986). The TS is primarily centered on a non-trivial exploration of all the solutions in a neighborhood. The components of the proposed TS algorithm are solution representation, neighborhood structure, tabu list, tabu tenure, aspiration criteria, intensification, diversification, and termination criteria.

One of the main components of the TS algorithm is neighborhood structure. To solve large-scale problems using local search is appropriate. In this way, for every move, several cities selected randomly and replacement with a determined number of nearest cities are reviewed. Ultimately, the best city is selected to substitute. This approach leads to better solution quality and reduce CPU time to reach the best solution (Basu & Ghosh, 2008). Tabu tenure parameter leads to the appropriate search in solution space and prevents algorithm to stop in local optima. In the proposed algorithm, tabu list is defined as two-dimensional array. When the swap is done, the tabu tenure parameter is placed in determined element, and other non-zero elements decrease a unit. It is possible to carry out the reverse of a move, one hope that the solution was sufficiently modified, so that it is improbable, but not impossible, to return to an already visited solution. Nevertheless, if such a situation arises, it is hoped that the tabu list would have changed; therefore the future trajectory of the search would change. The number of tabu moves must remain limited enough. For searching around the current solution, intensification mechanism is use. In the proposed algorithm, according to the number of swaps around the good solutions, neighborhood is searching properly (Glover & Laguna, 1993, 2002; Gendreau, 2003). In the proposed algorithm, the number of iterations without improvement considered as termination conditions. In the parameter setting, termination condition is tuned.

CONSTRUCTION PHASE

In the last phase of DIMMA, the proposed algorithms are implemented. For implementing, the java programming language is used to implement the proposed algorithms. The program was run on a personal computer with core 2 CPU at 2.66 GHz, 4 Gigabytes of Ram, and operating under Microsoft windows vista.

In the next step, the parameters of the proposed algorithms are tuned. We have used the DOE approach and Design-Expert software (Vaughn et al., 2000). In the next step, the performance of the proposed algorithms is analyzed.

Parameters Tuning Using DOE Approach

The parameters of the proposed algorithms are tuned using Design of Experiments (DOE) approach and Design-Expert statistical software (Vaughn et al., 2000). We can define an experiment as a test or series of tests in which purposeful changes are made to the input variables of a process or system so that we may observe and identify the reasons for changes that may be observed in the output response. DOE refers to the process of planning the experiment so that appropriate data that can be analyzed by statistical methods will be collected, resulting in valid and objective conclusions (Birattari, 2009).

The three basic principles of DOE are replication, randomization, and blocking. By replication, we mean a repetition of the basic experiment. By randomization, we mean that both the allocation of the experimental material and the order in which the individual runs or trials of the experiment are to be performed are randomly determined. Blocking is a design technique used to improve the precision with which comparisons among the factors of interest are made. Often blocking is used to reduce or eliminate the variability transmitted from nuisance factors; that is, factors that may influence the experimental response but in which we are not directly interested (Montgomery, 2005). The important parameters in DOE approach are response variable, factor, level, treatment and effect. The response variable is the measured variable of interest. In the analysis of metaheuristics, the typically measures are the solution quality and CPU time (Adenso, 2006). A factor is an independent variable manipulated in an experiment because it is thought to affect one or more of the response variables. The various values at which the factor is set are known as its levels. In metaheuristic performance analysis, the factors include both the metaheuristic tuning parameters and the most important problem characteristics (Birattari, 2002). A treatment is a specific combination of factor levels. The particular treatments will depend on the particular experiment design and on the ranges over which factors are varied. An effect is a change in the response variable due to a change in one or more factors (Ostle, 1963). Design of experiments is a tool that can be used to determine important parameters and interactions between them. Four-stages of DOE consist of screening and diagnosis of important factors, modeling, optimization (tuning), and assessment.

This methodology is called sequential experimentation, which is used to set the parameters in the DOE approach and has been used in this paper for the proposed algorithms (Montgomery, 2005; Ridge, 2007).

The first stage in the sequential experimentation approach is to determine all problem characteristics that affect at least one of the responses of interest. Without sufficient problem characteristics, the response surface models from later in the procedure will not make good predictions of performance on new instances.

The second stage involves a fractional factorial type design, analyzed with ANOVA. Its data are collected following the good practices for experiments with heuristic methods. At this stage, we have a model of each of the responses over the entire design space and this model has been confirmed to be accurate predictors of the actual metaheuristic. It is now possible to use this model for tuning the metaheuristic.

In the third stage, there are several possible optimization goals for tuning. We may wish to achieve a response with a given value (target value, maximum or minimum). Alternatively, we may wish that the response always fell within a given range (relative error less than 10%). More usually, we may wish to optimize several responses because of the heuristic compromise.

The assessment stage has two main aspects to the evaluation of recommended tuning parameter settings. Firstly, we wish to assess how well the recommended parameter settings perform in comparison to alternative settings. Secondly, we wish to determine how robust the recommended settings are when used with other problem characteristics (for more details see Ridge, 2007). These stages are implemented in Design-Expert statistical software.

To adjust the parameters of the proposed algorithms, four standard problems from TSPLIB website has been used (Reinelt, 1991). These problems based on the size and dimensions are classified into two groups. For each of problem groups, two blocks are considered. Problems ulysses22 and eil51 as first group block and problems gr202 and pcb442 as second group block are selected. For each of two groups, parameters tuning has been done separately.

In the proposed algorithms, solution quality and CPU time are considered as the response variables. For the proposed SA algorithm, final value of factors and their levels are shown in Table 2. Each block is considered with 16 treatments and main effects. The relative gap is derived from Equation (3). These parameters are fixed to solve the test problems.

$$relative\ gap = \frac{best\ so\ far\ tour\ length\ -\ optimal\ value}{optimal\ value}$$

(3)

For the proposed ACO algorithm, final value of factors and their levels are shown in Table 3. Each block is considered with 64 treatments and main effects. In the proposed GA algorithm, fac-

Table 2. Level factorial design and final values of parameters for the proposed SA algorithm

Factor	First group			Second group		
	Low	High	Final value	Low	High	Final value
initial temperature	8000	16000	15000	8000	18000	15000
cooling function	Geometric	Linear	Geometric	Geometric	Linear	Geometric
cooling rate	0.01	0.9999	0.9999	0.01	0.9999	0.9999
final temperature	0.01	0.0001	0.0001	0.01	0.0001	0.0001
candidate list rate	0.5	1	0.6	0.01	0.1	0.01

Table 3. Level factorial design and final values of factors for the proposed ACO algorithm

Factor	First group			Second group		
	Low	High	Final Value	Low	High	Final Value
α	5	20	5	2	20	3
β	5	25	5	5	25	10
q_0	0.01	0.5	0.5	0.7	1	0.3
local evaporate	0.01	0.5	0.01	0.08	0.3	0.3
global evaporate	0.01	0.5	0.01	0.08	0.3	0.09
number of ants	10	50	50	15	30	24
candidate list rate	0.01	1	0.8	0.01	0.05	0.05
iteration number	100	500	500	300	500	300
initial pheromone rate	1	3	1	1	2	1.5

Table 4. Level factorial design and final values of factors for the proposed GA algorithm

Factor	First group			Second group		
	Low	High	Final Value	Low	High	Final Value
population size	10000	15000	15000	2000	5000	5000
iteration number	500	1000	900	200	500	400
crossover rate	0.5	0.8	0.75	0.5	0.8	0.75
mutation rate	0.4	0.8	0.8	0.4	0.8	0.4
elitism parameter	0.05	0.2	0.05	0.05	0.2	0.13
candidate list rate	0.1	1	0.74	0.05	0.1	0.1

Table 5. Level factorial design and final values of factors for the proposed TS algorithm

Factor	First group			Second group		
	Low	High	Final Value	Low	High	Final Value
percent of swap	0.3	0.7	0.3	0.2	0.4	0.3
percent of candidate list	0.3	1	0.65	0.05	0.2	0.13
percent of diverse for first city	0.3	0.5	0.5	0.3	0.5	0.5
percent of diverse for second city	0.3	0.5	0.4	0.3	0.5	0.4
tabu tenure	15	35	15	100	200	150
termination condition	8000	30000	30000	1000	5000	5000
diversification condition	1000	5000	3000	1000	3000	2000

tors and their levels are shown in Table 4. Each block is considered with 64 treatments and main effects.

In the proposed TS algorithm, solution quality and CPU time are considered as the response variables. Final value of factors and their levels are shown in Table 5. Each block is considered with 16 treatments and main effects. These parameters are fixed to solve the test problems.

The Performance Analysis of the Proposed Algorithms

To evaluate the proposed algorithms, the standard problems with size of 16 to 1060 cities from TSPLIB website are used (Reinelt, 1991). The global optimal solutions of these problems are available in this website and are used to analyze the performance of the proposed algorithms. According to direct relationship between CPU time and problem size, the selected problems are divided into two groups. The first group includes the problem with less than or equal 100 cities and the second group includes more than 100 cities.

Each of the standard problems run ten times and the best and worst solution, the average of solutions, the average of CPU time, the average of relative gaps, and the relative gap of the best solution are achieved.

The computational results of the proposed algorithms for standard problems are shown in Tables 6 and 7. The averages of relative gaps, CPU time, and the best solution relative gap for the proposed SA algorithm are 10 percent, 22 seconds and 7 percent, respectively. In the ACO algorithm, the average of relative gap is 7 percent and CPU time is 50 seconds. In the proposed GA algorithm, the average of relative gap is 6 percent and CPU time is 190 seconds. In the proposed TS algorithm, the averages of relative gaps, CPU time, and the best solution relative gap are 11 percent, 300 seconds and 10 percent, respectively.

The last step of DIMMA is documentation that can be done by means of graphical tools. After solving the standard problems with the proposed algorithms, the average of relative gap, CPU time and best solution relative gap are compared on Figures 3, 4, and 5. As Figures are shown, when the dimensions of the problem become larger, the relative gap and CPU time are increased. Accord-

Table 6. Comparison of the proposed algorithms in best solution and average of solutions

Problems characteristics			Average of solutions				Best solution			
Problem	Number of cities	Global optima	SA	ACO	GA	TS	SA	ACO	GA	TS
ulysses16	16	6859	6859.00	6859.00	6859.00	6859.00	6859.00	6859.00	6859.00	6859.00
berlin52	52	7542	7914.39	7578.79	7542.00	7814.06	7544.32	7544.32	7542.00	7729.15
kroA100	100	21282	21923.69	21473.44	21740.55	22321.95	21356.24	21321.83	21576.72	22033.80
rd100	100	7910	8277.03	8091.45	8164.77	8485.83	8095.00	7934.84	7944.34	8342.82
a280	280	2579	2985.53	2766.75	2784.10	3008.23	2809.88	2726.45	2725.97	2952.23
gr666	666	294358	339243.80	331268.60	330680.40	353215.20	334523.00	329362.00	329108.00	351403.00
pr1002	1002	259045	310376.77	298397.40	296089.04	308546.29	302859.00	294381.00	293961.73	308532.99
u1060	1060	224094	268281.65	263901.04	255782.68	272768.72	261370.80	260785.00	250727.84	272410.35

Figure 3. The comparison of the proposed algorithms in the best solution relative gap

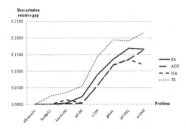

Figure 4. The comparison of the proposed algorithms in the average of relative gap

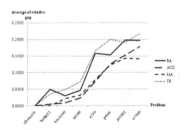

Figure 5. The comparison of the proposed algorithms in CPU time

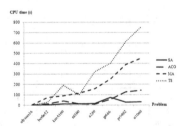

ing to the ten runs for each problem and calculate the total average, comparison is very accurate. Effectiveness and efficiency of the algorithms in solving various problems are evaluated by quality of solution and CPU time measures.

The performance analysis of the proposed algorithms for solving the standard problems is

Table 7. Comparison of the proposed algorithms in relative gap and CPU time

Problem	Best solution relative gap				Average of relative gap				CPU time (s)			
	SA	ACO	GA	TS	SA	ACO	GA	TS	SA	ACO	GA	TS
ulysses16	0.0000	0.0000	0.0000	0.0000	0.0000	0.0000	0.0000	0.0000	0.45	0.98	3.45	0.71
berlin52	0.0003	0.0003	0.0000	0.0248	0.0490	0.0050	0.0000	0.0360	3.73	11.09	72.75	26.89
kroA100	0.0035	0.0018	0.0138	0.0353	0.0300	0.0090	0.0210	0.0480	11.23	40.62	90.39	190.35
rd100	0.0234	0.0031	0.0043	0.0547	0.0460	0.0230	0.0320	0.0720	13.64	11.12	112.84	100.25
a280	0.0895	0.0572	0.0569	0.1447	0.1580	0.0730	0.0790	0.1660	16.2	10.30	157.40	320.40
gr666	0.1364	0.1189	0.1180	0.1938	0.1530	0.1250	0.1230	0.2000	75.55	57.14	250.26	400.84
pr1002	0.1691	0.1364	0.1347	0.1910	0.1980	0.1520	0.1430	0.1910	27.07	127.14	389.74	610.21
u1060	0.1663	0.1637	0.1188	0.2156	0.1970	0.1780	0.1410	0.2170	31.81	143.43	450.97	752.44
Average	**0.0736**	**0.0602**	**0.0558**	**0.1074**	**0.1040**	**0.0710**	**0.0670**	**0.1160**	**22.46**	**50.23**	**190.97**	**300.26**

Table 8. The Hamiltonian path of the proposed algorithms for 1071 Iranian cities

Algorithm	Average of path length (km)	Average of CPU time (s)	Worst path(km)	Best path(km)
SA	32906.9	45.76	33988	32233
ACO	32755.2	205.89	33253	32039
GA	33666.5	400.97	33696	33637
TS	33780.5	531.96	33826	33735

indicating the efficiency and effectiveness of algorithms in various problems.

The GA is better than other algorithms in comparing with the best solution relative gap and the average of relative gap. However, in comparing with CPU time, the SA algorithm is the best.

FINDING THE SHORTEST HAMILTONIAN PATH OF ALL IRANIAN CITIES

After solving test problems and evaluating the proposed algorithms, the shortest Hamiltonian path for 1071 Iranian cities is obtained. According to formal information, based on social-economic characteristics, these are the main cities of Iran. For this purpose, latitude and longitude of the cities are extracted (Iran roads map, 2008). Figure 6 is the two dimensional view of these cities. Then, this information is converted to appropriate for-

mat for the proposed algorithms. Longitude and latitudes are converted to geographical distances on the sphere with radius 6378.388 km and are stored on a matrix with dimensional 1071×1071.

The proposed algorithms run ten times and the averages of path length, the best and worst path (per km), and the time of each run (per seconds) are extracted (Table 8).

According to the number of cities, the proposed algorithms are implemented with parameters obtained for the second group problems in the tuning phase for each algorithm. These algorithms get distance matrix as input and output the best Hamiltonian path between 1071 cities. The relative superiority of the ACO algorithm to find the shortest Hamiltonian path for Iranian cities is evident. However, the SA algorithm has less CPU time. Figure 7 displays the best Hamiltonian path of the proposed ACO algorithm.

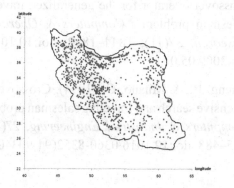

Figure 6. Two dimensional view of 1071 Iranian cities

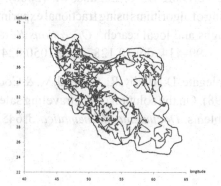

Figure 7. The best Hamiltonian path of the ACO algorithm for Iranian cities

CONCLUSION

In this paper, for the first time, two different types of metaheuristic algorithms, a single solution-based (SA and TS algorithms) and population solution-based algorithms (ACO and GA algorithms) were proposed to find shortest Hamiltonian path for 1071 Iranian cities and the results are compared with each other. The standard problems were used to evaluate the performance of the proposed algorithms. To adjust the best parameter values in the proposed algorithms, DOE method was used to find the most appropriate parameters. The computational results showed slight difference in the quality of solution obtained from the proposed algorithms. The SA algorithm solutions for standard problems are obtained in less time than the other algorithms, but the quality of solutions obtained by GA algorithm has a relative advantage. The quality of solution obtained by the ACO algorithm is better than other algorithms to find the shortest Hamiltonian path for 1071 Iranian cities, but it takes more CPU time.

For future researches, using other parameter tuning approaches such as dynamic and adaptive methods to achieve higher quality solutions and using statistical methods to evaluate the computational results of the metaheuristic algorithms are suggested.

REFERENCES

Adenso-Díaz, B., & Laguna, M. (2006). Fine-tuning of algorithms using fractional experimental designs and local search. *Operations Research*, *54*(1), 99–114. doi:10.1287/opre.1050.0243

Applegate, D., Bixby, R., Chvatal, V., & Cook, W. (1998). On the solution of the traveling salesman problems. *Documenta Mathematica*, *3*, 645–656.

Balaprakash, P., Birattari, M., Stützle, T., & Dorigo, M. (2010). Estimation-based metaheuristics for the probabilistic traveling salesman problem. *Computers & Operations Research*, *37*(11), 1939–1951. doi:10.1016/j.cor.2009.12.005

Balas, E. (1979). Disjunctive programming. *Annals of Discrete Mathematics*, *5*, 3–51. doi:10.1016/S0167-5060(08)70342-X

Balas, E. (1985). Disjunctive programming and a hierarchy of relaxations for discrete optimizations problems. *SIAM Journal on Algebraic and Discrete Methods*, *6*, 466–486. doi:10.1137/0606047

Basu, S., & Ghosh, D. (2008). *A review of the tabu search literature on traveling salesman problems*. Ahmedabad, India: Indian Institute of Management.

Bertsimas, D. (1988). *Probabilistic combinatorial optimization problems*. Unpublished doctoral dissertation, Massachusetts Institute of Technology, Cambridge, MA.

Birattari, M. (2009). *Tuning metaheuristics: A machine learning perspective*. Heidelberg, Germany: Springer-Verlag.

Birattari, M., Stuetzle, T., Paquete, L., & Varrentrapp, K. (2002). A racing algorithm for configuring metaheuristics, In *Proceedings of the Genetic and Evolutionary Computation Conference* (pp. 11-18).

Bontoux, B., Artigues, C., & Feillet, D. (2009). A memetic algorithm with a large neighborhood crossover operator for the generalized traveling salesman problem. *Computers & Operations Research*, *37*(11), 1844–1852. doi:10.1016/j.cor.2009.05.004

Cheng, R., & Mitsuo, G. (1994). Crossover on intensive search and traveling salesman problem. *Computers & Industrial Engineering*, *27*(1-4), 485–488. doi:10.1016/0360-8352(94)90340-9

Dantzig, G. B., Fulkerson, R., & Johnson, S. M. (1954). Solution of a large-scale traveling salesman problem. *Operations Research, 2*, 393–341. doi:10.1287/opre.2.4.393

Dorigo, M. (1992). *Optimization, learning and natural algorithms*. Unpublished doctoral dissertation, Dipartimento di Elettronica, Politecnico di Milano, Italy.

Dorigo, M., & Gambardella, M. (1997a). Ant colonies for the traveling salesman problem. *Bio Systems, 43*(2), 73–81. doi:10.1016/S0303-2647(97)01708-5

Dorigo, M., & Gambardella, M. (1997b). Ant colony system: A cooperative learning approach to the traveling salesman problem. *IEEE Transactions on Evolutionary Computation, 1*(1), 53–66. doi:10.1109/4235.585892

Dorigo, M., Maniezzo, V., & Colorni, A. (1996). An ant system: Optimization by a colony of cooperating agents. *IEEE Transactions on Systems, Man, and Cybernetics Part B, 26*(1), 29–41. doi:10.1109/3477.484436

Dreo, J., Petrowski, A., Siarry, P., & Taillard, E. (2006). *Metaheuristics for hard optimization*. Berlin, Germany: Springer-Verlag.

Eiben, A., & Smith, J. (2003). *Introduction to evolutionary computing*. Berlin, Germany: Springer-Verlag.

Garey, M. R., & Johnson, D. S. (1979). *Computers and intractability*. San Francisco, CA: Freeman and Co.

Gendreau, M. (2003). An introduction to tabu search . In Glover, F., & Kochenberger, G. A. (Eds.), *Handbook of metaheuristics* (pp. 37–54). Boston, MA: Kluwer Academic.

Ghoseari, K., & Sarhadi, H. (2009). Finding the shortest Iranian Hamiltonian tour by using combination of ant colony optimization and local research. *Transportation Research, 2*, 149–161.

Gitashenasi. (2008). *Geographical database of Iranian cities*. Tehran, Iran: Gitashenasi.

Glover, F. (1986). Future paths for integer programming and links to artificial intelligence. *Computers & Operations Research, 13*(5), 533–549. doi:10.1016/0305-0548(86)90048-1

Glover, F., & Laguna, M. (1997). *Tabu search*. Boston, MA: Kluwer Academic.

Glover, F., & Laguna, M. (2002). Tabu search . In Pardalos, P. M., & Resende, M. G. C. (Eds.), *Handbook of applied optimization* (pp. 194–208). New York, NY: Oxford University Press.

Gutin, G. (2003). Traveling salesman and related problems . In Gross, J., & Yellen, J. (Eds.), *Handbook of graph theory*. Boca Raton, FL: CRC Press.

Hahsler, M., & Hornik, K. (2007). Introduction to TSP - Infrastructure for the traveling salesperson problem. *Journal of Statistical Software, 23*(2), 1–21.

Hart, W., Krasnogor, N., & Smith, J. (2005). *Recent advances in memetic algorithms*. Berlin, Germany: Springer-Verlag. doi:10.1007/3-540-32363-5

Holland, H. (1975). *Adaption in natural and artificial systems*. Ann Arbor, MI: University of Michigan Press.

Jaillet, P. (1988). Apriori solution of a travelling salesman problem in which a random subset of the customers are visited. *Operations Research, 36*(6), 929–936. doi:10.1287/opre.36.6.929

Janaki, D., Sreenivas, T., & Ganapathy, K. (1996). Functional and parametric tests of integrated circuits. *Journal of Parallel and Distributed Computing, 37*(2), 207–212.

Johnson, S., & Papadimitriou, H. (1985). *The travelling salesman problem: A guided tour of combinatorial optimization*. New York, NY: John Wiley & Sons.

Kirkpatrick, S., Gelatt, C., & Vecchi, M. (1983). Optimization by simulated annealing. *Science, 220*, 671–680. doi:10.1126/science.220.4598.671

Lawler, E., Lenstra, J., Rinnooy, A., & Shmoys, D. (1985). *The traveling salesman problem: A guided tour of combinatorial optimization*. Chichester, UK: John Wiley & Sons.

Liu, Y. (2010). Different initial solution generators in genetic algorithms for solving the probabilistic traveling salesman problem. *Applied Mathematics and Computation, 216*(7), 125–137. doi:10.1016/j.amc.2010.01.021

Marinakis, Y., & Marinaki, M. (2010). A hybrid multi-swarm particle swarm optimization algorithm for the probabilistic traveling salesman problem. *Computers & Operations Research, 37*(3), 432–442. doi:10.1016/j.cor.2009.03.004

Montgomery, D. (2005). *Design and analysis of experiments* (6th ed.). New York, NY: John Wiley & Sons.

Moon, C., Kim, J., Choi, G., & Seo, Y. (2002). An efficient genetic algorithm for the traveling salesman problem with precedence constraints. *European Journal of Operational Research, 140*(3), 606–617. doi:10.1016/S0377-2217(01)00227-2

Moscato, P., & Cotta, C. (2005). A gentle introduction to memetic algorithms. *Operations Research & Management Science, 57*(2), 105–144.

Moscato, P., & Norman, M. (1992). A memtic approach for the traveling salesman problem implementation of a computational ecology for combinatorial optimization on message-passing systems . In Velaro, M., Onate, E., Jane, M., Larribo, J. L., & Suarez, B. (Eds.), *Parallel computing and transputer applications* (pp. 177–186). Amsterdam, The Netherlands: IOS Press.

Noon, C., & Bean, J. (1991). A lagrangian based approach for the asymmetric generalized traveling salesman problem. *Operations Research, 39*(4), 623–632. doi:10.1287/opre.39.4.623

Ostle, B. (1963). *Statistics in research* (2nd ed.). Ames, IA: Iowa State University Press.

Radharamanan, R., & Choi, L. (1986). A branch and bound algorithm for the travelling salesman and the transportation routing problems. *Computers & Industrial Engineering, 11*(1-4), 236–240. doi:10.1016/0360-8352(86)90085-9

Reinelt, G. (1991). TSPLIB: A traveling salesman problem library. *ORSA Journal on Computing, 3*, 376–384.

Ridge, E. (2007). *Design of experiments for the tuning of optimization algorithms*. Unpublished doctoral dissertation, Department of Computer Science, University of York, York, UK.

Sarubbi, J., Miranda, G., Luna, H., & Mateus, G. (2008). A branch and-cut algorithm for the multi commodity traveling salesman problem. In *Proceedings of the International Conference on Service Operations and Logistics, and Informatics*, Beijing, China.

Savelsbergh, M. (1985). Local search in routing problems with time windows. *Annals of Operations Research, 4*, 285–305. doi:10.1007/BF02022044

Tiankun, L., Chen, W., Zhen, X., & Zhang, Z. (2009). An improvement of the ant colony optimization algorithm for solving travelling salesman problem (TSP). In *Proceedings of the 5th International Conference on Wireless Communications, Networking and Mobile Computing*, Beijing, China.

Vaughn, N., Polnaszek, C., Smith, B., & Helseth, T. (2000). *Design-expert 6 user's guide*. Minneapolis, MN: Stat-Ease, Inc.

Yaghini, M., & Kazemzadeh, M. A. (2010). DIMMA: A design and implementation methodology for metaheuristic algorithms: A perspective from software development. *International Journal of Applied Metaheuristic Computing, 1*(4), 58–75.

Zamani, R., & Lau, K. (2010). Embedding learning capability in Lagrangian relaxation: An application to the travelling salesman problem. *European Journal of Operational Research, 201*(1), 82–88. doi:10.1016/j.ejor.2009.02.008

Zhang, J. (2009). Natural computation for the traveling salesman problem. In *Proceedings of the Second International Conference on Intelligent Computation Technology and Automation* (pp. 366-369).

This work was previously published in the International Journal of Applied Metaheuristic Computing, Volume 2, Issue 2, edited by Peng-Yeng Yin, pp. 74-92, copyright 2011 by IGI Publishing (an imprint of IGI Global).

Compilation of References

Abbass, H. A. (2001a). A monogenous MBO approach to satisfiability. In *Proceedings of the International Conference on Computational Intelligence for Modelling, Control and Automation, CIMCA'2001*, Las Vegas, NV.

Abbass, H. A. (2001b). Marriage in honey-bee optimization (MBO): a haplometrosis polygynous swarming approach. In *Proceedings of the Congress on Evolutionary Computation, CEC2001*, Seoul, Korea (pp. 207-214).

Abraham, A. (2003). Business intelligence from web usage mining. *Journal of Information & Knowledge Management, 2*, 375–390. doi:10.1142/S0219649203000565

Adenso-Diaz, B., Garcia-Carbajal, S., & Lozano, S. (2006). An empirical investigation on parallelization strategies for scatter search. *European Journal of Operational Research, 169*(2), 490–507. doi:10.1016/j.ejor.2004.08.011

Adenso-Díaz, B., & Laguna, M. (2006). Fine-tuning of algorithms using fractional experimental designs and local search. *Operations Research, 54*(1), 99–114. doi:10.1287/opre.1050.0243

Afentakis, P., & Gavish, B. (1986). Optimal Lot-Sizing for complex product structures. *Operations Research, 34*, 237–249. doi:10.1287/opre.34.2.237

Agrawal, R., Mannila, H., Srikant, R., Toivonen, H., & Verkamo, I. A. (1996). Fast discovery of association rules. In Fayyad, U. M., Piatetsky-Shapiro, G., Smyth, P., & Uthurusamy, R. (Eds.), *Advances in knowledge discovery and data mining* (pp. 369–376). Menlo Park, CA: AAAI.

Aha, D. W., & Bankert, R. L. (1996). A comparative evaluation of sequential feature selection algorithms. In Fisher, D., & Lenx, J.-H. (Eds.), *Artificial Intelligence and Statistics*. New York, NY: Springer.

Ahuja, R. K., Magnanti, T. L., & Oril, J. B. (1993). *Theory, algorithms, and application*. Upper Saddle River, NJ: Prentice Hall.

Aickelin, U., & Dowsland, K. (2003). An indirect genetic algorithm for a nurse scheduling problem. *Journal of Computers and Operations Research, 31*.

Akin, E., & Alatas, B. (2009). Multi-objective rule mining using a chaotic particle swarm optimization algorithm. *Knowledge-Based Systems, 22*(6), 455–460. doi:10.1016/j.knosys.2009.06.004

Al-Ani, A. (2005a). Feature subset selection using ant colony optimization. *International Journal of Computational Intelligence, 2*(1), 53–58.

Al-Ani, A. (2005b). Ant colony optimization for feature subset selection. *Transactions on Engineering. Computing and Technology, 4*, 35–38.

Alaykýran, K., Engin, O., & Döyen, A. (2007). Using ant colony optimization to solve the hybrid flow shop scheduling problems. *International Journal of Advanced Manufacturing Technology, 35*(5-6), 541–550. doi:10.1007/s00170-007-1048-2

Alba, E. (2005). *Parallel metaheuristics*. Hoboken, NJ: John Wiley & Sons. doi:10.1002/0471739383

Alvarez, A. M., Lez-Velarde, J. L. G., & De-alba, K. (2005). Scatter search for network design problem. *Annals of Operations Research, 138*, 159–178. doi:10.1007/s10479-005-2451-4

Ananda, R., Nirup, K., & Gang, Y. (1996). System operations advisor: A real-time decision support system for managing airline operations at United Airlines. *Interfaces, 26*(2), 50–58. doi:10.1287/inte.26.2.50

Anbil, R., Gelman, E., Patty, B., & Tanga, R. (1991). Recent advances in crew-pairing optimization at American Airlines. *Interfaces, 21*(1), 62–74. doi:10.1287/inte.21.1.62

Andersson, E., Housos, E., Kohl, N., & Wedelin, D. (1997). Crew pairing optimization. In Yu, G. (Ed.), *Operations research in the airline industry*. Boston, MA: Kluwer Academic.

Anghinolfi, D., & Paolucci, M. (2009). A new discrete particle swarm optimization approach for the single-machine total weighted tardiness scheduling problem with sequence-dependent setup times. *European Journal of Operational Research, 193*(1), 73–85. doi:10.1016/j.ejor.2007.10.044

Antosik, J. L. (1978). Automatic monthly crew assignment: A new approach. In *Proceedings of the 18th Airline Group International Federation of Operational Research Societies Symposium*.

Applegate, D., Bixby, R., Chvatal, V., & Cook, W. (1998). On the solution of the traveling salesman problems. *Documenta Mathematica, 3*, 645–656.

April, J., Better, M., Glover, F., Kelly, J., & Laguna, M. (2006). Enhancing business process management with simulation-optimization. In *Proceedings of the Winter Simulation Conference* (pp. 642-649).

Araya, S., Silva, M., & Weber, R. (2004). A methodology for web usage mining and its application to target group identification. *Fuzzy Sets and Systems*, 139–152. doi:10.1016/j.fss.2004.03.011

Arkin, E., Joneja, D., & Roundy, R. (1989). Computational complexity of uncapacitated multi-echelon, production planning problems. *Operations Research Letters, 8*, 61–66. doi:10.1016/0167-6377(89)90001-1

Audin, M. (2002). *Geometry*. Berlin, Germany: Springer.

Awad, M., & Latifur, K. (2009). *Design and implementation of data mining tools*. Boca Raton, FL: Auerbach Publications. doi:10.1201/9781420045918

Azimi, Z. N. (2005). Hybrid heuristics for examination timetabling problem. *Applied Mathematics and Computation, 163*(2), 705–733. doi:10.1016/j.amc.2003.10.061

Bacardit, J., & Butz, M. V. (2007). Data mining in learning classifier systems: comparing XCS with GAssist. In T. Kovacs, X. Llorà, K. Takadama, P. L. Lanzi, W. Stolzmann, & S. W. Wilson (Eds.), *Proceedings of the International Conference on Learning Classifier Systems* (LNCS 4399, pp. 282-290).

Back, T., & Schwefel, H. P. (1993). An overview of evolutionary algorithms for parameter optimization. *Evolutionary Computation, 1*(1). doi:10.1162/evco.1993.1.1.1

Bagyamani, J., & Thangavel, K. (2010). SIMBIC: Similarity based biclustering of expression data. *Communications in Information Processing and Management, 70*, 437–441. doi:10.1007/978-3-642-12214-9_73

Bahl, H., Ritzman, L., & Gupta, J. (1987). Determining lot sizes and ressource requirements: a review. *Operations Research, 35*(3), 329–345. doi:10.1287/opre.35.3.329

Balakrishnan, A., Magnanti, T. L., & Mirchandani, P. (1997). Annotated bibliographies in combinatorial optimization. In Dell'Amico, M., Maffioli, F., & Martello, S. (Eds.), *Network design* (pp. 311–334). New York, NY: John Wiley & Sons.

Balaprakash, P., Birattari, M., Stützle, T., & Dorigo, M. (2010). Estimation-based metaheuristics for the probabilistic traveling salesman problem. *Computers & Operations Research, 37*(11), 1939–1951. doi:10.1016/j.cor.2009.12.005

Balas, E. (1979). Disjunctive programming. *Annals of Discrete Mathematics, 5*, 3–51. doi:10.1016/S0167-5060(08)70342-X

Balas, E. (1985). Disjunctive programming and a hierarchy of relaxations for discrete optimizations problems. *SIAM Journal on Algebraic and Discrete Methods, 6*, 466–486. doi:10.1137/0606047

Balas, E., & Ho, A. (1980). Set covering algorithms using cutting planes, heuristics and sub-gradient optimization: A computational study. *Mathematical Programming Study, 12*, 37–60.

Ball, M. O., Golden, A., Assad, A., & Bodin, L. D. (1983). Planning for truck fleet size in the presence of a common-carrier option. *Decision Sciences, 14*, 103–120. doi:10.1111/j.1540-5915.1983.tb00172.x

Ball, M. O., Magnanti, T. L., Monma, C. L., & Nemhauser, G. L. (1995). *Network routing: Handbooks in operations research and management science* (*Vol. 8*). Amsterdam, The Netherlands: North Holland.

Ball, M., & Roberts, A. (1985). A graph partitioning approach to airline crew scheduling. *Transportation Science*, *19*(2), 107–126. doi:10.1287/trsc.19.2.107

Barchiesi, D. (2009). Adaptive non-uniforme, hyperellitist evolutionary method for the optimization of plasmonic biosensors. In *Proceedings of the International Conference on Computers & Industrial Engineering* (pp. 542–547).

Barchiesi, D., Macias, D., Belmar-Letellier, L., Labeke, D. V., de la Chapelle, M. L., Toury, T., Kremer, E., et al. (2008b). Plasmonics: Influence of the intermediate (or stick) layer on the efficiency of sensors. Applied Physics. B, Lasers and Optics, 93(1), 177–181

Barchiesi, D., Kremer, E., Mai, V., & Grosges, T. (2008a). A Poincare's approach for plasmonics: The plasmon localization. *Journal of Microscopy*, *229*(3), 525–532. doi:10.1111/j.1365-2818.2008.01938.x

Barnhart, C., Cohn, A., Johnson, E., Klabjan, D., Nemhauser, G., & Vance, P. (2003). Airline crew scheduling. In Hall, R. (Ed.), *Handbook of transportation science* (pp. 493–521). Boston, MA: Kluwer Academic.

Barnhart, C., Johnson, E. L., Nemhauser, G. L., Savelsbergh, M. W. P., & Vance, P. H. (1998). Branch-and-price: Column generation for solving huge integer programs. *Operations Research*, *46*, 316–329. doi:10.1287/opre.46.3.316

Barutt, J., & Hull, T. (1990). Airline crew scheduling: Supercomputers and algorithms. *SIAM News, 23*(6).

Basu, S., & Ghosh, D. (2008). *A review of the tabu search literature on traveling salesman problems*. Ahmedabad, India: Indian Institute of Management.

Bayardo, R. J., Jr., & Agrawal, R. (1999). Mining the most interesting rules. In *Proceedings of the 5th ACM SIGKDD International Conference on Knowledge Discovery and Data Mining* (pp. 145-153). New York, NY: ACM Press.

Baykasoglu, A., Ozbakor, L., & Tapkan, P. (2007). Artificial bee colony algorithm and its application to generalized assignment problem. In Chan, F. T. S., & Tiwari, M. K. (Eds.), *Swarm Intelligence, Focus on Ant and Particle Swarm Optimization* (pp. 113–144). Vienna, Austria: I-Tech Education and Publishing.

Beasley, J. E. (1990). OR-Library: Distributing test problems by electronic mail. *The Journal of the Operational Research Society, 41*, 1069–1072.

Bertsimas, D. (1988). *Probabilistic combinatorial optimization problems*. Unpublished doctoral dissertation, Massachusetts Institute of Technology, Cambridge, MA.

Besbes, W., Loukil, T., & Teghem, J. (2006). Using genetic algorithm in the multiprocessor flow shop to minimize the makespan. In. *Proceedings of the International Conference on Service Systems and Service Management, 2*, 1228–1233. doi:10.1109/ICSSSM.2006.320684

Better, M., Glover, F., & Laguna, M. (2007). Advances in Analytics: Integrating dynamic data mining with simulation optimization. *IBM Journal of Research and Development, 51*(3-4), 477–487. doi:10.1147/rd.513.0477

Bianchessi, N., & Righini, G. (2007). Heuristic algorithms for the vehicle routing problem with simultaneous pick-up and delivery. *Computers & Operations Research, 34*, 578–594. doi:10.1016/j.cor.2005.03.014

Bilgin, B., Ozcan, E., & Korkmaz, E. E. (2007). An experimental study on hyperheuristics and exam timetabling. In E. K. Burke & H. Rudova (Eds.), *Proceedings of the 6th International Conference on Practice and Theory of Automated Timetabling* (LNCS 3867, pp. 394-412).

Birattari, M., Stuetzle, T., Paquete, L., & Varrentrapp, K. (2002). A racing algorithm for configuring metaheuristics, In *Proceedings of the Genetic and Evolutionary Computation Conference* (pp. 11-18).

Birattari, M. (2009). *Tuning metaheuristics: A machine learning perspective*. Heidelberg, Germany: Springer-Verlag.

Bitran, G., & Yanasse, H. (1982). Computational complexity of the capacitated lot size problem. *Management Science, 46*(5), 724–738.

Blackburn, J., & Millen, R. (1982). Improved heuristics for multi-stage requirements planning systems. *Management Science*, *28*(1), 44–56. doi:10.1287/mnsc.28.1.44

Blazewicz, J., Ecker, K. H., Pesch, E., & Schmidt, G. (1996). *Scheduling Computer and Manufacturing Processes*. Berlin, Germany: Springer.

Bleuler, S., Prelic, A., & Zitzler, E. (2004). An EA framework for biclustering of gene expression data. In. *Proceedings of the Congress on Evolutionary Computation*, *1*, 166–173.

Blum, C., & Sampels, M. (2002). Ant colony optimization for FOP shop scheduling: a case study on different pheromone representations. In *Proceedings of the 2002 Congress on Evolutionary Computation (CEC'02)* (Vol. 2, pp. 1558-1563). Los Alamitos, CA: IEEE Computer Society Press.

Blum, C., Puchinger, J., Raidl, G. R., & Roli, A. (2011). Hybrid metaheuristics in combinatorial optimization: A survey. *Applied Soft Computing*, *11*, 4135–4151. doi:10.1016/j.asoc.2011.02.032

Bodin, L. D., Golden, B. L., Assad, A. A., & Ball, M. O. (1983). Routing and scheduling of Vehicles and crews. The state of the Art. *Computers & Operations Research*, *10*, 69–211.

Bolduc, M. C., Renaud, J., & Boctor, F. F. (2007). A heuristic for the routing and carrier selection problem. Short communication. *European Journal of Operational Research*, *183*, 926–932. doi:10.1016/j.ejor.2006.10.013

Bolduc, M. C., Renaud, J., Boctor, F. F., & Laporte, G. (2008). A perturbation metaheuristic for the vehicle routing problem with private fleet and common carriers. *The Journal of the Operational Research Society*, *59*, 776–787. doi:10.1057/palgrave.jors.2602390

Bontoux, B., Artigues, C., & Feillet, D. (2009). A memetic algorithm with a large neighborhood crossover operator for the generalized traveling salesman problem. *Computers & Operations Research*, *37*(11), 1844–1852. doi:10.1016/j.cor.2009.05.004

Bookbinder, J., & Koch, L. (1990). Production planning for mixed assembly/arborescent systems. *Journal of Operations Management*, *9*, 7–23. doi:10.1016/0272-6963(90)90143-2

Boschetti, M., Maniezzo, V., Roffilli, M., & Rohler, A. B. (2009). Matheuristics: Optimization, simulation and control. In *Proceedings of the 6th International Workshop on Hybrid Metaheuristics* (pp. 171-177).

Bosman, P. A. N., & Thierens, D. (2002). Multi-objective optimization with diversity preserving mixture-based iterated density estimation evolutionary algorithms. *International Journal of Approximate Reasoning*, *31*, 259–289. doi:10.1016/S0888-613X(02)00090-7

Bowman, V. J., & Glover, F. (1972). A note on zero-one integer and concave programming. *Operations Research*, *20*(1), 182–183. doi:10.1287/opre.20.1.182

Bozejko, W., & Wodecki, M. (2008). Parallel scatter search algorithm for the flow shop sequencing problem. In R. Wyrzykowski, J. Dongarra, K. Karczewski, & J. Wasniewski (Eds.), *Proceedings of the 7th International Conference on Parallel Processing and Applied Mathematics* (LNCS 4967, pp. 180-188).

Breedam, A. V. (1995). Vehicle routing: Bridging the gap between theory and practice. *Belgian Journal of Operations Research. Statistics and Computer Science*, *35*, 63–80.

Brusco, M. J., & Jacobs, L. W. (1993). A simulated annealing approach to the solution of flexible labor scheduling problems. *The Journal of the Operational Research Society*, *44*(12), 1191–1200.

Burke, E. K., Eckersley, A. J., McCollum, B., Petrovic, S., & Qu, R. (Eds.). (2003, August 13-16). Similarity measures for exam timetabling problems. In *Proceedings of the 1st Multidisciplinary Conference on Scheduling: Theory and Applications*, Nottingham, UK (pp. 120-136).

Burke, E. K., Elliman, D. G., Ford, P. H., & Weare, R. F. (1996). Examination timetabling in British universities: A survey. In E. K. Burke & P. Ross (Eds.), *Proceedings of the 1st International Conference on Practice and Theory of Automated Timetabling* (LNCS 1153, pp. 76-90).

Burke, E. K., Bykov, Y., Newall, J., & Petrovic, S. (2004). A time-predefined local search approach to exam timetabling problems. *IIE Transactions on Operations Engineering*, *36*(6), 509–528.

Byrne, J. (1988). A Preferential bidding system for technical aircrew (QANTAS AUSTRALIA). In *Proceedings of the 28th AGIFORS Symposium*, Cape Cod, MA.

Campos, V., Glover, F., Laguna, M., & Martí, R. (2001). An experimental evaluation of a scatter search for the linear ordering problem. *Journal of Global Optimization*, *21*, 397–414. doi:10.1023/A:1012793906010

Cantu-Paz, E. (2004). Feature subset selection, class separability, and genetic algorithms. In *Proceedings of the Genetic and Evolutionary Computation Conference* (pp. 959-970).

Cantu-Paz, E., Newsam, S., & Kamath, C. (2004). Feature selection in scientific application. In *Proceedings of the 2004 ACM SIGKDD International Conference on Knowledge Discovery and Data Mining* (pp. 788-793).

Caprara, A., Fischetti, M., Guida, P. L., Toth, P., & Vigo, D. (1999). Solution of large-scale railway crew planning problems: The Italian experience. In Wilson, N. H. M. (Ed.), *Computer-aided transit scheduling* (pp. 1–18). Berlin, Germany: Springer-Verlag.

Caprara, A., Fischetti, M., Guida, P. L., Toth, P., & Vigo, D. (1999b). Solution of large-scale railway crew planning problems: The Italian experience. In Wilson, N. (Ed.), *Computer-aided transit scheduling* (pp. 1–18). Berlin, Germany: Springer-Verlag.

Caprara, A., Fischetti, M., & Toth, P. (1999a). A heuristic algorithm for the set covering problem. *Operations Research*, *47*, 730–743. doi:10.1287/opre.47.5.730

Caprara, A., Toth, P., & Vigo, D. (1998). Modeling and solving the crew rostering problem. *Operations Research*, *46*, 820–830. doi:10.1287/opre.46.6.820

Cardoso, M. E., Salcedo, R. L., & de Azevedo, S. E. (1996). The simplex-simulated annealing approach to continuous non-linear optimization. *Computers & Chemical Engineering*, *20*(9), 1065. doi:10.1016/0098-1354(95)00221-9

Cardoso, M. E., Salcedo, R. L., Feyo de Azevedo, S., & Barbosa, D. (1997). A simulated annealing approach to the solution of MINLP problems. *Computers & Chemical Engineering*, *21*(12), 1349–1364. doi:10.1016/S0098-1354(97)00015-X

Carlier, J., & Néron, E. (2000). An exact method for solving the multi-processor flow-shop. *RAIRO: Recherche Operationnelle*, *34*(1), 1–25. doi:10.1051/ro:2000103

Carlisle, C., & Dozier, G. (2001). An off-the-shelf PSO. In *Proceedings of the Particle Swarm Optimization Workshop* (pp. 1-6).

Carraresi, P., Girardi, L., & Nonato, M. (1995). Network models, lagrangean relaxation and subgradients bundle approach in crew scheduling problems. In *Proceedings of the Sixth International Workshop on Computer-Aided Transit* (pp. 188-212).

Carter, C., & Ellram, L. (1998). Reverse logistics: A review of the literature and framework for future investigation. *Journal of Business Logistics*, *19*, 85–102.

Carter, M., & Johnson, D. G. (2001). Extended clique initialization in examination timetabling. *The Journal of the Operational Research Society*, *52*(5), 538–544. doi:10.1057/palgrave.jors.2601115

Carvalho, A. B., & Pozo, A. (2008). Non-ordered data mining rules through multi-objective particle swarm optimization: Dealing with numeric and discrete attributes. In *Proceedings of the International Conference on Hybrid Intelligent Systems* (pp. 495-500). Washington, DC: IEEE Computer Society.

Caserta, M., & Voß, S. (2009). Metaheuristics: Intelligent problem solving. In Maniezzo, V., Stützle, T., & Voß, S. (Eds.), *Matheuristics: Hybridizing metaheuristics and mathematical programming* (pp. 1–38). Berlin, Germany: Springer-Verlag.

Cendrillon, R., & Moonen, M. (2005). Iterative spectrum balancing for digital subscriber lines. In *Proceedings of the IEEE International Conference on Communications*.

Cendrillon, R., Moonen, M., Verlinden, J., Bostoen, T., & Yu, W. (2004, June). Optimal multi-user spectrum management for digital subscriber lines. In *Proceedings of the IEEE International Conference on Communications* (Vol. 1, pp. 1-5).

Cendrillon, R., Yu, W., Moonen, M., Verlinden, J., & Bostoen, T. (2006). Optimal multi-user spectrum management for digital subscriber lines. *IEEE Transactions on Communications*, *54*(5), 922–933. doi:10.1109/TCOMM.2006.873096

Chafekar, D., Xuan, J., & Rasheed, K. (2003). Constrained multi-objective optimization using steady state genetic algorithms. In E. Cantú-Paz, J. A. Foster, K. Deb, L. D. Davis, R. Roy, U.-M. O'Reilly et al. (Eds.), *Proceedings of the International Conference on Genetic and Evolutionary Computation, Part I* (LNCS 2723, p. 201).

Chakraborty, A., & Maka, H. (2005). Biclustering of gene expression data using genetic algorithm. In *Proceedings of the IEEE Symposium on Computational Intelligence in Bioinformatics and Computational Biology* (pp. 1-8).

Chanas, S., & Kuchta, D. (1996). Multiobjective programming in optimization of interval objective functions–A generalized approach. *European Journal of Operational Research, 94*(3), 594–598. doi:10.1016/0377-2217(95)00055-0

Chen, G., Liu, H., Yu, L., Wei, Q., & Zhang, X. (2006). A new approach to classification based on association rule mining. *Decision Support Systems, 42*, 674–689. doi:10.1016/j.dss.2005.03.005

Cheng, R., & Mitsuo, G. (1994). Crossover on intensive search and traveling salesman problem. *Computers & Industrial Engineering, 27*(1-4), 485–488. doi:10.1016/0360-8352(94)90340-9

Chen, J. F., & Wu, T. H. (2006). Vehicle routing problem with simultaneous deliveries and pickups. *The Journal of the Operational Research Society, 57*, 579–587. doi:10.1057/palgrave.jors.2602028

Chen, S. C., Lin, S. W., & Chou, S. Y. (2010). Enhancing the classification accuracy by scatter-search-based ensemble approach. *Applied Soft Computing, 11*(1).

Cheong, C. Y., Tan, K. C., & Veeravalli, B. (2009). A multi-objective evolutionary algorithm for examination timetabling. *Journal of Scheduling, 12*, 121–146. doi:10.1007/s10951-008-0085-5

Choi, E., & Tcha, D. W. (2007). A column generation approach to the heterogeneous fleet vehicle routing problem. *Computers & Operations Research, 34*, 2080–2095. doi:10.1016/j.cor.2005.08.002

Christofides, N., & Eilon, S. (1969). An algorithm for the vehicle dispatching problem. *Operations Research, 20*, 309–318. doi:10.1057/jors.1969.75

Christofides, N., Mingozzi, A., & Toth, P. (1979). The vehicle routing problem. In Christofides, N., Mingozzi, A., Toth, P., & Sandi, C. (Eds.), *Combinatorial optimization* (pp. 315–338). Chichester, UK: John Wiley & Sons.

Chu, C. W. (2005). A heuristic algorithm for the truckload and less-than-truckload problem. *European Journal of Operational Research, 165*, 657–667. doi:10.1016/j.ejor.2003.08.067

Cioffi, J., & Mohseni, M. (2004, March). Dynamic spectrum management - A methodology for providing significantly higher broadband capacity to the users. In *Proceedings of the 15th International Symposium on Services and Local Access*, Edinburgh, Scotland.

Cioffi, J., Jagannathan, S., et al. (2006, March). *Full Binder Level 3 DSM Capacity and Vectored DSL Reinforcement* (ANSI NIPP-NAI Contribution 2006-041). Las Vegas, NV: ANSI NIPP-NAI.

Cioffi, J., Jagannathan, S., Mohseni, M., & Ginis, G. (2007). CuPON: the Copper alternative to PON 100 Gb/s DSL networks. *IEEE Communications Magazine, 45*(6), 132–139. doi:10.1109/MCOM.2007.374437

Clark, A., & Scarf, H. (1960). Optimal policies for a multi-echelon inventory problem. *Management Science, 6*, 475–490. doi:10.1287/mnsc.6.4.475

Clarke, G., & Wright, J. W. (1964). Scheduling of vehicles from a central depot to a number of delivery points. *Operations Research, 12*, 568–581. doi:10.1287/opre.12.4.568

Clement, R., & Wren, A. (1995). Greedy genetic algorithms, optimizing mutations and bus driver scheduling. In Wilson, N. (Ed.), *Computer-aided transit scheduling* (pp. 213–235). Berlin, Germany: Springer-Verlag.

Clerc, M. (2009). *A method to improve standard PSO*. Retrieved January 29, 2010, from http://clerc.maurice.free.fr/pso/Design_efficient_PSO.pdf

Clerc, M. (2004). Discrete Particle Swarm Optimization, illustrated by the Traveling Salesman Problem. In *New Optimization Techniques in Engineering*. Berlin, Germany: Springer-Verlag.

Coello, C. A., & Lechuga, M. (2002). MOPSO: A proposal for multiple–objective particle swarm optimization. In *Proceedings of the 9th IEEE World Congress on Computational Intelligence*, Honolulu, HI (pp. 1051-1056). Washington, DC: IEEE Computer Society.

Coello, C. A., Pulido, G. T., & Lechuga, M. S. (2004). Handling multiple objectives with particle swarm optimization. *IEEE Transactions on Evolutionary Computation*, 8(3), 256–279. doi:10.1109/TEVC.2004.826067

Coenen, F., Leng, P., & Zhang, L. (2005). Threshold tuning for improved classification association rule mining. In T. B. Ho, D. Cheung, & H. Liu (Eds.), *Proceedings of the 9th Pacific-Asia Conference on Advances in Knowledge Discovery and Data Mining* (LNCS 3518, pp. 216-225).

Coenen, F., & Leng, P. (2007). The effect of threshold values on association rule based classification accuracy. *Data & Knowledge Engineering*, 345–689. doi:10.1016/j.datak.2006.02.005

Colorni, A., Dorigo, M., & Maniezzo, V. (1991). Distributed optimisation by ant colonies. In *Proceedings of ECAL91: First European Conference on Artificial Life* (pp. 134-142).

Comelli, M., Gourgand, M., & Lemoine, D. (2008). A review of tactical planning models. *Journal of Systems Science and Systems Engineering*, 18(2), 204–229. doi:10.1007/s11518-008-5076-8

Corberán, A., Fernández, E., Laguna, M., & Martí, R. (2002). Heuristic solutions to the problem of routing school buses with multiple objectives. *The Journal of the Operational Research Society*, 53, 427–435. doi:10.1057/palgrave.jors.2601324

Cordova, H., & van Biesen, L. (2010). *Using simplex bounded search for Optimum Spectrum Balancing in multiuser xDSL systems. Internal Report*. VUB.

Côté, P., Wong, T., & Sabourin, R. (2005). Application of a hybrid multi-objective evolutionary algorithm to the uncapacitated exam proximity problem. In E. Burke & M. Trick (Eds.), *Proceedings of the 5th International Conference on the Practice and Theory of Automated Timetabling* (LNCS 3616, pp. 294-312).

Crainic, T. G., & Gendreau, M. (2007). A scatter search heuristic for the fixed-charge capacitated network design problem. *Metaheuristics, Operations Research/Computer Science Interfaces Series, Part I*, 39, 25-40.

Crainic, T. G. (2000). Service network design in freight transportation. *European Journal of Operational Research*, 122, 272–288. doi:10.1016/S0377-2217(99)00233-7

Crainic, T. G., Dejax, P., & Delorme, L. (1989). Models for multimode location problem with interdepot balancing requirement. *Annals of Operations Research*, 18, 277–302. doi:10.1007/BF02097809

Crainic, T. G., & Gendreau, M. (2002). Cooperative parallel tabu search for capacitated network design. *Journal of Heuristics*, 8, 601–627. doi:10.1023/A:1020325926188

Crainic, T. G., Gendreau, M., & Farvolden, J. M. (2000). A Simplex-based tabu search method for capacitated network design. *INFORMS Journal on Computing*, 12, 223–236. doi:10.1287/ijoc.12.3.223.12638

Crainic, T. G., Gendreau, M., & Hernu, G. (2004). A slope scaling/Lagrangian perturbation heuristic with long-term memory for multicommodity capacitated fixed-charge network design. *Journal of Heuristics*, 10, 525–545. doi:10.1023/B:HEUR.0000045323.83583.bd

Crainic, T. G., Li, Y., & Toulouse, M. (2006). A first multilevel cooperative algorithm for capacitated multicommodity network design. *Computers & Operations Research*, 33, 2602–2622. doi:10.1016/j.cor.2005.07.015

Crainic, T. G., & Rousseau, J. M. (1987). The column generation principle and the airline crew scheduling problem. *INFOR*, 25, 136–151.

Crispim, J., & Brandão, J. (2005). Metaheuristics applied to mixed and simultaneous extensions of vehicle routing problems with backhauls. *The Journal of the Operational Research Society*, 56, 1296–1302. doi:10.1057/palgrave.jors.2601935

Curtis, S., Smith, B. M., & Wren, A. (1999). Forming bus driver schedules using constraint programming. In *Proceedings of the International Conference on the Practical Applications of Constraint Logic Programming* (pp. 239-254).

Curtis, S., Smith, B. M., & Wren, A. (2000). Constructing driver schedules using iterative repair. In *Proceedings of the 2ⁿᵈ International Conference on the Practical Applications of Constraint Technologies and Logic Programming.*

Dantzig, G. B., Fulkerson, R., & Johnson, S. M. (1954). Solution of a large-scale traveling salesman problem. *Operations Research, 2*, 393–341. doi:10.1287/opre.2.4.393

Dantzig, G. B., & Wolfe, P. (1960). Decomposition principles for linear programming. *Operations Research, 8*(1), 101–111. doi:10.1287/opre.8.1.101

Das, C., Maji, P., & Chattopadhyay, S. (2008). A novel biclustering algorithm for discovering value-coherent overlapping σ-Biclusters. *Advanced Computing and Communications*, 148-156.

Das, M., Roy, R., Dehuri, S., & Cho, S.-B. (2011). A new approach to associative classification based on binary multi-objective particle swarm optimization. *International Journal of Applied Metaheuristic Computing, 2*(2), 51–73. doi:doi:10.4018/jamc.2011040103

Davy, S., Barchiesi, D., Spajer, M., & Courjon, D. (1999). Spectroscopic study of resonant dielectric structures in near–field. *The European Physical Journal Applied Physics, 5*, 277–281. doi:10.1051/epjap:1999140

de la Iglesia, B., Reynolds, A., & Rayward-Smith, V. J. (2005). Developments on a multi-objective metaheuristic (MOMH) algorithm for finding interesting sets of classification rules. In *Proceedings of the Conference on Evolutionary Multi Criterion Optimization* (pp. 826-840).

de la Iglesia, B., Richards, G., Philpott, M. S., & Rayward-Smith, V. J. (2006). The application and effectiveness of a multi-objective metaheuristic algorithm for partial classification. *European Journal of Operational Research, 169*, 898–917. doi:10.1016/j.ejor.2004.08.025

Deb, K. (1995). *Optimization for engineering design: algorithms and examples.* New Delhi, India: Prentice Hall.

Deb, K. (2001). *Multiobjective optimization using evolutionary algorithms.* Chichester, UK: John Wiley & Sons.

Deb, K., Pratap, A., Agarwal, S., & Meyarivan, T. (2002). A fast and elitist multi-objective genetic algorithm: NSGA-II. *IEEE Transactions on Evolutionary Computation, 6*(2), 181–197. doi:10.1109/4235.996017

Dehuri, S., & Cho, S. B. (2009). Multi-criterion Pareto based particle swarm optimized polynomial neural network for classification: a review and state-of-the-art. *Journal of Computer Science Review, 3*, 19–40. doi:doi:10.1016/j.cosrev.2008.11.002

Dellaert, N., & Jeunet, J. (2000). Solving large unconstrained multilevel Lot-Sizing problems using a hybrid genetic algorithm. *International Journal of Production Research, 38*(5), 1083–1099. doi:10.1080/002075400189031

Dellaert, N., & Jeunet, J. (2002). Randomized multi-level Lot-Sizing heuristics for general product structures. *European Journal of Operational Research, 148*(1), 211–228. doi:10.1016/S0377-2217(02)00403-4

Dellaert, N., Jeunet, J., & Jonard, N. (2000). A genetic algorithm to solve the general multi-level lotsizing problem with time-varying costs. *International Journal of Production Economics, 68*, 241–257. doi:10.1016/S0925-5273(00)00084-0

Demsar, J. (2006). Statistical comparisons of classifiers over multiple datasets. *Journal of Machine Learning Research, 7*, 1–30.

Desaulniers, G., Desrosiers, J., Dumas, Y., Marc, S., Rioux, B., Solomon, M. M., & Soumis, F. (1997). Crew pairing at Air France. *European Journal of Operational Research, 97*(2), 245–259. doi:10.1016/S0377-2217(96)00195-6

Desaulniers, G., Desrosiers, J., Lasry, A., & Solomon, M. M. (1999). Crew pairing for a regional carrier. In Wilson, N. (Ed.), *Computer-aided transit scheduling* (pp. 19–41). Berlin, Germany: Springer-Verlag.

Desrochers, M., Gilbert, J., Sauve, M., & Soumis, F. (1992). CREW-OPT: Subproblem modeling in a column generation approach to urban crew scheduling. In Wilson, N. (Ed.), *Computer-aided transit scheduling.* Berlin, Germany: Springer-Verlag.

Desrochers, M., & Soumis, F. (1989). A column generation approach to the urban transit crew scheduling problem. *Transportation Science, 23*(1), 1–13. doi:10.1287/trsc.23.1.1

Desrochers, M., & Verhoog, T. W. (1991). A new heuristic for the fleet size and mix vehicle routing problem. *Computers & Operations Research, 18*(3), 263–274. doi:10.1016/0305-0548(91)90028-P

Dethloff, J. (2001). Vehicle routing and reverse logistics: The vehicle routing problem with simultaneous delivery and pick-up. *OR-Spektrum, 23*, 79–96. doi:10.1007/PL00013346

Diaby, M., & Ramesh, R. (1995). The Distribution Problem with Carrier Service: A Dual Based Penalty Approach. *ORSA Journal on Computing, 7*, 24–35.

Dillon, J. E., & Kontogiorgis, S. (1999). US Airways optimizes the scheduling of reserve flight crews. *Interfaces, 29*, 123–131. doi:10.1287/inte.29.5.123

Dorigo, M. (1992). *Optimization, learning and natural algorithms*. Unpublished doctoral dissertation, Dipartimento di Elettronica, Politecnico di Milano, Italy.

Dorigo, M., Maniezzo, V., & Colorni, A. (1996). Ant system: Optimization by a colony of cooperating agents. *IEEE Transactions on Systems, Man, and Cybernetics – Part B, 26*(1), 29-41.

Dorigo, M., & Gambardella, L. (1997). Ant Colony System: A Cooperative Learning Approach to the Traveling Salesman Problem. *IEEE Transactions on Evolutionary Computation, 1*(1), 53–66. doi:10.1109/4235.585892

Dorigo, M., & Gambardella, M. (1997a). Ant colonies for the traveling salesman problem. *Bio Systems, 43*(2), 73–81. doi:10.1016/S0303-2647(97)01708-5

Dorigo, M., & Gambardella, M. (1997b). Ant colony system: A cooperative learning approach to the traveling salesman problem. *IEEE Transactions on Evolutionary Computation, 1*(1), 53–66. doi:10.1109/4235.585892

Dorigo, M., Maniezzo, V., & Colorni, A. (1996). An ant system: Optimization by a colony of cooperating agents. *IEEE Transactions on Systems, Man, and Cybernetics Part B, 26*(1), 29–41. doi:10.1109/3477.484436

Dorigo, M., & Stützle, T. (2004). *Ant Colony Optimization*. Cambridge, MA: MIT Press.

Doumpos, M., & Pasiouras, F. (2005). Developing and testing models for replicating credit ratings: A multicriteria approach. *Computational Economics, 25*, 327–341. doi:10.1007/s10614-005-6412-4

Dreo, J., Petrowski, A., Siarry, P., & Taillard, E. (2006). *Metaheuristics for hard optimization*. Berlin, Germany: Springer-Verlag.

Drexl, A., & Kimms, A. (1997). Lot sizing and scheduling - Survey and extensions. *European Journal of Operational Research, 99*, 221–235. doi:10.1016/S0377-2217(97)00030-1

Drias, H., Sadeg, S., & Yahi, S. (2005). Cooperative bees swarm for solving the maximum weighted satisfiability problem. In *Proceedings of the IWAAN International Work Conference on Artificial and Natural Neural Networks* (LNCS 3512, pp. 318-325).

Duda, R. O., & Hart, P. E. (1973). *Pattern classification and scene analysis*. New York, NY: John Wiley & Sons.

Eberhart, R. C., & Kennedy, J. (1995). *A new optimizer using particle swarm theory*. Paper presented at the Sixth International Symposium on Micro Machine and Human Science.

Eiben, A., & Smith, J. (2003). *Introduction to evolutionary computing*. Berlin, Germany: Springer-Verlag.

Ekgasit, S., Thammacharoen, C., Yu, F., & Knoll, W. (2005). Influence of the metal film thickness on the sensitivity of surface plasmon resonance biosensors. *Applied Spectroscopy, 59*, 661–667. doi:10.1366/0003702053945994

El Moudani, W., & Mora-Camino, F. (2000). A fuzzy solution approach for the roster planning problem. In *Proceedings of the 9th IEEE International Conference on Fuzzy Systems*, San Antonio, TX.

Elias, S. E. G. (1966). *A mathematical model for optimizing the assignment of man and machine in public transport run cutting* ([]. Morgantown, WV: West Virginia University, Engineering Experiment Station.]. *Research Bulletin (Sun Chiwawitthaya Thang Thale Phuket)*, 81.

Engin, O., & Döyen, A. (2004). A new approach to solve hybrid flow shop scheduling problems by artificial immune system. *Future Generation Computer Systems, 20*(6), 1083–1095. doi:10.1016/j.future.2004.03.014

Euchi, J., & Chabchoub, H. (2009, July 6-9). Iterated Density Estimation with 2-opt local search for the vehicle routing problem with private fleet and common carrier. In *Proceedings of the International Conference on Computers & Industrial Engineering (CIE 2009)* (pp.1058-1063).

Euchi, J., & Chabchoub, H. (2010). A Hybrid Tabu Search to Solve the Heterogeneous Fixed Fleet Vehicle Routing Problem. *Logistics Research*, *2*(1), 3–11. doi:10.1007/s12159-010-0028-3

Feo, T., & Resende, M. G. C. (1995). Greedy randomized adaptive search procedures. *Journal of Global Optimization*, *2*, 1–27.

Fieldsend, J. E., & Singh, S. (2002). A multi-objective algorithm based upon particle swarm optimization, an efficient data structure and turbulence. In *Proceedings of the Workshop on Computational Intelligence*, Brimingham, UK (pp. 37-44).

Fikri, R., Grosges, T., & Barchiesi, D. (2004). Apertureless scanning near-field optical microscopy: Numerical Modeling of the lock-in detection. *Optics Communications, 238*(1-6), 15–23.

Fikri, R., Grosges, T., & Barchiesi, D. (2003). Apertureless scanning near-field optical microscopy: The need of the tip vibration modelling. *Optics Letters*, *28*(22), 2147–2149. doi:10.1364/OL.28.002147

Fischetti, M., Lodi, A., Martello, S., & Toth, P. (2001). A polyhedral approach to simplified crew and vehicle scheduling problems. *Management Science*, *47*(6), 833–850. doi:10.1287/mnsc.47.6.833.9810

Fischetti, M., Martello, S., & Toth, P. (1989). The fixed job schedule problem with working-time constraints. *Operations Research*, *37*(3), 395–403. doi:10.1287/opre.37.3.395

Fores, S., Proll, L. G., & Wren, A. (1999). An improved ILP system for driver scheduling. In Wilson, N. (Ed.), *Computer-aided transit scheduling* (pp. 43–62). Berlin, Germany: Springer-Verlag.

Forsyth, P., & Wren, A. (1997). *An ant system for bus driver scheduling* (Research Report No. 97.25). Leeds, UK: University of Leeds School of Computer Studies.

Frangioni, A., & Gendron, B. (2008). 0-1 reformulation of the multicommodity capacitated network design problem. *Discrete Applied Mathematics*, *157*, 1229–1241. doi:10.1016/j.dam.2008.04.022

Freling, R., Lentink, R. M., & Odijk, M. A. (2001). Scheduling train crews: A case study for the Dutch railways. In Voss, S., & Daduna, J. (Eds.), *Computer-aided scheduling of public transport* (pp. 153–165). Berlin, Germany: Springer-Verlag. doi:10.1007/978-3-642-56423-9_9

Gaballa, A. (1979). Planning callout reserve for aircraft delays. *Interfaces*, *9*, 78–86. doi:10.1287/inte.9.2pt2.78

Gajpal, J., & Abad, P. L. (2009). Multi-ant colony system (MACS) for a vehicle routing problem with backhauls. *European Journal of Operational Research*, *196*, 102–117. doi:10.1016/j.ejor.2008.02.025

Galvão, R. D., & Guimarães, J. (1990). The control of helicopter operations in the Brazilian oil industry: Issues in the design and implementation of a computerized system. *European Journal of Operational Research*, *49*, 266–270. doi:10.1016/0377-2217(90)90344-B

Ganesh, K., & Narendran, T. (2007). CLOVES: A cluster-and-search heuristic to solve the vehicle routing problem with delivery and pick-up. *European Journal of Operational Research*, *178*, 699–717. doi:10.1016/j.ejor.2006.01.037

Garcia-Lopez, F., Batista, B. M., & Moreno-Perez, J. A. (2003). Parallelization of the scatter search for the p-median problem. *Parallel Computing*, *29*, 575–589. doi:10.1016/S0167-8191(03)00043-7

Garcia-Lopez, F., Torres, M. G., Batista, B. M., Moreno-Perez, J. A., & Moreno-Vega, J. M. (2006). Solving features subset selection problem by a parallel scatter search. *European Journal of Operational Research*, *196*, 477–489. doi:10.1016/j.ejor.2004.08.010

Garey, M. R., & Johnson, D. S. (1979). *Computers and intractability*. San Francisco, CA: Freeman and Co.

Garey, M. R., & Johnson, D. S. (1979). *Computers and intractability: A guide to the theory of NP-completeness*. New York, NY: Freeman.

Garey, M. R., & Johnson, D. S. (1990). *Computers and intractability: A guide to the theory of NP-completeness*. Murray Hill, NJ: Bell Telephone Laboratories.

Gehring, H., & Homberger, J. (1999). A parallel hybrid evolutionary metaheuristic for the vehicle routing problem with time windows. In *Proceedings of the EUROGEN Conference on Evolutionary Algorithms in Engineering and Computer Science* (pp. 57-64).

Gendreau, M. (2003). An introduction to tabu search. In Glover, F., & Kochenberger, G. A. (Eds.), *Handbook of metaheuristics* (pp. 37–54). Boston, MA: Kluwer Academic.

Gendreau, M., Laporte, G., Musaraganyi, C., & Taillard, E. D. (1999). A tabu search heuristic for the heterogeneous fleet vehicle routing problem. *Computers & Operations Research*, *26*, 1153–1173. doi:10.1016/S0305-0548(98)00100-2

Gendron, B., & Crainic, T. G. (1994). *Relaxations for multicommodity capacitated network design problems* (Publication No. CRT-945). Montreal, QC, Canada: Centre de recherche sur les transports, Université de Montréal.

Gendron, B., Crainic, T. G., & Frangioni, A. (1998). Multicommodity capacitated network design. In Sanso, B., & Soriono, P. (Eds.), *Telecommunication network planning* (pp. 1–19). Boston, MA: Kluwer Academic.

Genin, P. (2003). *Planification tactique robuste avec usage d'un A.P.S. Proposition d'un mode de gestion par plan de référence*. Unpublished doctoral dissertation, Ecole supérieure des mines de Paris, France.

Ghamlouche, I., Crainic, T. G., & Gendreau, M. (2003). Cycle-based neighbourhoods for fixed-charge capacitated multicommodity network design. *Operations Research*, *51*, 655–667. doi:10.1287/opre.51.4.655.16098

Ghamlouche, I., Crainic, T. G., & Gendreau, M. (2004). Path relinking, cycle-based neighborhoods and capacitated multicommodity network design. *Annals of Operations Research*, *131*, 109–133. doi:10.1023/B:ANOR.0000039515.90453.1d

Ghoseari, K., & Sarhadi, H. (2009). Finding the shortest Iranian Hamiltonian tour by using combination of ant colony optimization and local research. *Transportation Research*, *2*, 149–161.

Giaferri, C., Hamon, J. P., & Lengline, J. G. (1982). Automatic monthly assignment of medium-haul cabin crew. In *Proceedings of the Airline Group International Federation of Operational Research Societies Symposium*, *22*, 69–95.

Giannopoulou, E. G. (2008). *Data mining in medical and biological research*. Rijeka, Croatia: INTECH.

Ginis, G., & Cioffi, J. M. (2002). Vectored transmission for digital subscriber line systems. *IEEE Journal on Selected Areas in Communications*, *20*(5), 1085–1104. doi:10.1109/JSAC.2002.1007389

Gitashenasi. (2008). *Geographical database of Iranian cities*. Tehran, Iran: Gitashenasi.

Glanert, W. (1984). A timetable approach to the assignment of pilots to rotations - Lufthansa. In *Proceedings of the 24th Airline Group International Federation of Operational Research Societies Symposium*.

Glover, F. (1989). Tabu Search I. *ORSA Journal on Computing, 1*(3), 190-206.

Glover, F. (1998). A template for scatter search and path relinking. In J. K. Hao, E. Lutton, E. Ronald, M. Schoenauer, & D. Snyers (Eds.), *Proceedings of the Third European Conference on Artificial Evolution* (LNCS 1363, pp. 3-54).

Glover, F. (1977). Heuristics for integer programming using surrogate constraints. *Decision Sciences, 8*, 156–166. doi:10.1111/j.1540-5915.1977.tb01074.x

Glover, F. (1986). Future paths for integer programming and links to artificial intelligence. *Computers & Operations Research*, *13*(5), 533–549. doi:10.1016/0305-0548(86)90048-1

Glover, F. (1989)... *Tabu Search - Part I. ORSA Journal on Computing*, *1*(3), 190–206.

Glover, F. (1990). Tabu Search II. *ORSA Journal on Computing*, *2*(1), 4–32.

Glover, F., & Kochenberger, G. A. (2003). *Handbook of metaheuristics*. Boston, MA: Kluwer Academic.

Glover, F., & Laguna, M. (1997). *Tabu search*. Boston, MA: Kluwer Academic.

Glover, F., & Laguna, M. (2002). Tabu search. In Pardalos, P. M., & Resende, M. G. C. (Eds.), *Handbook of applied optimization* (pp. 194–208). New York, NY: Oxford University Press.

Glover, F., Laguna, M., & Marti, R. (2000). Fundamentals of scatter search and path relinking. *Control and Cybernetics, 29*(3), 653–684.

Glover, F., Laguna, M., & Marti, R. (2000). Fundamentals of scatter search and path relinking. *Control and Cybernetics, 39*, 575–589.

Goldberg, D. E. (1989). *Genetic Algorithms in Search, Optimization, and Machine Learning*. Reading, MA: Addison-Wesley.

Golden, B.-L., Assad, A.-A., Levy, L., & Gheysens, F.-G. (1984). The fleet size and mix vehicle routing problem. *Computers & Operations Research, 11*(1), 49–66. doi:10.1016/0305-0548(84)90007-8

Golden, B., Wasil, E., Kelly, J., & Chao, I.-M. (1998). The impact of metaheuristics on solving the vehicle routing problem: algorithms, problem sets, and computational results. In Crainic, T., & Laporte, G. (Eds.), *Fleet management and logistics* (pp. 33–56). Boston, MA: Kluwer.

Gómez-Iglesias, A., Vega-Rodríguez, M. A., Castejon, F., Cardenas-Montes, M., & Morales-Ramos, E. (2010). Artificial bee colony inspired algorithm applied to fusion research in a grid computing environment. In *Proceedings of the 18th Euromicro Conference on Parallel, Distributed and Network-based Processing* (pp. 508-512).

Gourgand, M., Grangeon, N., & Norre, S. (1999). Metaheuristics for the deterministic hybrid flow shop problem. In *Proceedings of the International Conference on Industrial and Production management (IEPM'99)* (pp. 136-145).

Graves, S. (1981). In multi-stage production / inventory control systems: Theory and practice. *TIMS Studies in the Management Science, 16*, 95–110.

Grosan, C., Oltean, M., & Dumitrescu, D. (2003). Performance metrics for multi-objective optimization evolutionary algorithms. In *Proceedings of the Conference on Applied and Industrial Mathematics*, Oradea, Romania.

Groth, R. (1997). *Data mining: A hands-on approach for business professionals*. Upper Saddle River, NJ: Prentice Hall.

Guinet, A., Solomon, M., Kedia, P., & Dussa, A. (1996). A computational study of heuristics for two-stage flexible flowshops. *International Journal of Production Research, 34*, 1399–1415. doi:10.1080/00207549608904972

Gupta, J. (1988). Two-stage hybrid flowshop scheduling problem. *The Journal of the Operational Research Society, 39*, 359–364.

Gutierrez, F. J., Lerma-Rascon, M. M., Salgado-Garza, L. R., & Cantu, F. J. (2002). Biometrics and data mining: Comparison of data mining-based keystroke dynamics methods for identity verification. In C. A. Coello Coello, A. de Albornoz, L. E. Sucar, & O. C. Battistutti (Eds.), *Proceedings of the Second Mexican International Conference on Advances in Artificial Intelligence* (LNCS 2313, pp. 460-469).

Gutin, G. (2003). Traveling salesman and related problems. In Gross, J., & Yellen, J. (Eds.), *Handbook of graph theory*. Boca Raton, FL: CRC Press.

Hahsler, M., & Hornik, K. (2007). Introduction to TSP - Infrastructure for the traveling salesperson problem. *Journal of Statistical Software, 23*(2), 1–21.

Han, J., & Kamber, M. (2006). *Data mining concepts and techniques*. San Francisco, CA: Morgan Kaufmann.

Han, Y., Tang, J., Kaku, I., & Mu, L. (2009). Solving uncapacitated multilevel Lot-Sizing problems using a particle swarm optimization with flexible inertial weight. *Computers & Mathematics with Applications (Oxford, England), 57*(11-12), 1748–1755. doi:10.1016/j.camwa.2008.10.024

Haouari, M., Hidri, L., & Gharbi, A. (2006). Optimal scheduling of a two-stage hybrid flow shop. *Mathematical Methods of Operations Research, 64*(1), 107–124. doi:10.1007/s00186-006-0066-4

Hartigan, J. A. (1972). Direct clustering of a data matrix. *Journal of the American Statistical Association*, 123–129. doi:10.2307/2284710

Hart, W., Krasnogor, N., & Smith, J. (2005). *Recent advances in memetic algorithms*. Berlin, Germany: Springer-Verlag. doi:10.1007/3-540-32363-5

Haydon, B. (1973). Aircrew standby policies for long haul carriers. In *Proceedings of the 13th Airline Group International Federation of Operational Research Societies Symposium*.

Heinrich, C., & Schneeweiss, C. (1986). Multi-stage Lot-Sizing for general production systems. In Axsater, S., Schneeweiss, C., & Silver, E. (Eds.), *Multi-Stage Production Planning and Inventory Control*. Berlin, Germany: Springer Verlag.

Hoaa, X., Kirk, A., & Tabrizian, M. (2007). Towards integrated and sensitive surface plasmon resonance biosensors: A review of recent progress. *Biosensors & Bioelectronics, 23*, 151–160. doi:10.1016/j.bios.2007.07.001

Hoffman, K. L., & Padberg, M. (1993). Solving airline crew scheduling problems by branch-and-cut. *Management Science, 39*(6), 657–682. doi:10.1287/mnsc.39.6.657

Holden, N. F. (2007). A hybrid PSO/ACO algorithm for discovering classification rules in data mining. In *Proceedings of the GECCO Conference Companion on Genetic and Evolutionary Computation*, London, UK (pp. 2745-2750).

Holland, H. (1975). *Adaption in natural and artificial systems*. Ann Arbor, MI: University of Michigan Press.

Holland, J. H. (1992). *Adaptation in natural and artificial systems*. Cambridge, MA: MIT Press.

Holloway, S. (1997). *Straight and level: Practical airline economics*. London, UK: Ashgate.

Homola, J. (1997). On the sensitivity of surface Plasmon resonance sensors with spectral interrogation. *Sensors and Actuators. B, Chemical, 41*, 207–211. doi:10.1016/S0925-4005(97)80297-3

Horn, J., Nafploitis, N., & Goldberg, D. E. (1994). A niched Pareto genetic algorithm for multiobjective optimization. In *Proceedings of the First IEEE Conference on Evolutionary Computation*, Orlando, FL (pp. 82-87).

Hu, X., & Eberhart, R. C. (2002). Multiobjective optimization using dynamic neighborhood particle swarm optimization. In *Proceedings of the IEEE Congress Evolutionary Computation*, Honolulu, HI (pp. 1677–1681).

Hu, X., Lin, T. Y., & Jianchao, J. (2004). A new rough sets model based on database systems. *Fundamenta Informaticae*, 1-18.

Hu, X., Shi, Y., & Eberhart, R. (2004). Recent advances in particle swarm Evolutionary Computation. In *Proceedings of the 2004 Congress on Evolutionary Computation* (Vol. 1, pp. 90-97).

Huang, T., Kecman, V., & Kopriva, I. (2006). *Kernel based algorithms for mining huge data sets*. Berlin, Germany: Springer-Verlag.

Huap, S. H., Wulsin, L. R., Li, H., & Guo, J. (2009). Dimensionality reduction for knowledge discovery in medical claims database: Application to antidepressant medication utilization study. *Computer Methods and Programs in Biomedicine, 93*(2), 115–123. doi:10.1016/j.cmpb.2008.08.002

Ishibuchi, H., & Namba, S. (2004). Evolutionary multiobjective knowledge extraction for high dimentional pattern classification problems. In X. Yao, E. K. Burke, J. A. Lozano, J. Smith, J. J. Merelo-Guervós, J. A. Bullinaria et al. (Eds.), *Proceedings of the 8th International Conference on Parallel Problem Solving from Nature* (LNCS 3242, pp. 1123-1132).

Ishibuchi, H., & Nojima, Y. (2005). Accuracy-complexity tradeoff analysis by multiobjective rule selection. In *Proceedings of the Workshop on Computational Intelligence in Data Mining* (pp. 39-48).

Ishibuchi, H., Nakashima, T., & Murata, T. (109-133). Three-objective genetics-based machine learning for linguistic rule extraction. *Information Sciences*, 136.

Ishibuchi, H., Nozaki, K., Yamamoto, N., & Tanaka, H. (1995). Selecting fuzzy if-then rules for classification problems using genetic algorithms. *IEEE Transactions on Fuzzy Systems, 3*, 260–270. doi:10.1109/91.413232

Işıl, B., Güllü, K., & Kürşat, S. (2009). A multiobjective optimization framework for nano-antennas via normal boundary intersection (NBI) method. In *Proceedings of the IEEE AP-S International Symposium on Antennas and Propagation and USNC/URSI National Radio Science Meeting*, Charleston, SC.

Jaillet, P. (1988). Apriori solution of a travelling salesman problem in which a random subset of the customers are visited. *Operations Research, 36*(6), 929–936. doi:10.1287/opre.36.6.929

Jain, A., & Zongker, D. (1997). Feature selection: Evaluation, application, and small sample performance. *IEEE Transactions on Pattern Analysis and Machine Intelligence, 19*, 153–158. doi:10.1109/34.574797

Janaki, D., Sreenivas, T., & Ganapathy, K. (1996). Functional and parametric tests of integrated circuits. *Journal of Parallel and Distributed Computing, 37*(2), 207–212.

Jarboui, B., Damak, N., Siarry, P., & Rebai, A. (2007). A combinatorial particle swarm optimization for solving multi-mode resource-constrained project scheduling problems. *Journal of Applied Mathematics and Computation, 195*, 299–308. doi:doi:10.1016/j.amc.2007.04.096

Jaszkiewicz, A. (2000). *On the performance of multiple objective genetic local search on the 0/1 knapsack problem: A comparative experiment* (Working Paper No. RA002/2000). Poznan, Poland: Institute of Computer Science, Poznan University of Technology.

Jayaraman, V., & Ross, A. (2002). A simulated annealing methodology to distribution network design and management. *European Journal of Operational Research, 144*, 629–645. doi:10.1016/S0377-2217(02)00153-4

Jensen, R., & Shen, Q. (2004). Fuzzy-rough attribute reduction with application to web categorization. *Fuzzy Sets and Systems, 141*(3), 469–485. doi:10.1016/S0165-0114(03)00021-6

Jeunet, J., & Jonard, N. (2005). Single point stochastic search algorithms for the multi-level Lot-Sizing problem. *International Journal of Production Economics, 32*, 985–1006.

Jin, Z., Ohno, K., Ito, T., & Elmaghraby, S. (2002). Scheduling hybrid flowshops in printed circuit board assembly lines. *Production and Operations Management, 11*(1), 216–230.

Jin, Z., Yang, Z., & Ito, T. (2006). Metaheuristic algorithms for multistage hybrid flowshop scheduling problem. *International Journal of Production Economics, 100*, 322–334. doi:10.1016/j.ijpe.2004.12.025

Johnson, S. (1954). Optimal two- and three-stage production schedules with setup times included. *Naval Research Logistics Quarterly, 1*, 61–68. doi:10.1002/nav.3800010110

Johnson, S., & Papadimitriou, H. (1985). *The travelling salesman problem: A guided tour of combinatorial optimization.* New York, NY: John Wiley & Sons.

Jonker, J.-J., Piersma, N., & Van den Poel, D. (2004). Joint optimization of customer segmentation and marketing policy to maximize long-term profitability. *Expert Systems with Applications, 27*, 159–168. doi:10.1016/j.eswa.2004.01.010

Kamal, A., Manganaris, S., & Srikant, R. (1997). Partial classification using association rules. In *Proceedings of the AAAI Conference on Knowledge Discovery in Databases* (pp. 115-118).

Karaboga, D. (2005). *An idea based on honey bee swarm for numerical optimization* (Tech. Rep. No. TR06). Kayseri, Turkey: Erciyes University.

Karaboga, D., & Basturk, B. (2007). A powerful and efficient algorithm for numerical function optimization: Artificial Bee Colony (ABC) algorithm. *Journal of Global Optimization, 39*(3), 459–171. doi:10.1007/s10898-007-9149-x

Karaboga, D., & Basturk, B. (2008). On the performance of Artificial Bee Colony (ABC) algorithm. *Applied Soft Computing, 8*(1), 687–697. doi:10.1016/j.asoc.2007.05.007

Karaboga, D., & Ozturk, C. (2011). A novel clustering approach: Artificial Bee Colony (ABC) algorithm. *Applied Soft Computing, 11*(1), 652–657. doi:10.1016/j.asoc.2009.12.025

Katayama, N., Chen, M., & Kubo, M. (2009). Capacity scaling heuristic for the multicommodity capacitated network design problem. *Journal of Computational and Applied Mathematics, 232*, 90–101. doi:10.1016/j.cam.2008.10.055

Kaw, A. K. (2006). *Mechanics of composite materials.* Boca Raton, FL: CRC Press/Taylor and Francis.

Kaya, M. (2006). Multi-objective genetic algorithm based approaches for mining optimized fuzzy association rules. *Soft Computing, 10*, 578–586. doi:10.1007/s00500-005-0509-5

Kendall, G., & Mohd Hussin, N. (2005). A tabu search hyper-heuristic approach to the examination timetabling problem at the MARA University of Technology. In E. Burke & M. Trick (Eds.), *Proceedings of the 5th International Conference on the Practice and Theory of Automated Timetabling* (LNCS 3616, pp. 270-293).

Kennedy, J., & Eberhart, R. (1995). Particle swarm optimization. In *Proceedings of 1995 IEEE International Conference on Neural Networks* (pp. 1942-1948).

Kennedy, J., & Eberhart, R. (1997). A discrete binary version of the particle swarm algorithm. In *Proceedings of the International Conference on Systems, Man and Cybernetics* (Vol. 5, pp. 4104-4108).

Kennedy, J., & Mendes, R. (2002). Population structure and particle swarm performance. In *Proceedings of the IEEE Congress Evolutionary Computation,* Honolulu, HI (pp. 1671–1676).

Kennedy, J., & Eberhart, R. (1997). A discrete binary version of the particle swarm algorithm. *IEEE Transactions on Systems, Man, and Cybernetics, 5,* 4104–4108.

Khalouli, S., Ghedjati, F., & Hamzaoui, A. (2008). Method based on ant colony system for solving the hybrid flow shop scheduling problem. In *Proceedings of the 7th International Conference on Modelling, Optimization and SIMulation Systems (MOSIM'08)* (Vol. 2, pp. 1407-1416).

Kira, K., & Rendell, L. (1992). A practical approach to feature selection. In *Proceedings of the Ninth International Conference on Machine Learning,* Aberdeen, UK (pp. 249-256).

Kirkpatrick, S., Gelatt, C., & Vecchi, M. (1983). Optimization by simulated annealing. *Science, 220,* 671–680. doi:10.1126/science.220.4598.671

Klabjan, D., Johnson, E. L., Nemhauser, G. L., Gelman, E., & Ramaswamy, S. (2001). Airline crew scheduling with regularity. *Transportation Science, 35*(4), 359–374. doi:10.1287/trsc.35.4.359.10437

Klincewicz, J. G., Luss, H., & Pilcher, M. G. (1990). Fleet size planning when outside carrier services are available. *Transportation Science, 24,* 169–182. doi:10.1287/trsc.24.3.169

Kolomenskii, A., Gershon, P., & Schuessler, H. (1997). Sensitivity and detection limit of concentration and absorption measurements by laser-induced surface-plasmon resonance. *Applied Optics, 36,* 6539–6547. doi:10.1364/AO.36.006539

Koutsonikola, V. A., & Vakali, A. (2009). A fuzzy bi-clustering approach to correlate web users and pages. International. *Journal of Knowledge and Web Intelligence, 1*(1-2), 3–23. doi:10.1504/IJKWI.2009.027923

Kovacs, G., Groenwold, A. A., Jármai, K., & Farkas, J. (2004). Analysis and optimum design of fibre-reinforced composite structures. *Structural Multidisciplinary Optimization, 28*(2-3), 170–179. doi:10.1007/s00158-004-0425-9

Kretschman, E., & Raether, H. (1968). Radiative decay of nonradiative surface plasmons excited by light. *Zeitung Naturforschung A, 23,* 2135–2136.

Kroon, L., & Fischetti, M. (2001). Crew scheduling for Netherlands railways "destination-customer.". In Voss, S., & Daduna, J. (Eds.), *Computer-aided scheduling of public transport* (pp. 181–201). Berlin, Germany: Springer-Verlag. doi:10.1007/978-3-642-56423-9_11

Kuik, R., & Solomon, M. (1990). Multi-level Lot-Sizing problem: Evaluation of a simulated-annealing heuristic. *European Journal of Operational Research, 45*(1), 25–37. doi:10.1016/0377-2217(90)90153-3

Kwan, A. S. K., Kwan, R. S. K., Parker, M. E., & Wren, A. (1999). Producing train driver schedules under differing operating strategies. In Wilson, N. (Ed.), *Computer-aided transit scheduling* (pp. 129–154). Berlin, Germany: Springer-Verlag.

Laguna, M., & Martí, R. (2003). *Scatter search: Methodology and implementations.* Boston, MA: Kluwer Academic.

Laporte, G. (1992a). The Vehicle Routing Problem: An overview of exact and approximate algorithms. *European Journal of Operational Research, 59*(3), 345–358. doi:10.1016/0377-2217(92)90192-C

Laporte, G. (1992b). The traveling salesman problem: An overview of exact and approximate algorithms. *European Journal of Operational Research, 59*(2), 291–247. doi:10.1016/0377-2217(92)90138-Y

Larrañaga, P. (2002). A review on estimation of distribution algorithms. In Larrañaga, P., & Lozano, J. A. (Eds.), *Estimation of Distribution Algorithms. A New Tool for Evolutionary Computation* (pp. 80–90). Boston, MA: Kluwer.

Lasdon, L., Duarte, A., Glover, F., Laguna, M., & Marti, R. (2010). Adaptive memory programming for constrained global optimization. *Computers & Operations Research, 37,* 1500–1509. doi:10.1016/j.cor.2009.11.006

Laumanns, M., Thiele, L., Zitzler, E., Welzl, E., & Deb, K. (2002). *Running time analysis of a multi-objective evolutionary algorithm on a simple discrete optimization problem* (Tech. Rep. No. TIK-123). Zurich, Switzerland: Swiss Federal Institute of Technology.

Lawler, E., Lenstra, J., Rinnooy, A., & Shmoys, D. (1985). *The traveling salesman problem: A guided tour of combinatorial optimization.* Chichester, UK: John Wiley & Sons.

Layfield, C. J., Smith, B. M., & Wren, A. (1999). Bus relief point selection using constraint programming. In *Proceedings of the 1st International Conference on the Practical Applications of Constraint Technologies and Logic Programming.*

Lecaruyer, P., Canva, M., & Rolland, J. (2006). Metallic film optimization in a surface plasmon resonance biosensor by the extended Rouard method. *Applied Optics, 46*(12), 2361–2369. doi:10.1364/AO.46.002361

Lee, C.-H., & Fu, Y.-H. (2008). Web usage mining based on clustering of browsing. In *Proceedings of the Eighth International Conference on Intelligent Systems Design and Applications* (Vol. 1, pp. 281-286).

Lee, C. Y., & Yao, X. (2004). Evolutionary programming using mutations based on the Levy probability distribution. *IEEE Transactions on Evolutionary Computation, 8,* 1–13. doi:10.1109/TEVC.2003.816583

Lee, C., Cheng, T. C., & Lin, B. (1993). Minimizing the makespan in the 3-machine assembly-type flow-shop scheduling problem. *Management Science, 39*(5), 616–625. doi:10.1287/mnsc.39.5.616

Levine, D. (1996). Application of a hybrid genetic algorithm to airline crew-scheduling. *Computers & Operations Research, 23,* 547–558. doi:10.1016/0305-0548(95)00060-7

Li, W. H., & Pei, J. (2001). Accurate and efficient classification based on multiple class-association rules. In *Proceedings of the 1st IEEE International Conference on Data Mining* (pp. 369-376). Washington, DC: IEEE Computer Society.

Liang, J. J., Qin, A. K., Suganthan, P. N., & Baskar, S. (2006). Comprehensive learning particle swarm optimizer for global optimization of multimodal functions. *IEEE Transactions on Evolutionary Computation, 10*(3), 281–295. doi:10.1109/TEVC.2005.857610

Li, F., Golden, B. L., & Wasil, E. A. (2007). A record-to-record travel algorithm for solving the heterogeneous fleet vehicle routing problem. *Computers & Operations Research, 34,* 2734–2742. doi:10.1016/j.cor.2005.10.015

Li, H.-L., & Chen, M.-H. (2008). Induction of multiple criteria optimal classification rules for biological and medical data. *Computers in Biology and Medicine, 38,* 42–52. doi:10.1016/j.compbiomed.2007.07.006

Li, J. S., & Topor, R. (2002). Mining the optimal class association rule set. *Knowledge-Based Systems, 15,* 399–405. doi:10.1016/S0950-7051(02)00024-2

Li, L. (1994). Multilayer-coated diffraction gratings: Differential method of Chandezon et al. revisited. *Journal of the Optical Society of America. A, Optics, Image Science, and Vision, 11,* 2816–2828. doi:10.1364/JOSAA.11.002816

Linn, R., & Zhang, W. (1999). Hybrid flow shop scheduling: a survey. *Computers & Industrial Engineering, 37*(1-2), 57–61. doi:10.1016/S0360-8352(99)00023-6

Lin, S. (1965). Computer solutions of the traveling salesman problem. *The Bell System Technical Journal, 44,* 2245–2269.

Lin, S. (1965). Computer solutions of the traveling salesman problem. *The Bell System Technical Journal, 44,* 2245–2269.

Lin, S. W., & Chen, S. C. (2009). PSOLDA: A Particle swarm optimization approach for enhancing classification accurate rate of linear discriminant analysis. *Applied Soft Computing, 9,* 1008–1015. doi:10.1016/j.asoc.2009.01.001

Lin, S. W., Lee, Z. J., Chen, S. C., & Tseng, T. Y. (2008). Parameter determination of support vector machine and feature selection using simulated annealing approach. *Applied Soft Computing, 8*, 1505–1512. doi:10.1016/j.asoc.2007.10.012

Lin, S. W., Ying, K. C., Chen, S. C., & Lee, Z. J. (2008). Particle swarm optimization for parameter determination and feature selection of support vector machines. *Expert Systems with Applications, 35*, 1817–1824. doi:10.1016/j.eswa.2007.08.088

Liping, Z. (2006). A heuristic method for analyzing driver scheduling problem. *IEEE Transactions on Systems, Man, and Cybernetics. Part A, Systems and Humans, 36*(3).

Liping, Z. (2006). An heuristic method for analyzing driver scheduling problem. *IEEE Transactions on Systems, Man, and Cybernetics. Part A, Systems and Humans, 36*(3), 521–531. doi:10.1109/TSMCA.2005.853497

Liu, B., Hsu, W., & Ma, Y. (1998). Integrating classification and association rule mining. In *Proceedings of the 4th International Conference on Knowledge Discovery and Data Mining* (pp. 80-86).

Liu, H., Liu, L., & Zhang, H. (2011). A fast pruning redundant rule method using galois connection. *Applied Soft Computing, 11*, 130–137. doi:10.1016/j.asoc.2009.11.004

Liu, H., Sun, J., & Zhang, H. (2009). Post processing of associative classification rules using closed sets. *Expert Systems with Applications, 36*, 6659–6667. doi:10.1016/j.eswa.2008.08.046

Liu, Y. (2010). Different initial solution generators in genetic algorithms for solving the probabilistic traveling salesman problem. *Applied Mathematics and Computation, 216*(7), 125–137. doi:10.1016/j.amc.2010.01.021

Lofti, V., & Cerveny, R. (1991). A final-exam-scheduling package. *The Journal of the Operational Research Society, 42*, 205–216.

Lopez, F. G., Torres, M. G., Batista, B. M., Perez, J. A. M., & Moreno-Vega, J. M. (2006). Solving feature subset selection problem by a parallel scatter search. *European Journal of Operational Research, 169*, 477–489. doi:10.1016/j.ejor.2004.08.010

Lozano, J. A., Larrañaga, P., Inza, I., & Bengoetxea, E. (2006). *Towards a New Evolutionary Computation: Advances on Estimation of Distribution Algorithms (Studies in Fuzziness and Soft Computing)*. New York, NY: Springer-Verlag.

Madeira, S. C., & Oliveira, A. L. (2004). Biclustering algorithms for biological data analysis: A survey. *IEEE/ACM Transactions on Computational Biology and Bioinformatics, 1*, 24–45. doi:10.1109/TCBB.2004.2

Magnanti, T. L., & Wong, R. T. (1984). Network design and transportation planning: models and algorithms. *Transportation Science, 18*, 1–55. doi:10.1287/trsc.18.1.1

Majumdar, J., & Bhunia, A. K. (2006). Elitist genetic algorithm approach for assignment problem. *AMO- Advanced Modeling and Optimization, 8*(2), 135-149.

Majumdar, J., & Bhunia, A. K. (2010). Solving airline crew-scheduling problem with imprecise service time using genetic algorithm. *AMO - Advanced Modeling and Optimization, 12*(2).

Majumdar, J., & Bhunia, A. K. (2007). Elitist genetic algorithm for assignment problem with imprecise goal. *European Journal of Operational Research, 177*, 684–692. doi:10.1016/j.ejor.2005.11.034

Mangasarian, O. L. (1994). *Nonlinear programming* (p. 42). Philadelphia, PA: SIAM. doi:10.1137/1.9781611971255

Mansour, N., Tarhini, A., & Isahakian, V. (2003, July 14-18). Three-phase simulated annealing algorithms for exam scheduling. In *Proceedings of the ACS/IEEE International Conference on Computer Systems and Applications*, Tunis, Tunisia (p. 90).

Mansour, N., Isahakian, V., & Galayini, I. (2009). Scatter search technique for exam scheduling. *Applied Intelligence, 34*(2).

Maquera Sosa, N. G., Gandelman, D., & Sant'Anna, A. (2007). Logística inversa y ruteo de vehículos: Búsqueda dispersa aplicada al problema de ruteo de vehículos con colecta y entrega simultanea. In *Proceedings of the 1er Congreso de Logística y Gestión de la Cadena de Suministro*.

Marchettini, F. (1980). Automatic monthly cabin crew rostering procedure. In *Proceedings of the 20th AGIFORS Symposium*, New Delhi, India.

Marinaki, M., Marinakis, Y., & Zopounidis, C. (2007). Application of a genetic algorithm for the credit risk assessment problem. *Foundations of Computing and Decisions Sciences*, *32*(2), 139–152.

Marinaki, M., Marinakis, Y., & Zopounidis, C. (2009). Honey bees mating optimization algorithm for financial classification problems. *Applied Soft Computing*, *10*(3).

Marinakis, Y., Marinaki, M., & Matsatsinis, N. (2009a). A Hybrid discrete artificial bee colony – GRASP algorithm for clustering. In *Proceedings of the 39th International Conference on Computers and Industrial Engineering*, Troyes, France.

Marinakis, Y., Marinaki, M., & Matsatsinis, N. (2009b). A hybrid bumble bees mating optimization – GRASP algorithm for clustering. In E. Corchado, X. Wu, E. Oja, Á. Herrero, & B. Baruque (Eds.), *Proceedings of HAIS 2009* (LNAI 5572, pp. 549-556).

Marinakis, Y., & Marinaki, M. (2010). A hybrid multi-swarm particle swarm optimization algorithm for the probabilistic traveling salesman problem. *Computers & Operations Research*, *37*(3), 432–442. doi:10.1016/j.cor.2009.03.004

Marinakis, Y., Marinaki, M., Doumpos, M., Matsatsinis, N., & Zopounidis, C. (2008). Optimization of nearest neighbor classifiers via metaheuristic algorithms for credit risk assessment. *Journal of Global Optimization*, *42*, 279–293. doi:10.1007/s10898-007-9242-1

Marinakis, Y., Marinaki, M., Doumpos, M., & Zopounidis, C. (2009). Ant colony and particle swarm optimization for financial classification problems. *Expert Systems with Applications*, *36*(7), 10604–10611. doi:10.1016/j.eswa.2009.02.055

Marsten, R. E., Muller, M. R., & Killion, C. L. (1979). Crew planning at flying tiger: A successful application of integer programming. *Management Science*, *25*(12), 1175–1183. doi:10.1287/mnsc.25.12.1175

Marsten, R., & Shepardson, F. (1981). Exact solution of crew scheduling problems using the set partitioning model: Recent successful applications. *Networks*, *11*, 165–177. doi:10.1002/net.3230110208

Martello, S., & Toth, P. (1986). A heuristic approach to the bus driver scheduling problem. *European Journal of Operational Research*, *24*(1), 106–117. doi:10.1016/0377-2217(86)90016-0

Martello, S., & Toth, P. (1990). *Knapsack problem: algorithms and computer implementation*. Chichester, UK: John Wiley & Sons.

Martí, R., Laguna, M., & Glover, F. (2006). Principles of scatter search. *European Journal of Operational Research*, *169*, 359–372. doi:10.1016/j.ejor.2004.08.004

McCollum, B. (2007) A perspective on bridging the gap between research and practice in university timetabling. In E. K. Burke & H. Rudova (Eds.), *Proceedings of the 6th International Conference on the Practice and Theory of Automated Timetabling* (LNCS 3867, pp. 3-23).

McGill, R., Tukey, J. W., & Wayne, A. (1978). Variations of box PlotsAuthor(s): LarsenSource. *The American Statistician*, *32*(1), 12–16. doi:10.2307/2683468

Mendes, R., Kennedy, J., & Neves, J. (2004). The fully informed particle swarm: Simpler maybe better. *IEEE Transactions on Evolutionary Computation*, *8*, 204–210. doi:10.1109/TEVC.2004.826074

Michalewicz, Z. (1992). *Genetic Algorithms + Data Structure = Evolution Programs*. New York, NY: Springer-Verlag.

Mingozzi, A., Boschetti, M. A., Ricciardelli, S., & Bianco, L. (1999). A set partitioning approach to the crew scheduling problem. *Operations Research*, *47*(6), 873–888. doi:10.1287/opre.47.6.873

Min, H. (1989). The multiple vehicle routing problem with simultaneous delivery and pickup points. *Transportation Research*, *23A*, 377–386.

Minoux, M. (1989). Network synthesis and optimum network design problems: Models, solution methods and applications. *Networks*, *19*, 313–360. doi:10.1002/net.3230190305

Minoux, M. (2001). Discrete cost multicommodity network optimization problems and exact solution methods. *Annals of Operations Research*, *106*, 19–46. doi:10.1023/A:1014554606793

Minoux, M. (2004). Polynomial approximation schemes and exact algorithms for optimum curve segmentation problems. *Discrete Applied Mathematics*, *144*, 158–172. doi:10.1016/j.dam.2004.05.003

Mitchell, J. S. (1977). Goal programming for scheduling reserve pilot availability. In *Proceedings of the Airline Group International Federation of Operational Research Societies Symposium*.

Mitchell, M. (1996). *An introduction to genetic algorithms*. Cambridge, MA: MIT Press.

Montané, F. A., & Galvão, R. D. (2006). A tabu search algorithm for the vehicle routing problem with simultaneous pick-up and delivery service. *Computers & Operations Research*, *33*, 595–619. doi:10.1016/j.cor.2004.07.009

Montgomery, D. (2005). *Design and analysis of experiments* (6th ed.). New York, NY: John Wiley & Sons.

Moon, C., Kim, J., Choi, G., & Seo, Y. (2002). An efficient genetic algorithm for the traveling salesman problem with precedence constraints. *European Journal of Operational Research*, *140*(3), 606–617. doi:10.1016/S0377-2217(01)00227-2

Morishita, S. (1998). On classification and regression. In *Proceedings of the First International Conference on Discovery Science* (pp. 40-57).

Moscato, P., & Cotta, C. (2005). A gentle introduction to memetic algorithms. *Operations Research & Management Science*, *57*(2), 105–144.

Moscato, P., & Norman, M. (1992). A memtic approach for the traveling salesman problem implementation of a computational ecology for combinatorial optimization on message-passing systems. In Velaro, M., Onate, E., Jane, M., Larribo, J. L., & Suarez, B. (Eds.), *Parallel computing and transputer applications* (pp. 177–186). Amsterdam, The Netherlands: IOS Press.

Mosheiov, G. (1994). The traveling salesman problem with pick-up and delivery. *European Journal of Operational Research*, *79*, 299–310. doi:10.1016/0377-2217(94)90360-3

Mostaghim, S., & Teich, J. (2003). Strategies for finding good local guides in multiobjective particle swarm optimization. *IEEE Swarm Intelligence Symposium*, 26-33.

Moudani, W., & Mora-Camino, F. (2000). A fuzzy solution approach for the rostering planning problem. In *Proceedings of the 9th IEEE International Conference on Fuzzy Systems*, San Antonio, TX.

Moudani, W., Cosenza, C. A., De-Coligny, M., & Mora-Camino, F. (2001). A bi-criterion approach for the airlines crew rostering problem. In *Proceedings of the First International Conference on Evolutionary Multi-Criterion Optimization*, Zurich, Switzerland.

Moudani, W., & Mora-Camino, F. (2011). Management of bus driver duties using data mining. *International Journal of Applied Metaheuristic Computing*, *2*(2).

Mühlenbein, H., & Mahnig, T. (1999). FDA – a scalable evolutionary algorithm for the optimization of additively decomposed functions. *Evolutionary Computation*, *7*(4), 353–376. doi:10.1162/evco.1999.7.4.353

Nagy, G., & Salhi, S. (2005). Heuristic algorithms for single and multiple depot vehicle routing problems with pickups and deliveries. *European Journal of Operational Research*, *162*, 126–141. doi:10.1016/j.ejor.2002.11.003

Naik, G. N., Krishna Murty, A. V., & Gopalakrishnan, S. (2005). A failure mechanism based failure theory for laminated composites including the effect of shear stress. *Composite Structures*, *69*(2), 219–227. doi:10.1016/j.compstruct.2004.06.014

Narendra, P. M., & Fukunaga, K. (1977). A branch and bound algorithm for feature subset selection. *IEEE Transactions on Computers*, *26*(9), 917–922. doi:10.1109/TC.1977.1674939

Nawaz, M., Enscore, E., & Ham, I. (1983). A heuristic algorithm for the m-machine, n-job flowshop sequencing problem. *Omega*, *11*, 91–95. doi:10.1016/0305-0483(83)90088-9

Neff, H., Zong, W., Lima, A., Borre, M., & Holzhuter, G. (2006). Optical properties and instrumental performance of thin gold films near the surface plasmon resonance. *Thin Solid Films*, *496*, 688–697. doi:10.1016/j.tsf.2005.08.226

Nelder, J. A., & Mead, R. (1965). A simplex for function minimization. *The Computer Journal*, *7*, 308–313.

Nemhauser, G. L. (1966). *Introduction to dynamic programming*. New York, NY: John Wiley & Sons.

Néron, E., Baptiste, P., & Gupta, J. (2001). Solving an hybrid flow shop problem using energetic reasoning and global operations. *Omega, 29*, 501–511. doi:10.1016/S0305-0483(01)00040-8

Nguyen, S., & Kachitvichyanukul, V. (2010). Movement strategies for multi-objective particle swarm optimization. *International Journal of Applied Metaheuristic Computing, 1*(3), 59–79. doi:10.4018/jamc.2010070105

Noon, C., & Bean, J. (1991). A lagrangian based approach for the asymmetric generalized traveling salesman problem. *Operations Research, 39*(4), 623–632. doi:10.1287/opre.39.4.623

Nowicki, E., & Smutnicki, C. (1998). The flow shop with parallel machines: A tabu search approach. *European Journal of Operational Research, 106*(2-3), 226–253. doi:10.1016/S0377-2217(97)00260-9

Omkar, S. N., Khandelwal, R., Yathindra, S., Naik, G. N., & Gopalakrishnan, S. (2008). Artificial immune system for multi-objective design optimization of composite structures. *Engineering Applications of Artificial Intelligence, 21*(8), 1416–1429. doi:10.1016/j.engappai.2008.01.002

Omkar, S. N., Mudigere, D., Naik, G. N., & Gopalakrishnan, S. (2008). Vector evaluated particle swarm optimization (VEPSO) for multi-objective design optimization of composite structures. *Computers & Structures, 86*(1-2), 1–14. doi:10.1016/j.compstruc.2007.06.004

Omkar, S. N., & Senthilnath, J. (2009). Artificial bee colony for classification of acoustic emission signal. *International Journal of Aerospace Innovations, 1*(3), 129–143. doi:10.1260/175722509789685865

Omkar, S. N., Senthilnath, J., Khandelwal, R., Narayana Naik, G., & Gopalakrishnan, S. (2011). Artificial Bee Colony (ABC) for multi-objective design optimization of composite structures. *Applied Soft Computing, 11*(1), 489–499. doi:10.1016/j.asoc.2009.12.008

Orlowska, E. (1998). *Incomplete information: Rough set analysis*. Heidelberg, Germany: Physica-Verlag.

Osman, I. H., & Salhi, S. (1996). Local search strategies for the VFMP. In Rayward-Smith, V. J., Osman, I. H., Reeves, C. R., & Smith, G. D. (Eds.), *Modern Heuristic Search Methods* (pp. 131–153). New York, NY: Wiley.

Ostle, B. (1963). *Statistics in research* (2nd ed.). Ames, IA: Iowa State University Press.

Ozcan, E., Mısır, M., Ochoa, G., & Burke, E. K. (2010). A reinforcement learning – great-deluge hyper-heuristic for examination timetabling. *International Journal of Applied Metaheuristic Computing, 1*(1), 39–59. doi:10.4018/jamc.2010102603

Ozdemir, H. T., & Mohan, C. K. (2001). Flight graph genetic algorithm for crew scheduling in airlines. *Information Sciences, 133*, 165–173. doi:10.1016/S0020-0255(01)00083-4

Pagnot, T., Barchiesi, D., Labeke, D. V., & Pieralli, C. (1997). Use of a SNOM architecture to study fluorescence and energy transfer near a metal. *Optics Letters, 22*, 120–122. doi:10.1364/OL.22.000120

Papandriopoulos, J., & Evans, J. S. (2009). SCALE: A low-complexity distributed protocol for spectrum balancing in multiuser DSL networks. *IEEE Transactions on Information Theory, 55*(8). doi:10.1109/TIT.2009.2023751

Parpinelli, R. S., Lopes, H. S., & Freitas, A. A. (2002). An ant colony algorithm for classification rule discovery. In H. Abbas, R. Sarker, & C. Newton (Eds.). *Data mining: A heuristic approach* (pp. 191-208).

Pawlak, Z. (1982). Rough sets. *International Journal of Computer and Information Sciences, 11*(5), 341–356. doi:10.1007/BF01001956

Pawlak, Z. (1991). *Rough sets: Theoretical aspects and reasoning about data*. Boston, MA: Kluwer Academic.

Pedrycz, W., Park, B. J., & Pizzi, N. J. (2009). Identifying core sets of discriminatory features using particle swarm optimization. *Expert Systems with Applications, 36*, 4610–4616. doi:10.1016/j.eswa.2008.05.017

Pei, J., Upadhyaya, S. J., Farooq, F., & Govidaraju, V. (2004). Data mining for intrusion detection: Techniques, applications, and systems. In *Proceedings of the 20th International Conference on Data Engineering* (p. 877). Washington, DC: IEEE Computer Society.

Pelikan, M., Goldberg, D. E., & Lobo, F. G. (2002). A survey of optimization by building and using probabilistic models. *Computational Optimization and Applications, 21*(1), 5–20. doi:10.1023/A:1013500812258

Peters, J. F., & Skowron, A. (2004). *Transactions on rough sets 1*. Berlin, Germany: Springer-Verlag.

Pham, D. T., Kog, E., Ghanbarzadeh, A., Otri, S., Rahim, S., & Zaidi, M. (2006). The bees algorithm - A novel tool for complex optimization problems. In *IPROMS 2006: Proceedings of the 2nd International Virtual Conference on Intelligent Production Machines and Systems*.

Pillay, N., & Banzhaf, W. (2009). A study of heuristic combinations for hyper-heuristic systems for the uncapacitated examination timetabling problem. *European Journal of Operational Research, 197*(2), 482–491. doi:10.1016/j.ejor.2008.07.023

Pitakaso, R., Almeder, C., Doener, K. F., & Hartl, R. (2007). A max-min ant system for unconstrained multi-level Lot-Sizing problems. *Computers & Operations Research, 34*, 2533–2552. doi:10.1016/j.cor.2005.09.022

Portmann, M., Vignier, A., Dardilhac, D., & Dezalay, D. (1998). Branch and bound crossed with GA to solve hybrid flowshops. *European Journal of Operational Research, 107*(2), 389–400. doi:10.1016/S0377-2217(97)00333-0

Press, W. H., & Teukolsky, S. A. (1991). Simulated annealing optimization over continuous spaces. *Computers in Physics, 5*(4), 426.

Psaraftis, H. (1983). K-interchange procedures for local search in a precedence-constrained routing problem. *European Journal of Operational Research, 13*, 391–402. doi:10.1016/0377-2217(83)90099-1

Punj, G., & Stewart, D. W. (1983). Cluster analysis in marketing research: Review and suggestions for application. *JMR, Journal of Marketing Research, 20*, 134–148. doi:10.2307/3151680

Qi, X., Bard, J., & Yu, G. (2004). Class scheduling for pilot training. *Operations Research, 52*(1), 148–162. doi:10.1287/opre.1030.0076

Qu, R., & Burke, E. K. (2009). Hybridisation within a graph based hyperheuristic framework for university timetabling problems. *The Journal of the Operational Research Society, 60*, 1273–1285. doi:10.1057/jors.2008.102

Qu, R., Burke, E. K., McCollum, B., Merlot, L. T. G., & Lee, S. Y. (2009). A survey of search methodologies and automated system development for examination timetabling. *Journal of Scheduling, 12*, 55–89. doi:10.1007/s10951-008-0077-5

Radharamanan, R., & Choi, L. (1986). A branch and bound algorithm for the travelling salesman and the transportation routing problems. *Computers & Industrial Engineering, 11*(1-4), 236–240. doi:10.1016/0360-8352(86)90085-9

Rajasethupathy, K., Scime, A., Rajasethupathy, K. S., & Murray, G. R. (2009). Finding "persistent rules": Combining association and classification results. *Expert Systems with Applications, 36*, 6019–6024. doi:10.1016/j.eswa.2008.06.090

Raquel, C., Prospero, C., & Naval, J. (2005). An effective use of crowding distance in multiobjective particle swarm optimization. In *Proceedings of the Conference on Genetic and Evolutionary Computation*, Washington, DC (pp. 257-264). New York, NY: ACM Press.

Rathipriya, R., Thangavel, K., & Bagyamani, J. (2011). Evolutionary biclustering of clickstream data. *International Journal Computer Science, 8*, 32–38.

Ratnaweera, A., Halgamuge, S. K., & Watson, H. C. (2004). Self-organizing hirarchical particle swarm optimizer with time varying acceleration coefficients. *IEEE Transactions on Evolutionary Computation, 8*(3), 240–255. doi:10.1109/TEVC.2004.826071

Reddy, J. N. (1997). *Mechanics of laminated composite plates*. Boca Raton, FL: CRC Press/Taylor and Francis.

Rego, C. (2000). *Scatter search for vehicle routing problem*. Paper presented at the INFORMS National Meeting, Salt Lake City, UT.

Reinelt, G. (1991). TSPLIB: A traveling salesman problem library. *ORSA Journal on Computing, 3*, 376–384.

Riane, F., Artiba, A., & Elmaghraby, S. (1998). A hybrid three stage flow-shop problem: efficient heuristics to minimize makespan. *European Journal of Operational Research, 109*, 321–329. doi:10.1016/S0377-2217(98)00060-5

Ridge, E. (2007). *Design of experiments for the tuning of optimization algorithms* (Unpublished doctoral dissertation). University of York, York, UK.

Rizk, N., & Martel, A. (2001). *Supply chain flow planning methods: a review of the Lot-Sizing literature*. Quebec City, QC, Canada: Université Laval.

Rochat, Y., & Taillard, E. D. (1995). Probabilistic diversification and intensification in local search for vehicle routing. *Journal of Heuristics, 1*, 147–167. doi:10.1007/BF02430370

Rokach, L. (2008). Genetic algorithm-based feature set partitioning for classification problems. *Pattern Recognition, 41*, 1676–1700. doi:10.1016/j.patcog.2007.10.013

Rosenberger, J. M. (2001). *Topics in airline operations*. Unpublished doctoral dissertation, Georgia Institute of Technology, Atlanta, GA.

Rubin, J. (1973). A technique for the solution of massive set covering problems, with application to airline crew scheduling. *Transportation Science, 7*(1), 34–48. doi:10.1287/trsc.7.1.34

Ruiz, R., & Vázquez-Rodríguez, J. (2010). The hybrid flow shop scheduling problem. *European Journal of Operational Research, 205*(1), 1–18. doi:10.1016/j.ejor.2009.09.024

Russell, R. A., & Chiang, W. (2006). Scatter search for the vehicle routing problem with time windows. *European Journal of Operational Research, 169*, 606–622. doi:10.1016/j.ejor.2004.08.018

Ryan, D. M., & Foster, B. A. (1981). An integer programming approach to scheduling. In *Proceedings of the International Conference on Computer Scheduling of Public Transport* (pp. 269-280).

Ryan, D. M. (1992). A solution of massive generalized set partitioning problems in aircrew rostering. *The Journal of the Operational Research Society, 43*(5), 459–467.

Salhi, S., & Nagy, G. (1999). A cluster insertion heuristic for single and multiple depot vehicle routing problems with backhauling. *The Journal of the Operational Research Society, 50*, 1034–1042.

Santos, D., Hunsucker, J., & Deal, D. (1996). An evaluation of sequencing heuristics in flow shops with multiple processors. *Computers & Industrial Engineering, 30*, 681–691. doi:10.1016/0360-8352(95)00184-0

Sarra, D. (1988). The automatic assignment model (Saturn - Alitalia). In *Proceedings of the 28th Airline Group International Federation of Operational Research Societies Symposium*.

Sarubbi, J., Miranda, G., Luna, H., & Mateus, G. (2008). A branch and-cut algorithm for the multi commodity traveling salesman problem. In *Proceedings of the International Conference on Service Operations and Logistics, and Informatics*, Beijing, China.

Sasisekharan, R., Seshadri, V., & Weiss, S. M. (1996). Data mining and forecasting in large-scale telecommunication networks. *IEEE Expert Intelligent Systems and their Applications, 11*(1), 37-43.

Savelsbergh, M. (1985). Local search in routing problems with time windows. *Annals of Operations Research, 4*, 285–305. doi:10.1007/BF02022044

Schaefer, A. J., Johnson, E. L., Kleywegt, A. J., & Nemhauser, G. L. (2005). Airline crew scheduling under uncertainty. *Transportation Science, 39*(3), 340–348. doi:10.1287/trsc.1040.0091

Schaerf, A. (1999). A survey of automated timetabling. *Artificial Intelligence Review, 13*, 87–127. doi:10.1023/A:1006576209967

Schaffer, J. D. (1984). *Multi objective optimization with vector evaluated genetic algorithms*. Unpublished doctoral dissertation, Vanderbilt University, Nashville, TN.

Scheuerer, S., & Wendolsky, R. (2006). A scatter search heuristic for the capacitated clustering problem. *European Journal of Operational Research, 169*, 533–547. doi:10.1016/j.ejor.2004.08.014

Schoen, F. (1991). Stochastic techniques for global optimization: a survey of recent advances. *Journal of Global Optimization, 1*, 207. doi:10.1007/BF00119932

Schwefel, H. (1995). *Evolution & Optimum Seeking*. New York, NY: John Wiley & Sons.

Sengoku, H., & Yoshihara, I. (1998). A fast TPS solver using GA on JAVA. In *Proceedings of the 3rd International Symposium on Artificial Life and Robotics* (pp. 283-288).

Shao, L. S., & Fu, G. X. (2008). Disaster prediction of coal mine gas based on data mining. *Journal of Coal Science and Engineering, 14*(3), 458–463. doi:10.1007/s12404-008-0099-9

Shelokar, P. S., Jayaraman, V. K., & Kulkarni, B. D. (2004). An ant colony classifier system: application to some process engineering problems. *Computers & Chemical Engineering, 28*, 1577–1584. doi:10.1016/j.compchemeng.2003.12.004

Shi, Y., & Eberhart, R. (1998). A modified particle swarm optimizer. In *Proceedings of 1998 IEEE World Congress on Computational Intelligence* (pp. 69-73).

Shi, Y., & Eberhart, R. (1998). A modified particle swarm optimizer. In *Proceedings of the Evolutionary Computation Conference* (pp. 69-73).

Shi, Y., & Eberhart, R. (1999). Empirical study of paticle swarm otpimization. In *Proceedings of the IEEE World Congress on Evolutionary Computation* (pp. 6-9). Washington, DC: IEEE Computer Society.

Shina, H. W., & Sohnb, S. Y. (2004). Segmentation of stock trading customers according to potential value. *Expert Systems with Applications, 27*, 27–33. doi:10.1016/j.eswa.2003.12.002

Shultmann, F., Zumkeller, M., & Rentz, O. (2006). Modeling reverse logistic task within closed-loop supply chains: An example from the automotive industry. *European Journal of Operational Research, 171*, 1033–1050. doi:10.1016/j.ejor.2005.01.016

Siedlecki, W., & Sklansky, J. (1988). On automatic feature selection. *International Journal of Pattern Recognition and Artificial Intelligence, 2*(2), 197–220. doi:10.1142/S0218001488000145

Simpson, R. W. (1969). *Scheduling and routing models for airline systems* (Tech. Rep. No. R-68-3). Cambridge, MA: MIT Press.

Smith, B. M., Layfield, C. J., & Wren, A. (2000). A constraint programming preprocessor for a bus driver scheduling system. In *Proceedings of the Workshop on Constraint Programming and Large Scale Discrete Optimization* (pp. 131-150).

Smith, B. M., Layfield, C. J., & Wren, A. (2000). A constraint programming preprocessor for a bus driver scheduling system. In Freuder, E., & Wallace, R. (Eds.), *Constraint programming and large scale discrete optimization* (*Vol. 57*). Providence, RI: American Mathematical Society.

Smith, B. M., & Wren, A. (1988). A bus crew scheduling system using a set covering formulation. *Transportation Research, 22A*, 97–108.

Sohoni, M., Bailey, G., Martin, K., Carter, H., & Johnson, E. (2003). Delta optimizes continuing-qualification-training schedules for pilots. *Interfaces, 33*, 57–70. doi:10.1287/inte.33.5.57.19253

Sohoni, M., Johnson, E., & Bailey, T. (2004). Long-range reserve crew manpower planning. *Management Science, 50*(6), 724–739. doi:10.1287/mnsc.1030.0141

Solomon, M. M. (1987). Algorithms for the vehicle routing and scheduling problems with time window constraints. *Operations Research, 35*, 254–265. doi:10.1287/opre.35.2.254

Song, K. B., Chung, S. T., Ginis, G., & Cioffi, J. M. (2002). Dynamic spectrum management for next-generation DSL systems. *IEEE Communications Magazine, 40*(10), 101–109. doi:10.1109/MCOM.2002.1039864

Sosa, N. G. M., Galvão, R. D., & Gandelman, D. A. (2007). Algoritmo de busca dispersa aplicado ao problema clássico de roteamento de vehiculos. *Pesquisa Operacional, 27*, 293–310. doi:10.1590/S0101-74382007000200006

Srinivasa Rao, R., Narasimham, S. V. L., & Ramalingaraju, M. (2008). Optimization of distribution network configuration for loss reduction using artificial bee colony algorithm. *International Journal of Electrical Power and Energy Systems Engineering, 1*(2).

Srinivas, N., & Deb, K. (1995). Multiobjective function optimization using nondominated sorting genetic algorithms. *Evolutionary Computation, 2*(3), 221–248. doi:10.1162/evco.1994.2.3.221

Srivastava, J., Cooley, R., Deshpande, M., & Tan, P. N. (2000). Web usage mining: Discovery and applications of usage patterns from web data. *SIGKDD Explorations*, *1*(2), 12–23. doi:10.1145/846183.846188

Steinzen, I., Gintner, V., Suhl, L., & Kliewer, N. (2010). A time-space network approach for the integrated vehicle- and crew-scheduling problem with multiple depots. *Journal Transportation Science*, *44*(3).

Stützle, T., & Hoos, H. (1997). Improvement in the ant system: introducing min-max ant system. In *Proceedings of the International Conference on Artificial Neuronal Networks and Genetic Algorithms* (pp. 266-274).

Taillard, E. D. (1999). A heuristic column generation method for the heterogeneous fleet VRP. *RAIRO*, *33*(1), 1–14. doi:10.1051/ro:1999101

Tang, O. (2004). Simulated annealing in lot sizing problem. *International Journal of Production Economics*, *88*(2), 173–181. doi:10.1016/j.ijpe.2003.11.006

Tarantilis, C., Kiranoudis, C., & Vassiliadis, V. (2004). A threshold accepting metaheuristic for the heterogeneous fixed fleet vehicle routing problem. *European Journal of Operational Research*, *152*, 148–158. doi:10.1016/S0377-2217(02)00669-0

Tasgetiren, M. F., Sevkli, M., Liang, Y., & Gencyilmaz, G. (2004). Particle Swarm Optimization Algorithm for Permutation Flowshop Sequencing Problem. In M. Dorigo et al. (Eds.), *Proceedings of ANTS 2004* (LNCS 3172, pp. 382-389).

Teodorovic, D., & Dell'Orco, M. (2005). Bee colony optimization - A cooperative learning approach to complex transportation problems. In *Advanced OR and AI Methods in Transportation* (pp. 51-60).

Thabtah, F. A., & Cowling, P. I. (2007). A greedy classification algorithm based on association rule. *Applied Soft Computing*, *7*, 1102–1111. doi:10.1016/j.asoc.2006.10.008

Thabtah, F., Cowling, P. I., & Hammoud, S. (2006). Improving rule sorting, predictive accuracy, and training time in associative classification. *Expert Systems with Applications*, *31*, 414–426. doi:10.1016/j.eswa.2005.09.039

Thompson, J., & Dowsland, K. (1998). A robust simulated annealing based examination timetabling system. *Computers & Operations Research*, *25*, 637–648. doi:10.1016/S0305-0548(97)00101-9

Tiankun, L., Chen, W., Zhen, X., & Zhang, Z. (2009). An improvement of the ant colony optimization algorithm for solving travelling salesman problem (TSP). In *Proceedings of the 5th International Conference on Wireless Communications, Networking and Mobile Computing*, Beijing, China.

Toth, P., & Vigo, D. (2002). Models relaxations and exact approaches for the capacitated vehicle routing problem. *Discrete Applied Mathematics*, *123*, 487–512. doi:10.1016/S0166-218X(01)00351-1

Toth, P., & Vigo, D. (2002). *The vehicle routing problem: SIAM monographs on discrete, mathematics and applications*. Philadelphia, PA: Society for Industrial & Applied Science.

Treacy, W. S., & Cavey, M. (2000). Credit risk rating systems at large US banks. *Journal of Banking & Finance*, *24*, 167–201. doi:10.1016/S0378-4266(99)00056-4

Tsai, C. Y., & Chiu, C. C. (2004). A purchase-based market segmentation methodology. *Expert Systems with Applications*, *27*, 265–276. doi:10.1016/j.eswa.2004.02.005

Tsiaflakis, P., & Moonen, M. (2008, April). Low-complexity dynamic spectrum management algorithms for digital subscribre lines. In *Proceedings of the IEEE International Conference on Acoustics, Speech, and Signal Processing*, Las Vegas, NV (pp. 2769-2772).

Tsiaflakis, P., Diehl, M., & Moonen, M. (2008). Distributed spectrum management algorithms for multiuser DSL networks. *IEEE Transactions on Signal Processing*, *56*(10). doi:10.1109/TSP.2008.927460

Tsiaflakis, P., Vangorp, J., Moonen, M., Verlinden, J., Van Acker, K., & Cendrillon, R. (2005). An efficient lagrange multiplier search algorithm for optimal spectrum balancing in crosstalk dominated xDSL systems. *IEEE Journal on Selected Areas in Communications*, *4*, 4.

UCI. (2010). *Machine learning repository*. Retrieved from http://archive.ics.uci.edu/ml

Ugray, Z., Lasdon, L., Plummer, J., & Bussieck, M. (2009). Dynamic filters and randomized drivers for the multi-start global optimization algorithm MSNLP. *Optimization Methods and Software, 24,* 635–656. doi:10.1080/10556780902912389

Vanier, F. (2009). *World broadband statistics: Q1 2009.* Retrieved from http://www.point-topic.com

Vaughn, N., Polnaszek, C., Smith, B., & Helseth, T. (2000). *Design-expert 6 user's guide.* Minneapolis, MN: Stat-Ease, Inc.

Veenhuis, C. (2006). Advanced Meta-PSO. In *Proceedings of the IEEE Sixth International Conference on Hybrid Intelligent Systems* (pp. 54–59).

Venkata Rao, R., & Pawar, P. J. (2010). Parameter optimization of a multi-pass milling process using non-traditional optimization algorithms. *Applied Soft Computing, 10,* 445–456. doi:10.1016/j.asoc.2009.08.007

Veral, E., & LaForge, R. (1985). The performance of a simple incremental Lot-Sizing rule in a multilevel inventory environment. *Decision Sciences, 16,* 57–72. doi:10.1111/j.1540-5915.1985.tb01475.x

Vercellis, C. (2009). *Business intelligence: Data mining and optimization for decision making.* New York, NY: John Wiley & Sons.

Vignier, A., Commandeur, C., & Proust, P. (1997). New lower bound for the hybrid flowshop scheduling problem. In *Proceedings of IEEE Sixth International Conference on Emerging Technologies and Factory Automation (ETFA 97)* (pp. 446-451).

Volgenant, T., & Jonker, R. (1987). On some generalizations of the traveling salesman problem. *The Journal of the Operational Research Society, 38,* 1073–1079.

Vollmann, T. E., Berry, D. W., & Whybark, D. C. (1997). *Manufacturing planning and control Systems* (4th ed.). New York, NY: McGraw-Hill.

Voros, J. (2002). On the relaxation of multi-level dynamic Lot-Sizing models. *International Journal of Production Economics, 77*(1), 53–61. doi:10.1016/S0925-5273(01)00202-X

Wagner, H., & Whitin, T. (1958). Dynamic version of the economic lot size model. *Management Science, 5,* 89–96. doi:10.1287/mnsc.5.1.89

Ward, R. E., & Deibel, L. E. (1972, November). *The advancement of computerized assignment of transit operators to vehicles through programming techniques.* Paper presented at the Joint National Meeting of the Operations Research Society of America, Atlantic City, NJ.

Wassan, N. A., Wassan, A. H., & Nagy, G. (2008). A reactive tabu search algorithm for the vehicle routing problem with simultaneous pickups and deliveries. *Journal of Combinatorial Optimization, 15*(4), 368–386. doi:10.1007/s10878-007-9090-4

Wedde, H. F., Farooq, M., & Zhang, Y. (2004). BeeHive: An efficient fault-tolerant routing algorithm inspired by honey bee behavior. In M. Dorigo (Ed.), *Ant colony optimization and swarm intelligence* (LNCS 3172, pp. 83-94).

Weir, J. D. (2002). *A three phase approach to solving the bidline generation problem with an emphasis on mitigating pilot fatigue through circadian rule enforcement.* Unpublished doctoral dissertation, Georgia Institute of Technology, Atlanta, GA.

Wong, S., & Li, C. S. (Eds.). (2006). *Life science data mining.* Singapore: World Scientific.

Wren, A. (1999). Scheduling, timetabling and rostering - a special relationship? In E. K. Burke & P. M. Ross (Eds.), *Proceedings of the International Conference on Practice and Theory of Automated Timetabling* (LNCS 1153, pp. 46-75).

Wren, A., Kwan, R. S. K., & Parker, M. E. (1994). Scheduling of rail driver duties. *Computers in railways–IV, 2,* 81-89.

Wren, A., & Rousseau, J. M. (1995). Bus driver scheduling: An overview. In Wilson, N. (Ed.), *Computer-aided transit scheduling* (pp. 174–187). Berlin, Germany: Springer-Verlag.

Wren, A., & Wren, D. O. (1995). A genetic algorithm for public transport driver scheduling. *Computers & Operations Research, 22.*

Xu, G., Zong, Y., Dolog, P., & Zhang, Y. (2010). Co-clustering analysis of weblogs using bipartite spectral projection approach. In R. Setchi, I. Jordanov, R. J. Howlett, & L. C. Jain (Eds.), *Proceedings of the 14ᵗʰ International Conference on Knowledge-Based and Intelligent Information and Engineering Systems* (LNCS 6278, pp. 398-407).

Yaghini, M., & Kazemzadeh, M. A. (2010). DIMM A: A design and implementation methodology for metaheuristic algorithms: A perspective from software development. *International Journal of Applied Metaheuristic Computing, 1*(4), 58–75.

Yagmahan, B., & Yenisey, M. (2010). multi-objective ant colony system algorithm for flow shop scheduling problem. *Expert Systems with Applications, 37,* 1361–1368. doi:10.1016/j.eswa.2009.06.105

Yang, X. S. (2005). Engineering optimizations via nature-inspired virtual bee algorithms. In J. M. Yang & J. R. Alvarez (Eds.), *Proceedings of IWINAC 2005* (LNCS 3562, pp. 317-323).

Yao, X. (1999). *Evolutionary computation: Theory and applications.* Singapore: World Scientific.

Yoshihara, I., & Sengoku, H. (2000). Scheduling bus driver's services based on genetic algorithm. In *Proceedings of the International Conference on Artificial Intelligence in Science and Technology* (pp. 62-67).

Yu, W., Lui, R., & Cendrillon, R. (2004, December). Dual optimization methods for multiuser orthogonal frequency-division multiplex systems. In *Proceedings of the IEEE International Conference on Global Communications* (Vol. 1, pp. 225-229).

Yusta, S. C. (2009). Different metaheuristic strategies to solve the feature selection problem. *Pattern Recognition Letters, 30,* 525–534. doi:10.1016/j.patrec.2008.11.012

Yu, W., Ginis, G., & Cioffi, J. (2002). Distributed multiuser power control for digital subscriber lines. *IEEE Journal on Selected Areas in Communications, 20*(5), 1105–1115. doi:10.1109/JSAC.2002.1007390

Zachariadis, E., Tarantilis, C., & Kiranoudis, C. (2009). A hybrid metaheuristic algorithm for the vehicle routing problem with simultaneous delivery and pick-up service. *Expert Systems with Applications, 36*(2), 1070–1081. doi:10.1016/j.eswa.2007.11.005

Zamani, R., & Lau, K. (2010). Embedding learning capability in Lagrangian relaxation: An application to the travelling salesman problem. *European Journal of Operational Research, 201*(1), 82–88. doi:10.1016/j.ejor.2009.02.008

Zangwill, W. (1966). A deterministic multiproduct, multifacility production and inventory model. *Operations Research,* 486–507. doi:10.1287/opre.14.3.486

Zeghal, F. M., & Minoux, M. (2006). Modeling and solving a crew assignment problem in air transportation. *European Journal of Operational Research, 175*(1), 187–209. doi:10.1016/j.ejor.2004.11.028

Zhang, J. (2009). Natural computation for the traveling salesman problem. In *Proceedings of the Second International Conference on Intelligent Computation Technology and Automation* (pp. 366-369).

Zhan, Z.-H., Zhang, J., Li, Y., & Chung, H.-S. (2009). Adaptive Particle Swarm Optimization. *IEEE Transactions on Systems, Man, and Cybernetics. Part B, Cybernetics, 6*(39), 1362–1381. doi:10.1109/TSMCB.2009.2015956

Zhao, X. (2008). Convergent analysis on evolutionary algorithm with non-uniform mutation. In *Evolutionary Computation* (pp. 940–944).

Zhao, X., Cao, X.-S., & Hu, Z.-C. (2007). Evolutionary programming based on non-uniform mutation. *Applied Mathematics and Computation, 192,* 1–10. doi:10.1016/j.amc.2006.06.107

Zhong, N., & Skowron, A. (2001). A rough set-based knowledge discovery process. *International Journal of Applied Mathematics and Computer Science, 11*(3), 603–619.

Zimmermann, H. J. (1996). *Fuzzy set theory and its applications.* Boston, MA: Kluwer Academic.

Zitzler, E., & Thiele, L. (1998). *An evolutionary algorithm for multi-objective optimization: the strength pareto approach* (Tech. Rep. No. TIK-43). Zurich, Switzerland: Swiss Federal Institute of Technology.

Zitzler, E., Deb, K., & Thiele, L. (1999). *Comparison of multiobjective evolutionary algorithms: empirical results.* Zurich, Switzerland: Swiss Federal Institute of Technology.

Zitzler, E., Deb, K., & Thiele, L. (2000). Comparison of multiobjective evolutionary algorithms: Empirical results. *Evolutionary Computation*, *8*(2), 173–195. doi:10.1162/106365600568202

Zong, Y., Xu, G., Dolog, P., & Zhang, Y. (2010). Co-clustering for weblogs in semantic space. In L. Chen, P. Triantafillou, & T. Suel (Eds.), *Proceedings of 11th International Conference on Web Information Systems Engineering* (LNCS 6488, pp. 120-127).

About the Contributors

Peng-Yeng Yin received his BS, MS, and PhD degrees in computer science from National Chiao Tung University (Hsinchu, Taiwan). From 1993 to 1994, he was a visiting scholar in the Department of Electrical Engineering at the University of Maryland (College Park, MD, USA) and in the Department of Radiology at Georgetown University (Washington D.C., USA). In 2000, he was a visiting professor in the Visualization and Intelligent Systems Laboratory (VISLab) in the Department of Electrical Engineering at the University of California (Riverside, USA). From 2006 to 2007, he was a visiting professor at Leeds School of Business, University of Colorado. From 2001 to 2003, he was a professor in the Department of Computer Science and Engineering, Ming Chuan University (Taoyuan, Taiwan). Since 2003, he has been a professor of the Department of information Management, National Chi Nan University (Nantou, Taiwan). Dr. Yin received the Overseas Research Fellowship from Ministry of Education (1993) and the Overseas Research Fellowship from National Science Council (2000). He has received the best paper award from the Image Processing and Pattern Recognition Society of Taiwan. He is a member of the Phi Tau Phi Scholastic Honor Society and listed in *Who's Who in the World*, *Who's Who in Science and Engineering*, and *Who's Who in Asia*. Dr. Yin has published more than 100 academic articles in reputable journals and conferences including IEEE Trans. on Pattern Analysis and Machine Intelligence, IEEE Trans. on Knowledge and Data Engineering, IEEE Trans. on Education, Pattern Recognition, Annals of Operations Research, IEEE International Conference on Computer Vision, etc. He has been on the editorial board of the *International Journal of Advanced Robotic Systems*, the *Open Artificial Intelligence Journal*, the *Open Artificial Intelligence Letters*, the *Open Artificial Intelligence Reviews* and served as a program committee member in many international conferences. His current research interests include artificial intelligence, evolutionary computation, metaheuristics, pattern recognition, content-based image retrieval, relevance feedback, machine learning, computational intelligence, operations research, and computational biology.

* * *

Rahim Akhavan is a faculty member of School of Railway Engineering, Kermanshah University of Technology. I studied Bachelor of Science in industrial engineering and Master of Science program in railway engineering at Iran University of Science and Technology. My research interests are transportation planning, optimization with metaheuristics, operations research, multicommodity network design problems and optimization in transportation models. I write a book in the field of metaheuristics in Persian, a journal paper in IJAMC in the field of network design, and an under review journal paper in the field of network design and metaheuristics. In addition, I presented 2 conference papers in the field of solving combined car blocking and train makeup planning with Ant Colony Optimization, and Simulated annealing for network design.

J. Bagyamani is working as Associate Prof. & Head, Department of Computer Science, Government Arts College, Dharmapuri and currently pursuing her PhD in the Department of Computer Science, Periyar University, Salem as a Teacher fellow under UGC XI Plan FIP Scheme. She has published 6 papers in reputable International Journals and received two best paper awards in International Conferences. Her areas of interest are data mining, biclustering of Gene Expression Data using heuristic and meta-heuristic techniques and web mining.

Dominique Barchiesi is a Full Professor of theoretical physics, applied mathematics and statistics at the University of Technology of Troyes. He conducts active research in numerical modelling, optimization and advanced methods with application to engineering of nanotechnologies and plasmonics. The result of his research has been published in over one hundred eighty-five articles, conferences, and book chapters since 1993, in the fields of optics, electromagnetism, signal processing, operational research and finite element methods.

Habib Chabchoub is a professor at the University of Sfax. Currently he is the director of the Institut des Hautes Etudes Commerciales of Sfax. He obtained his PhD administration sciences from the Laval University in Canada. He occupied various administrative positions in several academic institutions in Tunisia. His main research interests are multi-criteria decision aid, vehicle routing, logistics and transport, mathematical programming, and so on. He published more than 30 papers in international journals and more than 60 communications. He is the director of 50 theses. He was the founder of the Superior Institute of Industrial Management in Sfax and different training programs of masters and doctorates. He is member in several Tunisian and foreign associations. Besides, he was the responsible of sessions in the national and international conferences and referee for the national and international journals.

Madhabananda Das is a professor and Dean in School of Computer Engineering, KIIT University, Bhubaneswar, Orissa. He received his B. Tech degree in electronics and telecommunication engineering and M. Tech degree in electronic systems and communication from Sambalpur University, Orissa, India, in 1981 and 1985 respectively. He received his Ph.D. in Computer Engineering in 2010 from KIIT University, Bhubaneswar, Orissa. His research interests include Swarm intelligence, Evolutionary Computation, Fuzzy set theory, Neural Network and Rule Mining. He has already published about 10 research papers in International Conferences and as book chapters.

Sung-Bae Cho received the Ph.D. degrees in computer science from KAIST (Korea Advanced Institute of Science and Technology), Taejeon, Korea, in 1993. He was an Invited Researcher of Human Information Processing Research Laboratories at ATR (Advanced Telecommunications Research) Institute, Kyoto, Japan from 1993 to 1995, and a Visiting Scholar at University of New South Wales, Canberra, Australia in 1998. He was also a Visiting Professor at University of British Columbia, Vancouver, Canada from 2005 to 2006. Since 1995, he has been a Professor in the Department of Computer Science, Yonsei University. His research interests include neural networks, pattern recognition, intelligent man-machine interfaces, evolutionary computation, and artificial life. He is a Senior Member of IEEE and a Member of the Korea Information Science Society, the IEEE Computer Society, the IEEE Systems, Man, and Cybernetics Society, and the Computational Intelligence Society.

Hernan X. Cordova was born in Guayaquil, Ecuador. His current research interests are broadband technologies and the mathematical optimization of practical problems.

Satchidananda Dehuri is a Reader and Head in P.G. Department of Information and Communication Technology, Fakir Mohan University, Vyasa Vihar, Balasore, Orissa. He received his M.Sc. degree in Mathematics from Sambalpur University, Orissa in 1998, and the M.Tech. and Ph.D. degrees in Computer Science from Utkal University, Vani Vihar, Orissa in 2001 and 2006, respectively. He completed his Post Doctoral Research in Soft Computing Laboratory, Yonsei University, Seoul, Korea under the BOYSCAST Fellowship Program of DST, Govt. of India. In 2010 he received Young Scientist Award in Engineering and Technology for the year 2008 from Odisha Vigyan Academy, Department of Science and Technology, Govt. of Odisha. He was at the Center for Theoretical Studies, Indian Institute of Technology Kharagpur as a Visiting Scholar in 2002. During May-June 2006 he was a Visiting Scientist at the Center for Soft Computing Research, Indian Statistical Institute, Kolkata. His research interests include Evolutionary Computation, Neural Networks, Pattern Recognition, Data Warehousing and Mining, Object Oriented Programming and its Applications and Bioinformatics. He has already published about 100 research papers in reputed journals and referred conferences, has published three text books for undergraduate and Post graduate students and edited five books, and is acting as an editorial member of various journals. He chaired the sessions of various International Conferences.

Laurent Deroussi, PhD, is assistant professor in Computer Science, at the University of Clermont-Ferrand. His research interests are in the areas of production systems, supply chains and complex systems. Some of his main contributions concern the field of flexible manufacting systems, for which he focuses on off-line scheduling and design problems. He also works in the field of metaheuristics, stochastic algorithms, hybrid algorithms and optimization-simulation coupling models.

Abraham Duarte is an Associate Professor in the Computer Science Department at the Rey Juan Carlos University (Madrid, Spain). He received his doctoral degree in Computer Sciences from the Rey Juan Carlos University. His research is devoted to the development of models and solution methods based on meta-heuristics for combinatorial optimization and decision problems under uncertainty. He has published more than 30 papers in prestigious scientific journals and conference proceedings such us European Journal of Operational Research, INFORMS Journal on Computing, Computational Optimization and Applications or Computers & Operations Research. Dr Duarte is reviewer of the Journal of Heuristic, Journal of Mathematical Modeling and Algorithms, INFORMS Journal on Computing, Applied Soft Computing, European Journal of Operational Research and Soft Computing. He is also member of the program committee of the conferences MAEB, HIS, ISDA or MHIPL.

Jalel Euchi is a Ph.D. student at the Department of Quantitative Methods, University of Sfax, Tunisia also he prepared his thesis in the form of co-supervised at the University of Havre in France. He received his Bachelor of Operational Research degree and Master degree in Operational Research and Production Management from the Department of Quantitative Methods of the University of Sfax, Tunisia. His primary research interests are in complex vehicle routing problems, heuristics and meta-heuristics algorithms to solve the NP-Hard problems, computational operations research, logistics research and supply chains. He participated to different international conferences and he was the corresponding author in several published papers.

Dan Abensur Gandelman is a PhD student at Universidade Federal do Rio de Janeiro, studying Industrial Engineering. He also has a Master Degree in Industrial Engineering and he is graduated in Electronic Engineering, both at Universidade Federal do Rio de Janeiro.

Fatima Ghedjati is an Associate Professor in IUT of Reims, University of Reims Champagne-Ardenne, France. She received her MS degree from the University of Nancy I and PhD (1994) in computer science from the University of Paris 6, France. Her research interests are scheduling, combinatorial optimization, Multi-criteria decision, heuristics, meta-heuristics (Genetic Algorithms, Tabu Search, Ant Colony Optimization).

Laurence Giraud-Moreau, PhD, is an Associate Professor of numerical mechanics at the University of Technology of Troyes. She conducts actives research on numerical modelling, optimization and remeshing methods with applications to forming processes.

Fred Glover is the Chief Technology Officer in charge of algorithmic design and strategic planning initiatives for OptTek Systems, Inc., and holds the title of Distinguished Professor, Emeritus, at the University of Colorado, Boulder. He has authored or co-authored more than 400 published articles and eight books in the fields of mathematical optimization, computer science and artificial intelligence, and is the originator of the optimization search procedure called Tabu Search (Adaptive Memory Programming), for which Google returns more than a million results. Fred Glover is the recipient of the von Neumann Theory Prize, the highest honor of the INFORMS society, and is an elected member of the U. S. National Academy of Engineering. His numerous other awards and honorary fellowships include those from the AAAS, the NATO Division of Scientific Affairs, INFORMS, DSI, USDCA, ERI, AACSB, Alpha Iota Delta and the Miller Institute for Basic Research in Science.

Thomas Grosges, PhD, is an Associate Professor of numerical and theoretical physics, and applied mathematics at the University of Technology of Troyes. Dr. Grosges conducts active research on numerical developments related to modelling the interaction of matter and light and to the optimization of numerical solvers in computer sciences and their applications. His work was published in a number of physics, applied physics, engineering and computer science academic journals, conference publications and books.

Abdelaziz Hamzaoui is currently a Professor and the Director of the IUT of Troyes, University of Reims Champagne-Ardenne, France. He received his MS and PhD degrees from University of Reims Champagne-Ardenne both in Electrical and Control Engineering in 1989 and 1992 respectively. He obtained the Diploma of Habilitation to Direct Research (HDR) from the University of Reims Champagne-Ardenne in Electrical and Control Engineering in 2004. His research interests include intelligent control, fuzzy control, robust adaptive control, power system and renewable energy.

Mohammad Karimi is an MSc. Student at the Department of Rail Transportation Engineering, School of Railway Engineering, Iran University of Science and Technology. His research interests are metaheuristics optimization methods, multicommodity network design problems, and optimization in rail transportation problems such as train formation problem and car blocking. He has some published papers in the field of network design and metaheuristics.

Sameh Kessentini received an engineering degree and a Master of Science degree from Tunisia Polytechnic School. She is currently working toward the PhD degree from the University of Technology of Troyes. Her current research interests are numerical and optimization methods with application to engineering of nanotechnologies and plasmonic.

Safa Khalouli received the MS degree in Systems' Optimization and Security (OSS) from the University of Technology Troyes-France in 2005 and the Ph.D degree in computer science from the University of Reims-Champagne-Ardenne-France, in 2010. Her current research interests include combinatorial optimization, scheduling, metaheuristics, heuristics.

Manuel Laguna is the Media One Professor of Management Science and Senior Associate Dean at the Leeds School of Business of the University of Colorado Boulder. He started his career at the University of Colorado in 1990, after receiving master's (1987) and doctoral (1990) degrees in Operations Research and Industrial Engineering from the University of Texas at Austin. He has done extensive research in the interface between computer science, artificial intelligence and operations research to develop solution methods for practical problems in operations-management areas such as logistics and supply chains, telecommunications, decision-making under uncertainty and optimization of simulated systems. Dr. Laguna has more than one hundred publications, including more than sixty articles in academic journals and four books. He is editor-in-chief of the *Journal of Heuristics*, is in the international advisory board of the *Journal of the Operational Research Society* and has been guest editor of the *Annals of Operations Research* and the *European Journal of Operational Research*.

Marc Lamy de la Chapelle, PhD, is Full Professor at the University of Paris 13. He is specialised in the vibrational spectroscopies and more especially on the Surface Enhanced Raman Scattering and related topics such as plasmon resonance and optical properties of nanoparticles. He is currently working on the applications of these techniques to characterization of biological structures and on the use of nanotechnologies for the development of biosensor.

Leon Lasdon received his Ph.D. in Systems Engineering from Case Institute of Technology in 1964. He taught in the Operations Research Department at Case from 1964 to 1977, when he joined the McCombs School of Business at The University of Texas at Austin. He holds the David Bruton Jr. Chair in Business Decision Support Systems in the Information, Risk, and Operations Management Department. Prof. Lasdon is an active contributor to nonlinear programming algorithms and software. He is co-author (with Dan Fylstra) of the Microsoft Excel Solver. His OQNLP and MSNLP multistart solvers for smooth nonconvex optimization are available within GAMS and TOMLAB. His LSGRG2 nonlinear optimizer is available within the Frontline Systems Premium Excel Solver and the multistart systems, and is also widely used in process control. He is the author or co-author of over 120 refereed journal articles and three books. Recent papers are available at www.utexas.edu/courses/lasdon (link to "papers").

David Lemoine is assistant professor in Computer Science, at the Ecole des Mines de Nantes in France. His research interests include issues related to tactical planning optimization (lot-sizing models ...): mathematical programming (mathematical models, lagrangean relaxations ...), approximated methods (heuristics, metaheuristics ...) and simulation. He also works on the links between tactical and operational points of view by taking into account, for instance, maintenance consideration into tactical planning process.

Nashat Mansour is a Professor of Computer Science and the Assistant-Dean of the School of Arts and Sciences at the Lebanese American University. He received B.E. and M.Eng.Sc. degrees in Electrical Engineering from the University of New South Wales, Australia and M.S. in Computer Engineering and Ph.D. in Computer Science from Syracuse University, USA. He is also the Director of the Software Institute at the Lebanese American University and the currnet President of the Arab Computer Society. Dr. Mansour has served on national and regional advisory committees entrusted to develop strategies and policies for promoting R&D in Computing and IT for economic and social development. He conducts research in the areas of: application of metaheuristics and data mining to real-world problems, biomedical informatics, and search based software engineering.

Gladys Maquera graduated in Mathematics Education from the Peruvian university union (Peru), a Master in Electrical Engineering from the University of Campinas (Brazil), PhD in Production Engineering from Federal University of Rio de Janeiro (Brazil) and post doctorate from the University Federal Fluminense (Brazil). He has experience in Production Engineering with emphasis in Operations Research and the area of computer science in the area of Artificial Intelligence. He currently develops research projects, development and innovation, He is Director of Research at the Peruvian university union Juliaca branch and principal investigator of the National Council for Science, Technology and Innovation of Peru (CONCYTEC).

Magdalene Marinaki, PhD, was born in Chania, Greece. She received a Diploma in Production Engineering and Management from the Technical University of Crete, Greece and a PhD from the same University, in 2002. Currently she serves as a permanent staff (research associate) in the Technical University of Crete, Chania, Greece. Her research interests focus on computational methods in optimization problems, on Dynamic Optimization, on Nature Inspired Methods, on Supply Chain Management, on Metaheuristics algorithms on Data Mining and on Optimal Control. She is the author of two books and more than forty articles in international scientific journals and books. She has more than thirty presentations in international and national scientific conferences. She has worked as a researcher in more than 10 European projects.

Yannis Marinakis, PhD, was born in Chania, Greece, in 1976. He received a Diploma in Production Engineering and Management from the Technical University of Crete, Greece, in 1999 and a PhD, from the same University, in 2005. He is currently a Lecturer in the Technical University of Crete, Chania, Greece. His research interests focus on computational methods in optimization problems, on Nature Inspired Methods, on Supply Chain Management and on Metaheuristics algorithms on Data Mining. He teaches the following courses: Combinatorial Optimization, Game Theory, Design and Optimization in Supply Chain Management, Heuristic and Metaheuristic Algorithms (postgraduate) and Optimization of Large Scale Systems (postgraduate). He is the author of two books and more than fifty articles in international scientific journals and books (Computers and Operations Research, European Journal of Operational Research, Journal of Global Optimization, Computational Optimization and Applications, Journal of Combinatorial Optimization, Expert Systems with Applications, Annals of Operations Research, Applied Soft Computing). He has more than thirty five presentations in international and national scientific conferences.

Rafael Martí is Professor in the Statistics and Operations Research Department at the University of Valencia, Spain. His teaching includes courses on Operations Management in Business, Statistics in Social Sciences, Mathematical Programming for Math majors and Management Science at the Masters and Doctoral level. His research interest focuses on the development of metaheuristics for hard optimization problems. He is co-author of several books (e.g., "Scatter Search" Kluwer 2003 and "The linear ordering problem" Springer 2010) and is currently Area Editor in the Journal of Heuristics and Associate Editor in the Mathematical Programming Computation and the International Journal of Metaheuristics; he has published more than 50 JCR-indexed journal papers.

Nikolaos Matsatsinis, PhD, is a Professor of Information and Intelligent Decision Support Systems at the Department of Production Engineering and Management of the Technical University of Crete, Greece. He is Chairman of the Department, Director of Decision Support Systems Lab and President of the Hellenic Operational Research Society (HELORS). He has contributed as scientific or project coordinator on over of forty scientific projects. He is the author or co-author of sixteen books and over of seventy five articles in international scientific journals and books. His research interests fall into the areas of Decision Support Systems, Artificial Intelligent and Multi-Agent Systems, Heuristics and Metaheuristics Algorithms, e-Marketing, Multicriteria Decision Analysis, Group Decision Support Systems, Workflow Management Systems. Prof. Matsatsinis is member of the Institute of IEEE; AAAI; IFIP; CEPIS; INFORMS; IFORS; EURO; EU/ME; EFITA.

Mohsen Momeni is an MSc. Student in Rail Transportation Engineering, School of Railway Engineering, Iran University of Science and Technology. His research interests are metaheuristics optimization methods, parameter tuning of metaheuristics, multicommodity network design problems, and optimization in rail transportation problems such as train formation problem and car blocking. He has some published papers in the field of network design and metaheuristics.

Walid Moudani received his Master diploma degree and Doctorate degree in computer science from Institute National Polytechnique of Toulouse (INPT) and Laboratory for Analysis and Architecture of Systems LAAS-CNRS in 1997 and 2001, respectively. He is Associate Professor in Lebanese University since 2001. He is member of the International Society on Multi-Criteria Decision-Making (MCDM). His main research area is in the development and design of new methods and systems towards the solution of planning and operations problems. His main contribution in operations research is in proposing a reactive approach for air transportation operations planning (crew, fleet assignment and flight scheduling) and decision making approach for revenue management. He published more than 20 scientific papers in the areas of optimization, image recognition, compression and analysis, data warehouse, data mining and its applications. He is currently with the Group of Bioinformatics and Modeling in the research center of AZM for research in biotechnology and its applications in the Lebanese University. Director at Centre National des Arts et Métiers - CNAM, Tripoli - Lebanon Center, from 2005 until 2009.

Félix Mora-Camino was an Aeronautical Engineer from ENSICA, Toulouse, 1974, Docteur Ingénieur in Automatic Control from Institute National des Sciences Appliquées de Toulouse-INSAT, 1977, Docteur of Sciences in Operations Research from Université Paul Sabatier-UPS, 1987. He has been a Professor of Automatic Control and Operations Research at the Air Transportation Department of ENAC, from 1989 until today. Félix Mora-Camin is the Head of the LARA (Automation and Operations Research

Laboratory) of the Air Transport Department at ENAC, Associate researcher at LAAS du CNRS, Fault Detection and Diagnostic Group, from 1992 until today. His main research subjects include Fault detection and identification, fault tolerant flight control systems.

Mrunmaya Mudigere received the BE degree in mechanical engineering from the Viswesvarayya Technological University in 2009. Currently he is working as a project assistant in the Department of Aerospace Engineering at the Indian Institute of Science, Bangalore.

G. Narayana Naik received the BE degree in Civil Engineering from the Jawaharlal Nehru National College of Engineering, Shimoga in 1989, the M.Tech. degree in Civil Engineering from Indian Institute of Technology, Madras in 1993 and Ph.D. degree in Aerospace Engineering at the Indian Institute of Science, Bangalore in 2007. He has joined the Department of Aerospace Engineering at the Indian Institute of Science, Bangalore, where he is currently working as Principal Research Scientist. His research interests are in the area of Nano-composites, Solid Mechanics, Theory of Elasticity, Analysis and Design of Structures, Composite structures, Computational Mechanics, Damage Mechanisms, Smart Materials and Structures, Genetic Algorithms, etc.

S.N.Omkar received the BE degree in mechanical engineering from the University Viswesvarayya College of Engineering in 1985, the MSc (Engg) degree in aerospace engineering from Indian Institute of Science in 1992 and the PhD degree in aerospace engineering from the Indian Institute of Science, Bangalore, in 1999. He joined the Department of Aerospace Engineering at the Indian Institute of Science, Bangalore, where is currently a principal research scientist. His research interests include helicopter dynamics, biologically inspired computational techniques, fuzzy logic and parallel computing.

Kiran Patil received the BE degree in mechanical engineering from the Viswesvarayya Technological University in 2005, the M.Tech. degree in Mechanical Engineering from Viswesvarayya Technological University in 2007. Currently he is working as a project associate in the Department of Aerospace Engineering at the Indian Institute of Science, Bangalore.

John Plummer is Senior Lecturer of Quantitative Methods in the McCoy College of Business, Department of Computer Information Systems and Quantitative Methods at Texas State University, San Marcos Texas. He received his PhD degree from the Business School, University of Texas at Austin in 1984, with earlier MBA and BS in Chemical Engineering degrees from Texas A&M. His research interests include implementation and refinement of nonlinear programming algorithms and software, interfaces to algebraic modeling systems, and multi-start heuristics for global optimization. He is co-author of the OQNLP and MSNLP multistart Solvers included in the GAMS modeling language.

Mohadeseh Rahbar is an MSc. Student at the Department of Rail Transportation Engineering, School of Railway Engineering, Iran University of Science and Technology. Her research interests are metaheuristics optimization methods, exact optimization methods, multicommodity network design problems, and optimization in rail transportation problems. She has some published papers in the field of network design and metaheuristics.

R. Rathipriya is working as Assistant Professor in Periyar University, Salem, India. She received her Bachelor of Science and Master of Science degrees in Computer Science from the Periyar University. Currently, she is pursuing her PhD in Bharathiyar University, India. Her research interests are in several areas of data mining, web mining, Optimization techniques and Bio-Informatics.

Cesar Rego is a Professor at the School of Business of the University of Mississippi. He received a MSc in Operations Research and Systems Engineering from the School of Technology of the University of Lisbon, and a Ph.D. in Computer Science from the University of Versailles. His research focuses on mathematical optimization, computer science, and artificial intelligence. Dr. Rego's innovations in the field of metaheuristics include the invention of the Relaxation Adaptive Memory Programming (RAMP) approach for solving complex optimization problems.

Rahul Roy is a M. Tech student in School of Computer Engineering, KIIT University, Bhubaneswar, Orissa, India. He completed his Integrated M.Sc. degree in Computer science from Assam University, Silchar, India, in 2009. He received gold medal and Ananya Devi endowment award for securing First class First position in M.Sc. His research interest includes Evolutionary Computation and Rule mining.

Annibal Parracho Sant'Anna is a Professor at Universidade Federal Fluminense and has a Ph.D. in Statistics by the University of California, Berkeley and M. Sc in Mathematics by IMPA/CNPq, he graduated in Mathematics and in Economics at Universidade Federal do Ro de Janeiro. He developed Pos-doctoral studies at the University of New South Wales. He has been the Head of the Institute of Mathematics and of the Laboratory of Statistics of Universidade Federal do Rio de Janeiro. He was until recently the President of the Brazilian Society of Operations Research.

Mohammadreza Sarmadi is an MSc. Student in Rail Transportation Engineering, School of Railway Engineering, Iran University of Science and Technology. His research interests are metaheuristics optimization methods, parameter tuning of metaheuristics, multicommodity network design problems, and optimization in rail transportation problems. He has some published papers in the field of network design and metaheuristics.

Ghia Sleiman-Haidar is a graduate of the Lebanese American University where she received a B.S. and M.S. degrees in Computer Science. Her research interests are in exam timetabling and parallelism.

K. Thangavel is presently the Prof. & Head, Department of Computer Science Periyar University, Salem. He has completed his PhD in Gandhigram Rural University. His areas of interest are data mining, image processing, mobile computing and rough set theory and Optimization Techniques. He is reviewer of reputable journals. He has received Young Scientist Award 2009 from Tamilnadu State Council for Science and Technology.

Leo Van Biesen was born in Elsene, Belgium, on August 31, 1955. He received the degree in electro-mechanical engineer and the Ph.D. degree from Vrije Universiteit Brussel (VUB), Brussels, Belgium, in 1978 and 1983, respectively. He is currently a Full Senior Professor of fundamental electricity, electrical measurement techniques, signal theory, computer-controlled measurement systems, telecommunication, underwater acoustics, and Geographical Information Systems for sustainable development of environments with VUB. His current interests are signal theory, modern spectral estimators, time-domain reflectometry, wireless local loops, xDSL technologies, underwater acoustics, and expert systems for intelligent instrumentation. Prof. Van Biesen was the President of the International Measurement Confederation (IMEKO) until September 2006. He is also a member of the Board of the European Telecommunication Engineers Federation (FITCE), Belgium, and the International Scientific Radio Union (URSI), Belgium. He was the Chairman of IMEKO TC-7 from 1994 to 2000, the President Elect of IMEKO from 2000 to 2003, and the Liaison Officer between the IEEE and IMEKO.

Masoud Yaghini is an Assistant Professor at the Department of Rail Transportation Engineering, School of Railway Engineering, Iran University of Science and Technology. His research interests include data mining, optimization, metaheuristic algorithms, and application of data mining and optimization techniques in rail transportation planning. He published several books and papers in the field of data mining, metaheuristics, and rail transportation planning.

Adnan Yassine is a professor at the University of Havre- France. Currently he is the director of the Laboratory of Mathematics Application of Havre. His teaching and research focus on numerical analysis (numerical algorithms, scientific computing, mathematical programming), convex analysis and non-convex optimization convex and no convex (differentiable and no differentiable), global optimization, and financial optimization Operations Research (combinatorial optimization, optimization in networks, scheduling and graph theory). She works on several scientific fields such as: applications to interior point methods for nonlinear optimization problems, genetic algorithms in scheduling, operations research applied to problems in logistics and industrial, economic and financial optimization, quadratic programming into binary variables, the integer linear programming, polyhedral methods, meta-heuristics, relaxation and Lagrangian decomposition and linearization.

Constantin Zopounidis, PhD, is Professor of Financial Management and Operations Research, at the Department of Production Engineering and Management at Technical University of Crete (Greece). He is Editor-in-Chief in the following journals: Operational Research: An International Journal (Springer), International Journal of Multicriteria Decision Making (Interscience),The Journal of Financial Decision Making (Klidarithmos) and The Journal of Computation Optimization in Economics and Finance (Nova Publishers). He is also Associate Editor in New Mathematics and Natural Computation (World Scientific), Optimization Letters (Springer), International Journal of Banking, Accounting and Finance (Inderscience), International Journal of Data Analysis Techniques and Strategies (Inderscience) and European Journal of Operational Research (Elsevier). In recognition of his scientific work, his research has been awarded and among all he has been attributed in 1996, the Gold Medal and Diploma of Social and Human Sciences from the MOISIL International Foundation, for his research in multicriteria intelligent decision support systems and their application to the scientific world of financial management and credit risk assessment, in 2000, the Best Interdisciplinary Research Paper Award, from the Decision Sciences Institute and in 2009 he was a winner of the highly commended paper awarded from the Emerald Literati Network. He has published 60 books in international publishing companies and more than 400 papers are appearing in international scientific journals on finance, accounting and operations research, fields that his research interests fall into. He has accomplished many lectures as Visiting Professor in many European Universities.

Index